CLIMATE CHANGE: A WICKED PROBLEM

Under one cover, Frank Incropera provides a comprehensive, objective, and critical assessment of all issues germane to the climate change debate: science, technology options, economic ramifications, cultural and behavioral issues, the influence of special interests and public policy, geopolitics, and ethical dimensions. The underlying science is treated in depth, but in an approachable and accessible manner. A strong case is made for the reality of anthropogenic climate change, while confronting the range of issues that remain uncertain and deconstructing opposing views. Incropera assesses the strengths and weaknesses of technology options for mitigating the effects of climate change, analyzes nontechnical factors – economic, cultural, and political – and provides an in-depth treatment of ethical implications. This book is intended for those wishing to become fully informed about climate change and is designed to provide the reader with a firm foundation for drawing his or her own conclusions.

Frank P. Incropera is Clifford and Evelyn Brosey Professor of Mechanical Engineering at the University of Notre Dame, where he also served as Dean of Engineering from 1998 until 2006. He spent a majority of his career at Purdue University, and among his many honors he has received the American Society of Engineering Education (ASEE) Ralph Coats Roe Award for excellence in teaching (1982), the ASEE George Westinghouse Award for contributions to education (1983), the American Society of Mechanical Engineers (ASME) Heat Transfer Memorial Award (1988), the Melville Medal for the best original paper published by ASME (1988), and the Worcester Reed Warner Medal of ASME (1995). He received the Senior Scientist Award from the Alexander von Humboldt Foundation of the Federal Republic of Germany in 1988 and in 1996 was elected to the U.S. National Academy of Engineering. In 2001, he was named by the Institute for Scientific Information as one of the 100 most frequently cited engineering researchers in the world. He is a Fellow of ASME and the American Association for the Advancement of Science (AAAS). Professor Incropera has had a long-standing interest in transport phenomena and in recent years has turned his attention to the broad range of technical and nontechnical issues associated with transition to a sustainable energy future.

CAMBRIDGE
UNIVERSITY PRESS

32 Avenue of the Americas, New York, NY 10013-2473, USA

Cambridge University Press is part of the University of Cambridge.

It furthers the University's mission by disseminating knowledge in the pursuit of education, learning, and research at the highest international levels of excellence.

www.cambridge.org
Information on this title: www.cambridge.org/9781107521131

© Frank P. Incropera 2016

This publication is in copyright. Subject to statutory exception and to the provisions of relevant collective licensing agreements, no reproduction of any part may take place without the written permission of Cambridge University Press.

First published 2016

Printed in the United States of America

A catalog record for this publication is available from the British Library.

Library of Congress Cataloging in Publication Data
Incropera, Frank P.
Climate change: a wicked problem: complexity and uncertainty
at the intersection of science, economics, politics, and human
behavior / Frank P. Incropera, University of Notre Dame.
    pages   cm
Includes bibliographical references and index.
ISBN 978-1-107-10907-0 (hardback) – ISBN 978-1-107-52113-1 (pbk.)
1. Climatic changes – Environmental aspects.   2. Climatic changes – Social
aspects.   3. Greenhouse effect, Atmospheric.   4. Energy consumption –
Environmental aspects.   I. Title.
QC903.I475   2016
363.738′74–dc23   2015016109

ISBN 978-1-107-10907-0 Hardback
ISBN 978-1-107-52113-1 Paperback

Cambridge University Press has no responsibility for the persistence or accuracy of URLs for external or third-party Internet Web sites referred to in this publication and does not guarantee that any content on such Web sites is, or will remain, accurate or appropriate.

Climate Change:
A Wicked Problem

Complexity and Uncertainty at the Intersection of Science, Economics, Politics, and Human Behavior

FRANK P. INCROPERA

University of Notre Dame

CAMBRIDGE
UNIVERSITY PRESS

To Andrea, who would rather I had spent my time in other ways, and to our grandchildren, Wally, Michael, Mallory, Brandon, and Kyle. May they have the wisdom and resilience to meet the challenges that await them.

Contents

Foreword

Frank Incropera has done a masterful job of making the case for taking action now to reduce the impact of greenhouse gases attributable to human activity. He does it in a way that even the most ardent skeptics will have to acknowledge is a persuasive and balanced case that respects counterarguments but engages them thoroughly and convincingly.

I have had the pleasure of knowing Dr. Incropera for almost twenty years. I first met him when he agreed to leave Purdue University and move north to take the helm as Dean of the College of Engineering at the University of Notre Dame. As a member of the Advisory Council of the College, I watched as Dr. Incropera put in place the building blocks of a great engineering school. And while many of those building blocks – nanotechnology, bioengineering, advanced control systems, supersonic wind tunnels, and the like – were beyond the ken of this nuclear submarine engineer, I was delighted to discover that the dean and I shared a passion for all facets of the production and use of energy for the benefit of society.

Over the years, I and several other members of the Advisory Council were asked to speak to his classes about the practical aspects of the energy system in America and around the world. You see, Dr. Incropera is more than just a superb engineer. As is amply demonstrated in his book, he has a gift for taking the most complex of technical subjects and making them understandable to casual readers. As one who almost flunked quantum mechanics, I can attest that this is a real gift. Beyond the technical, Dr. Incropera is always interested in the whole range of considerations for any complex energy system. Obviously, a sustainable system must be founded on good science. That is just the starting point. Economics, the environment, history, politics, tax policy, and socioeconomic issues all have roles to play in understanding why we have the energy mix we have today. More importantly, he believes that unless you clearly understand the role each of these dimensions plays, it will be impossible to craft a solution to this "wicked problem."

As a teacher, Dr. Incropera's approach was to expose his students to multiple points of view to allow them to understand that there is no silver bullet. He would seek out executives from large energy companies, oil companies, venture capitalist firms, research organizations, and equipment manufacturers so his class could complement the sound technical base he was giving them with the real-world context in which they would have to operate. That philosophy is reflected in this book.

Let me illustrate his approach by using my own experience as a lecturer in his classes. I have been a senior executive at a company that was almost bankrupted building a nuclear power plant (GHG emissions free, I might add), a company that was one of the largest coal burners in the United States in a region where manufacturing demanded vast quantities of cheap electricity, and a company that has one of the cleanest generation footprints in the United States. Dr. Incropera had great intellectual curiosity about the driving forces behind the decisions that the companies, policy makers, and investors made in crafting the various regional energy mixes. And he relentlessly tried to extract from each of his guests the essence of what drove the decisions we made in hopes of imparting that knowledge to his students. In many ways, this book does what he did in his classes, and it works.

Dr. Incropera's intellectual curiosity is manifest throughout this book. In the early chapters, he provides that sound technical base he provided to his students. By cataloging a clear compilation of the growing body of climate change science, he lays out an indisputable case for action. What distinguishes his work is that he is refreshingly honest where the evidence is contradictory. He acknowledges that some theories are unprovable in our lifetimes or even in the lifetime of several generations. But he brings that common sense that we come to admire in our everyday lives that at some point the evidence points us in the direction of doing something. That sense of urgency is enhanced by his discussions of the consequences of inaction.

His research doesn't stop there. He proceeds to do a deep dive into all of the other factors that either got us to where we are or will make it difficult to get us out. His book thoughtfully covers potential mitigation strategies, public policy options, and the history of efforts to craft national and international solutions.

Several aspects of this book should make it required reading for every public policy student, legislator, regulator, and business decision maker in the energy space. By acknowledging the legitimate economic and reliability benefits of fossil fuels, he makes the dialogue less of a personal attack. His discussion in Chapter 9 of the complex human behavioral issues suggests he believes that people generally don't engage in irrational activities. Today's energy mix was developed to address real human needs. In fact, the reason that demand for fossil fuels continues to grow in developing nations is that they are effective at addressing those fundamental needs. And while acknowledging that goes a long way in getting people to focus on less harmful alternatives to accomplish the same goals, he doesn't let people off the hook for sloppy intellectual analysis.

The other insight in the book is that just because we can't fix all of the problems doesn't mean we should do nothing. I was involved in the efforts from 2008 through 2010 to craft comprehensive climate change legislation. The Waxman-Markey Bill passed the U.S. House in 2009 with a razor-thin margin. Its complexities, however, doomed it in the Senate, where a bipartisan effort to make the legislation more palatable stalled long enough to allow unrelated intervening developments to torpedo the initiative. Since then, Congress has been unable to accomplish anything on climate change. I believe part of that is the legacy of an overly complex Waxman-Markey Bill that no one wants to revisit. Dr. Incropera's solution would be to not give up, just because a coordinated national or global effort is highly unlikely. His recommendation is to work hard on what we can do now – conversion to natural gas, battery storage to advance solar and wind resources, more nuclear, and more electric vehicles. In one sense, that is exactly what is happening. Many states are working those edges with renewable mandates, energy efficiency programs, electric vehicle incentives, and the like. The federal government's regulators continue to make their mark through their regulatory push to enact more aggressive vehicle fuel efficiency standards and greenhouse gas rules that will likely keep the states as our incubators for thoughtful solutions.

This state of play leaves much national policy work undone. But Dr. Incropera thoughtfully addresses what policy paths would enhance existing greenhouse gas reduction strategies. For all who are taking his advice not to wait until it is too late, I would recommend that they read this book to strengthen their resolve and give them insights into the issues they must engage in order to succeed.

Anthony F. Earley Jr.
Chairman, CEO, and President
PG&E Corporation

Foreword

I have known Frank Incropera for more than thirty years. We both served as the chair of our respective mechanical engineering departments and then later in other leadership positions. In our respective roles, we often discussed both scientific and academic issues. While we did not always agree, I always found him to be an incredibly insightful individual who has a unique way in which he views the world and the environment around us.

As the author of what is perhaps the most highly regarded textbook on the subject of heat and mass transfer, he is an internationally renowned authority and someone whose opinion I greatly respect and friendship I greatly value. As the Clifford and Evelyn Brosey Professor of Mechanical Engineering at the University of Notre Dame, Professor Incropera continues to address important problems from a unique "problem-solving" perspective and has a deep knowledge and passion for trying to understand the fundamental basis of issues from an engineering perspective. In keeping with his past record, this is exactly what he has done in this latest publication, *Climate Change: A Wicked Problem*.

Who is this book written for? For anyone who is earnestly seeking to increase the breadth of his or her knowledge and understanding of one of the twenty-first century's most contentious issues. Professor Incropera expresses his views and opinions forthrightly throughout, yet his work addresses the issue from the perspective of an engineer, a scientist, an educator, and a pragmatist. As a fair-minded arbiter, he wades deep into the science, but his book also covers numerous related aspects of the issue from the perspective that most interested non-scientists will find accessible.

Professor Incropera dons several hats in his latest book; whether he's momentarily discussing climate change from the standpoint of business, history, ethics, politics, or economics, the scientist/engineer is always standing nearby, ready to lay down facts and figures as the foundation for potential solutions. Throughout, he changes hats seamlessly, looking at challenges through the eyes of different stakeholders, but always returning to his engineering and scientific base, where he has a standing that few have in this often fevered discussion. He has reached a conclusion, summarized succinctly at the end of Chapter 5: "With regard to scientific matters, the debate is all but over. The Earth will continue to warm due to human forcings, and manifestations of warming will become more pronounced." Whether the reader agrees or disagrees, there is much to recommend in the chapters that follow.

In Chapter 6, for example, Professor Incropera offers potential solutions in a discussion of "Mitigation, Adaptation, and Geoengineering." By Chapter 11, he posits a call to action that resonates strongly at the Georgia Institute of Technology, a world-class technological university where sustainability has long been part of our ethos. Ever the pragmatist, Professor Incropera makes a practical, compelling case for what can be done in a time when what perhaps should be done has too many hurdles to clear.

"Mitigation is not an option; it is a necessity, a cornerstone of efforts to deal with global warming," Professor Incropera writes, before going on to say, "Simply put, energy efficiency must be driven into every facet of human activity." The goal is achievable. Forward-thinking businesses have already incorporated that credo into their operations because they understand that it's wise from both economic and environmental perspectives.

At Georgia Tech, sustainability, of which energy efficiency is a vital part, is woven throughout our curriculum and drives a good deal of the research conducted by our faculty, staff, and students. From my personal perspective, I have to ask, "What happens if we are wrong and climate change is not the result of human activity? What is the resulting penalty for the actions proposed in this book?" I recognize that trade-offs, some controversial, will be necessary to bring Professor Incropera's recommendations to fruition, but the long-term payoffs will be significant economically and environmentally, and will accrue to the generations that follow.

Whether the reader is a climate change novice or an experienced hand, a scientist or an interested layman, *Climate Change: A Wicked Problem* provides invaluable information and insights with which to intelligently engage in shaping the future of this monumental challenge. It will challenge your thinking regardless of your starting point.

G.P. "Bud" Peterson, President
Georgia Institute of Technology
Atlanta, Georgia

Foreword

There is something innately human about caring for other humans and caring for our planet as a whole. Yet, on an issue such as climate change, which arguably could affect all humans and the planet, we are surrounded by controversies, conflicts, and debates in our social discourse. Why? It is because among the many issues that we all face today, climate change spans perhaps the most number of dimensions: scientific, economic, social, ethical, religious, and political. Confronted with this magnitude of complexity, we often grasp only a few facets that we can individually fathom and ignore the other dimensions, epitomizing the story of the blind men and an elephant. It is, therefore, not surprising to find people from different (and even similar) backgrounds talking past each other. Clarity is elusive. In such moments of confusion we need someone to simplify, distill, and connect the dots for us. This remarkable book by Frank Incropera does exactly that – it offers the most balanced, unbiased, and holistic view of this highly complex landscape.

The book starts with energy, which is the lifeblood of our modern life and our economy. When we flip a light switch, drive to our neighborhood grocery store, or do a Google search, we unwittingly receive the benefits of 250 years of industrial revolution that started with the steam engine running on coal to modern computers powered increasingly by natural gas. And this industrial revolution has been largely about how we sourced, distributed, and used energy. It was and continues to be predominantly based on fossil energy. Burning fossil fuels emits carbon dioxide in the atmosphere, which has been claimed to be the key culprit behind global warming.

Dr. Incropera explains with utmost clarity what we know, what is the uncertainty in our knowledge, and what we don't know from the scientific viewpoint. How does our climate work? How much carbon dioxide have we emitted so far? How long does it last in the atmosphere? Are there other sources of carbon dioxide? Are there other greenhouse gases and what influence do they have compared to carbon dioxide? What are the feedback mechanisms and what are the tipping points? How close are we to them? Dr. Incropera systematically helps the reader navigate through these difficult topics in the most uncomplicated and undemanding manner. He explains the difference between global warming and climate change, and why we are much more certain about the former and know much less about the latter. He addresses the issue of whether extreme weather events are connected to global warming and what the uncertainties are in our knowledge, what we can claim and what we cannot claim.

While it is important to understand the root causes, Dr. Incropera devotes a large section of the book to what we can do about it. A piece of this focuses on what we can do to mitigate global warming, how we ought to adapt to it, and what the risks are associated with geoengineering. But this is not just a technical issue, because technology is connected to economics; after all, energy is a commodity that all citizens in a modern economy use. It needs public policy because our choices of energy affect everyone else. Should there be financial incentives, should there be regulation, and if so, how much? Dr. Incropera even takes on the politics of climate change in the most nonpartisan manner, first delving into global politics and then into the debates in federal, state, and local governments. He extracts for the reader the underlying gist of the political debates and why the politics have come to the present state of affairs. Energy and climate invariably involve industry and the corporate world. Dr. Incropera describes how sound business policies can be mutually inclusive to environmental protection, and what the corporate world has (and has not) done so far to address climate change.

It is well known that the debate on climate change has dissenting opinions, and most books present either one side or the other. What is unique about Dr. Incropera's book is that in addition to the conventional wisdom on climate change, he devotes time to present the dissenting opinions as well. With unusual clarity and balance, he offers the arguments and distills them for the reader and dissects the knowns, unknowns, and uncertainties in these arguments, respecting the reader's prerogative to make up their own mind. There is a certain human element to this debate that becomes emotional and personal, which Dr. Incorpera presents with extraordinary clarity.

If the predictions of global warming turn out to be correct, it will affect the world as a whole: the close to 10 billion people, businesses, nations, and ecosystems. This makes it a human issue of extraordinary proportion, warranting critical understanding of topics such as ethics, social justice, and religion. The world is replete with diversity of thought and philosophies on such matters, with deep cultural and historical underpinnings. Unlike science, there is no right or wrong answer. It is, therefore, rare to find books that deal with both science and such aspects of human nature and connect the dots between them. Dr. Incropera dares to take this on, tapping into the innate humanity among us, and does exceptionally well in confronting this difficult juxtaposition of science and humanities. He characteristically ends the book with an action plan that not only touches on technology and public policy but also integrates personal and social values.

As we enter a period of intense global and national discourse on this important topic, the timing of this book could not be better. People worldwide need to read Dr. Incropera's book to map out for themselves the panoramic view of this multidimensional complex issue.

Arun Majumdar
Jay Precourt Professor, Stanford University
Former Vice President for Energy, Google
Founding Director, US Advanced Research Projects
Agency – Energy (ARPA-E)
Former U.S. Acting Undersecretary of Energy,
Department of Energy

Preface

To state the obvious, climate change is an environmental problem. But it has features that distinguish it from other well-publicized problems. For one, competing agents create uncertainty in linkages between cause and effect. With other environmental problems, deleterious effects are due solely to anthropogenic activities. There are no other agents. It is well known that automotive and power plant emissions such as carbon monoxide, sulfur dioxide, and particulates adversely affect human health and the environment. The same can be said of refrigerants that deplete stratospheric ozone. However, for climate change, anthropogenic agents associated with emissions of greenhouse gases such as carbon dioxide are superimposed on natural causes, and distinguishing between their effects is not a trivial matter.

Another distinguishing feature deals with time scales. For other environmental problems, adverse effects are near-term, if not immediate. In contrast, significant inertia is associated with the long residence time of atmospheric greenhouse gases and the slow rate at which equilibrium is achieved between the Earth's atmosphere and oceans. The effects of today's emissions are not felt today but over time, and it may be decades before they are unequivocally revealed. Lastly, unlike many environmental problems, climate change is not a regional or national problem. It is global.

The foregoing features make climate change a uniquely challenging environmental problem. But there's more. Four decades ago, Rittel and Webber (1973) introduced the notion of a *wicked problem*. Wicked problems are inherently societal problems, and in pluralistic societies with diverse interests and traditions there is seldom consensus on the nature of the problem, much less its solution. A wicked problem has many stakeholders, and any attempt at a solution has multiple consequences as its implications ripple across the many affected parties. Whether the solution is *right* or *wrong* is not judged by absolute or objective standards but by the interests and values of the stakeholders. Climate change, or more specifically anthropogenic climate change, is a prototypical wicked problem.

By integrating a comprehensive set of relevant factors, this book is intended to inform the climate change debate in ways that recognize existing uncertainties and tensions, as well as limitations to achieving timely and meaningful solutions. Drawing on the most recent literature, the underlying science is treated in depth. A strong case is made for the reality of anthropogenic climate change, but not without confronting the range of issues that remain uncertain and deconstructing opposing views. A pragmatic approach is taken on options for mitigating the effects of climate change. Strengths and weaknesses of the options are identified, including limitations that render some options problematic and measures that must be taken to facilitate substantive contributions by others. Assessments include economic considerations, targets of opportunity for innovation, and barriers imposed by special interests, politics, and human behavior.

Although climate change is a comparatively new aspect of a longer-standing relationship between energy and the environment, the two are inextricable. In the 1950s and 1960s there was growing concern for the effects of fossil and nuclear fuels on air, water, and soil pollution, but it was not until the 1990s that climate change began to receive serious attention. Since then, there has been growing recognition that the issue is central to any assessment of energy options. Linkages between energy and climate change are addressed throughout this book, with the view that multiple options must be pursued to reduce the use of fossil fuels, but that economic realities preclude an abrupt withdrawal. Like it or not, the world is awash in fossil fuels, and they will continue to be used, possibly throughout the century. The challenge is to reduce consumption in ways that do not impair the global economy while significantly reducing the threat of climate change.

Issues contributing to the complexity of climate change are treated in eleven chapters. Because global warming is strongly tied to energy utilization, Chapter 1 provides an introduction to the different forms and uses of energy, the importance of energy to economic development, and the impact of energy utilization on the environment. Energy, Economics, and Environment (three big Es) are joined at the hip. It is not good enough to judge an energy portfolio exclusively in terms of its environmental impact, nor is it sufficient to judge it solely on the basis of economic considerations. The need for integration, by its very nature, mandates compromise.

Chapters 2 through 5 consider the scientific origins of global warming. A 2011 poll of the American public revealed that only 44% believed in the scientific basis for anthropogenic warming, down from 75% in 2001 (Harris, 2011). The remaining 56% were evenly divided between nonbelievers and those who simply weren't sure. A more recent survey of twenty nations – developed and developing – revealed that the United States is not alone in questioning the anthropogenic origins of climate change (IPSOS, 2014). Respondents were asked the following question: *To what extent do you agree or disagree* (that) *the climate change we are currently seeing is a **natural** phenomenon that happens from time to time?* Among Americans, Indians, and Chinese, approximately 50% agreed with the statement, and even in Great Britain (48%) and Germany (39%) there was significant agreement. Recognition that there is in fact a problem begins with the underlying science. What can we say with certainty about natural and anthropogenic agents of warming? What don't we know?

Chapter 2 deals with *natural drivers* of climate change, features of the global energy balance, and aspects of radiation propagation in the Earth's atmosphere that can alter the balance. Chapter 3 deals with *anthropogenic drivers* of warming and climate change. Greenhouse gases are identified and characterized in terms of relevant parameters, and emission trajectories – past and future – are provided. Chapter 4 deals with the extent to which warming has occurred, the contribution of anthropogenic agents, and prospects for future warming. Uncertainties and contentious issues are examined, including the most recent hiatus in the temperature record. Chapter 5 considers the effects of warming and climate change on humankind and the natural world. From rising sea levels to extreme weather events, evidence points to significant effects on the built and natural environments, water resources, food production, and human health and security.

A cautionary note! This book provides a comprehensive assessment of global warming and climate change, one that addresses all relevant factors – scientific and otherwise. It is also written with readers of varied backgrounds – scientific and nonscientific – in mind. That said, Chapters 3 and 4 and portions of Chapter 5 may be tough sledding for those disinclined to deal with scientific details. If you fit that description, I encourage you to make the effort. It will provide you with the *state of the science* circa 2014, including results that deconstruct efforts to dismiss the science. If you wish, the chapter footnotes can be ignored without loss of key material and arguments.

Three lines of defense against the effects of climate change are assessed in Chapter 6. The first line is to *mitigate* the effects by reducing factors that contribute to warming. Chapter 6 provides a comprehensive and critical assessment of technology options. What can be done to reduce greenhouse gas emissions and stabilize atmospheric concentrations at acceptable levels by transitioning to carbon-free sources of energy? Attention is focused on important and problematic issues such as the transition from coal to natural gas for power generation, implementation of carbon capture and sequestration, the role of nuclear power, and the pace at which the use of renewables can be increased. However, economic factors loom large in choosing policy options for decarbonizing the world's energy portfolio, and it is likely that the adverse effects of climate change will not be sufficiently dampened by mitigation measures. Enter the second line of defense – *adaptation* – which involves measures taken to increase the *resilience* of humans, their artifacts, and the environment to climate change. But such measures may also prove insufficient if greenhouse gas emissions continue on their current trajectory. That leaves *geoengineering*, measures of last resort designed to alter the climate system in ways that negate the effects of greenhouse gas emissions.

By focusing on technological options for dealing with climate change, Chapter 6 plays a central role in the book's narrative. That said, even as an engineer, I am not sanguine about prospects for technology doing it alone. It is a necessary part of the solution, but it may not be sufficient. Governments – national, state, and local – must play a role, and public policy options are discussed in Chapter 7. Other chapters of the book, particularly 1, 10, and 11, consider the need for a sea change in cultural norms.

Since climate change is a global problem, a critical question is whether consensus and cooperation can be reached on appropriate solutions. Achieving cooperation on a global scale is the most important requirement for dealing with the problem. Chapter 8 deals with the politics of climate change. From the United Nations to the governing bodies of the world's nations to legislative groups within states of each nation and to cities and towns within the states, politics strongly influences what is, or isn't, done to address the problem. In effect, Chapter 8 is a history lesson on the politics of climate change and, moving forward, a primer on difficulties associated with achieving consensus.

Discussion of climate politics continues in Chapter 9 with consideration of the strategies used by special interests in the United States to thwart mitigation measures. But barriers to dealing with climate change are not entirely technical, economic, or political, and Chapter 9 also considers cultural conditions that influence decisions. In the face of scientific complexity and uncertainty, what cultural and behavioral factors cause some to dismiss the problem and others to push for solutions?

To this point, climate change is treated as a juxtaposition of the three _Es_ (energy, economy and the environment). But there remains one more _E_, namely the _ethics_ of climate change. If there are ethical implications to ignoring warming and climate change, what are the moral frameworks and religious traditions that inform these implications? In Chapter 10 these dimensions are explored in search of moral guidance. Although ethical theories are not without ambiguity, Aristotelian (virtue) ethics provides moral clarity consistent with a significant body of religious doctrine.

In a final analysis, what can be said about the problem of climate change and possible solutions? Chapter 11 makes it clear that the problem is real, serious, and must be addressed. It calls for an aggressive approach to reducing greenhouse gas emissions by amplifying public education and accelerating implementation of energy efficiency and conservation measures along with the use of carbon-free sources of energy and lifestyle changes. However, recognizing economic, social, and political realities, the reduction of atmospheric greenhouse gas concentrations to acceptable levels is highly unlikely. Adaptation must therefore combine with mitigation as integral pieces of the world's climate change strategy.

In recent years, I have had many opportunities to speak on climate change, often to nonscientific or technical audiences. I have also engaged in less formal conversations with friends and colleagues. These interactions have left me with several perceptions: (1) there is general (public) awareness of climate change; (2) to the extent that they exist, opinions on whether it is or is not a problem are often strongly held; and (3) many people, well educated or not, have a limited knowledge of the matter, despite many articles and books on the subject. It is these perceptions that provided the impetus for this book.

The book has several distinguishing features, beginning with its comprehensive and critical assessment of all issues germane to the climate change debate – the underlying science, technology options for mitigating the effects of climate change, economic ramifications of the options, cultural and behavioral issues, the influence of special interests and public policy, geopolitical issues, and the ethical implications of climate change. Contents of the book are true to the title labeling climate change as a _wicked problem_. The book also reflects what I bring to the table as an engineer. As well as extensive involvement with energy sciences and technologies in education and research, I have had many opportunities to engage with thought leaders in energy sectors of the business community, including electric utilities; producers of fossil, nuclear, and renewable energy; and the transportation sector. These industries play a significant role in shaping energy policy – in the United States and elsewhere – and determining what actually gets done to reduce emissions. Their views on climate change span a broad spectrum, but a common thread is their sensitivity to costs and returns on capital investments. This sensitivity has sharpened my perspective on differences between what should be done and what can be done and has no doubt contributed to pragmatic elements of the book.

By integrating the many facets of climate change that inform debate on the subject, my goal is to provide the reader a foundation for shaping personal views, assessing new findings, and contributing to future discourse. The science will continue to evolve, as new data emerge and more is learned about energy exchange between and within the Earth's oceans, land, and atmosphere. Technologies that lend themselves to mitigating or dealing with the effects of climate change will advance, affecting the economic viability of specific measures. New knowledge and technologies will influence our perceptions of risk and our sense of urgency. Actions taken or not taken will continue to be influenced by the shifting sands of public opinion and geopolitics. *Your engagement with the issues matters.*

Acknowledgments

The broad scope of this book presented several challenges. Although my background as an engineer specializing in energy sciences and technologies equipped me to deal with the scientific and technological aspects of climate change, could I do justice to other relevant issues, from economics and public policy to ethics and human behavior? Could all issues be integrated in a readable and compelling manner? To whatever extent I've succeeded, I'm indebted to the assistance of others: To Steve Batill and Bill Colonis, who read early versions of the manuscript and identified weaknesses. To Marjorie Pfeill, who read what I thought was a satisfactory version of the chapter on ethics and suggested ways to make it better. To Anthony F. Earley Jr., Michael O'Sullivan, and Pat Eilers, who broadened and deepened my understanding of corporate perspectives on energy options, particularly economic implications. To my colleagues in Notre Dame's Mendoza College of Business – especially Leo Burke, Tom Frecka, and Jessica McManus – who provided many opportunities for me to engage them and their students in discussions of energy and climate change. To the book's editor, Matt Lloyd, whose involvement during the final stages of writing added value to the end product. And, to the many students – especially Felipe Witchger and Michael Della Penna – who have taken my course on Energy Technology and Policy and have stretched my thinking on the subject.

Abbreviations

ACC	Abrupt climate change
ACESA	American Clean Energy and Security Act
AIS	Antarctic ice sheet
ALEC	American Legislative Exchange Council
AMOC	Atlantic Meridional Overturning Circulation
AOGCM	Atmosphere-Ocean General Circulation Model
API	American Petroleum Institute
AR 3,4,5	Third, Fourth, and Fifth Assessment Reports of the IPCC
CAA	U.S. Clean Air Act
CAFE	Corporate average fuel economy standards
CC	Carbon content of a fuel (mass of carbon per mass of fuel, kg-C/kg-fuel)
CCOC	Cost of carbon capture ($/t-$CO_2$)
CCS	Carbon capture and sequestration
CCSP	Climate Change Science Program
CDM	Clean development mechanism of the Kyoto Protocol
CDR	A form of geoengineering involving carbon dioxide removal from the atmosphere to reduce its contribution to global warming
CEI	Competitive Enterprise Institute
CFC	Chlorofluorocarbon
CI	Carbon intensity of a fuel (carbon emissions per energy consumption, kg-C/GJ-fuel or kg-CO_2/MWh_e)
CIA	U.S. Central Intelligence Agency
COEJL	Coalition on the Environment and Jewish Life
COP	(United Nations) Conference of Parties
CPP	EPA Clean Power Plan

CRU	Climate Research Unit of the University of East Anglia
DOD	U.S. Department of Defense
DOT	U.S. Department of Transportation
EAIS	East Antarctic Ice Sheet
EC	European Commission
EI	Energy intensity (ratio of a nation's energy consumption to its GDP, MJ/$)
EIA	U.S. Energy Information Administration
ENSO	El Niño Southern Oscillation
EOR	Enhanced oil recovery
EOS	Earth orbiting system
EP	European Parliament
EPA	U.S. Environmental Protection Agency
ETS	EU emissions trading scheme
EU	European Union
EV	Electric vehicle
EWE	Extreme weather event
FCCC	(United Nations) Framework Convention on Climate Change
FEMA	U.S. Federal Emergency Management Administration
FIT	Feed-in tariff
GAO	U.S. Government Accountability Office
GCC	Global climate change; Global Climate Coalition
GDP	Gross domestic product
GHG	Greenhouse gas
GIS	Greenland ice sheet
GISS	NASA Goddard Institute of Space Sciences
GWP	Gross world product or global warming potential
HadCRU	UK Hadley Centre and Climate Research Unit of the University of East Anglia
HCFC	Hydrochlorofluorocarbon
HDI	Human development index
HFC	Hydrofluorocarbon
HHV	Higher heating value (energy per unit mass, kJ/kg) of a fuel
HV	Heating value (energy per unit mass, kJ/kg) of a fuel
ICE	Internal combustion engine
IEA	International Energy Agency
IGCC	Integrated gasification and combined cycle
IMF	International Monetary Fund
IPCC	Intergovernmental Panel on Climate Change

ITC	Investment tax credit
KP	Kyoto Protocol
LCA	Life cycle assessment (of GHG emissions)
LCOE	Levelized cost of supplying electricity to the grid
LDV	Light-duty vehicle (car, SUV, light truck, or minivan)
LHV	Lower heating value (energy per unit mass, kJ/kg) of a fuel
LNG	Liquefied natural gas
NAE	National Association of Evangelicals; U.S. National Academy of Engineering
NASA	U.S. National Aeronautics and Space Administration
NCAT45	Category 4 and 5 hurricanes and typhoons
NFIP	U.S. National Flood Insurance Protection Program
NGCC	Natural gas combined cycle power plant
NIMBY	Not in my backyard
NOAA	U.S. National Oceanic and Atmospheric Administration
NOX	Oxides of nitrogen (also NO_x)
NPCC	New York Panel on Climate Change
NRC	U.S. National Research Council
NSF	U.S. National Science Foundation
NSPS	New source performance standards of the EPA
NYT	*New York Times*
OECD	Organisation for Economic Co-operation and Development (thirty-four developed nations, largely from the EU and North America, as well as Australia, Chile, Iceland, Israel, Japan, Korea, New Zealand, Switzerland, and Turkey)
OFC	Oxy-fuel combustion
OPEC	Organization of Petroleum Exporting Countries
PC	Pulverized coal
PFC	Perfluorinated chemical
PHEV	Plug-in hybrid electric vehicle
POC	Products of combustion
ppb	Concentration of a species in a mixture expressed in parts per billion by volume (ratio of number of molecules to molecules of dry air)
ppm	Concentration of a species in a mixture expressed in parts per million by volume (ratio of number of molecules to molecules of dry air)
PPP	Purchasing power parity

ppt	Concentration of a species in a mixture expressed in parts per trillion by volume (ratio of number of molecules to molecules of dry air)
PTC	Production tax credit
PV	Solar photovoltaic power
RCP	Representative Concentration Pathway
REC	Renewable energy credit
REDD	Reducing emissions from deforestation and degradation in developing nations
RF	Radiative forcing (change in the net downward radiative flux at the tropopause due to changes in a parameter that affects the transfer of short-wave solar or long-wave terrestrial radiation to or from the Earth)
RGGI	Regional Greenhouse Gas Initiative
RPS	Renewable portfolio standard
SCPC	Supercritical pulverized coal (power plant)
SRES	UN Special Report on Emission Scenarios
SRM	A form of geoengineering involving solar radiation management by increasing reflection to counter the warming effect of GHG emissions
SST	Sea surface temperature
SUV	Sport utility vehicle
TCEP	Texas Clean Energy Project
Tcm	Trillion cubic meters (at standard temperature and pressure)
THC	Thermohaline circulation
TOA	Top (outer edge) of the Earth's atmosphere
TPES	Total primary energy supply
TSI	Total solar irradiance (measured in W/m^2)
TVA	Tennessee Valley Authority
UN	United Nations
UNEP	United Nations Environmental Programme
USCCB	United States Conference of Catholic Bishops
USCPC	Ultra-supercritical pulverized coal (power plant)
WAIS	West Antarctic ice sheet
WDPM	Papal World Day of Peace Message
WMO	World Meteorological Organization
WSJ	*Wall Street Journal*

CHAPTER 1

Energy, economics, and climate change

We can't engage in a serious examination of climate change without considering its strong ties to energy. More than any other factor, anthropogenic contributions to climate depend on how energy is produced and used.

Over his illustrious career, Richard Smalley (1943–2005), a Nobel Laureate and pioneer in the field of nanoscience and technology, was invited to give many lectures on his work. However, in the last few years of his life, he felt compelled to use the lectures as a vehicle for sharing his concerns about the world's energy future. In one of his slides he presented his views on humanity's top ten problems of the next fifty years. His list included food, water, the environment, poverty, war, disease, education, democracy, and population. While we might attach different weights to the significance of each concern, we would probably agree that all are to be taken seriously. However, for Smalley, there was no equivocation on what belonged at the top of the list. Meeting the world's energy needs was paramount and linked, to varying degree, with the other nine.

1.1 Energy: an indispensable resource

It would be difficult to overstate the importance of energy to the well-being of humankind. It is the resource that sustains all life and economic activity. It enables the production and distribution of all manner of goods and services, as well as human mobility on the ground and in the air. It is absolutely essential to achieving an acceptable standard of living, and in the words of Paul Roberts (2004, p. 6), "Access to energy has emerged as the overwhelming imperative of the twenty-first century."

While preindustrial societies functioned entirely on energy derived from the Sun, the Industrial Revolution marked a transition to the use of fossil fuels and by the mid-twentieth century to nuclear energy. It is difficult to appreciate the enormity of today's global energy supply chain. In 2013, humankind consumed approximately 505 quadrillion (505,000 trillion) British thermal units (Btu) of energy, or simply 505 quads (BP, 2014a).[1] The amount is staggering, and trillions of dollars are spent annually to produce and distribute this energy. That said, about one-third of the world's population still live in or near poverty, with many lacking the energy required to meet the most basic human needs. Movement of people from poverty to a decent standard of living, along with a growing world population, guarantees continued growth in the demand for energy. But generally energy production and consumption do not occur without adverse environmental effects, and some forms of energy are more benign than others.

1.2 Energy 101: a taxonomy

Forms of energy are diverse, and any taxonomy should include a distinction between primary sources of energy and energy carriers. Primary sources can be characterized as renewable or nonrenewable and as carbon-free or carbonaceous. A primary source of energy is simply one that exists naturally. In contrast, an energy carrier does not exist in a natural form and can only be produced by converting the energy associated with a primary source.

There are two major energy carriers: electricity and hydrogen. Electricity has been vital to human advancement for more than a century and will become even more important in the years ahead. Although hydrogen is, at best, a bit player in today's energy supply chain, it could one day play a more prominent role. But for human consumption, electricity and hydrogen are not inherent gifts of nature. Some artifact of human innovation must be used to convert a primary energy source to electrical energy or hydrogen.

Primary sources of energy are highlighted in Figure 1.1. Once used, a nonrenewable source of energy is not replenished. It is simply depleted. One can think of these sources as stored within the Earth and consisting of fossil and nuclear fuels. There is only so much, and when a nonrenewable resource is used, it reduces the amount left in storage. Continued withdrawal leads to depletion or to a point where reserves are so diminished that further withdrawal is impractical.

Fossil fuels (coal, natural gas, and petroleum) consist of hydrocarbon molecules, and the chemical energy associated with the bonds between carbon and hydrogen atoms can be released by chemical reactions,

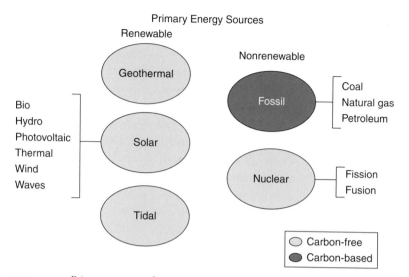

FIGURE 1.1. Primary sources of energy: renewable and nonrenewable, carbon-free and carbonaceous.

typically by burning the fuels. The chemical energy of the fuel is converted to thermal energy of the high-temperature products of combustion, which can then be used for space heating in homes and commercial buildings, for process heat in factories, to power automobiles and aircraft, and to produce electricity. Fossil fuels have several desirable attributes. They have large energy densities (energy content per unit mass or volume); they are abundant; and they can be produced and supplied to the consumer at comparatively low costs. Not surprisingly, they are widely used, and the global infrastructure and capital investments associated with producing, distributing, and using the fuels are enormous. Consider the vast array of oil and gas wells; coal mines; supertankers, pipelines, and freight trains; space and process heating systems; automobiles, trucks, boats, and aircraft; and electric power plants.

Fossil fuels have sustained economic growth since the eighteenth century and will remain important well into the twenty-first century. But there is a downside to burning the fuels. The products of combustion, which are discharged to the atmosphere, include constituents that contribute to atmospheric pollution and global warming. The challenge is one of using the fuels in an environmentally benign fashion.

Nuclear energy is highly concentrated and is also nonrenewable. It can be released by means of a fission or fusion reaction. Fission reactions entail splitting heavy atomic nuclei such as uranium-235 into fragments, thereby releasing large amounts of energy that can be harnessed for useful purposes. Fission reactors are widely used to produce electricity, but not without environmental issues related to reactor safety and disposal of radioactive wastes produced by the fission process.

A fusion reaction combines lighter nuclei to form a heavier nucleus, again with the release of a large amount of energy. Fusion is an attractive target for two reasons. Light nuclei required to fuel the reaction are abundant in the world's oceans, providing a nearly inexhaustible supply, and products of the reaction are benign. However, despite the billions of dollars that have been spent – and continue to be spent – on attempts to contain a fusion reaction, commercialization of the process is far from imminent. Even if the reaction can be sustained, the engineering problems associated with developing a viable power production system would be enormous and costly. Fusion technology is in some sense a Holy Grail, but one that is not likely to be achieved in this century, if ever.

Although fossil and nuclear fuels are both nonrenewable, they differ in one important way. Because fossil fuels are carbonaceous, their products of combustion include carbon dioxide, the largest contributor to global warming. In contrast, nuclear fuels are carbon-free.

Renewable forms of energy are also carbon-free and for all practical purposes can never be depleted. Geothermal energy is derived from energy that was stored within the Earth during its formation and energy that is continuously released by nuclear (fission) reactions. High temperatures within the Earth's core and mantle provide the driving potential for conduction of thermal energy to the Earth's crust, where pressurized steam or hot water are generated at depths accessible to drilling from the Earth's surface. Once accessed, thermal energy associated with the steam or hot water can be used for space and process heat or for power generation. Although geothermal energy is being harnessed throughout the world, its contribution to global energy consumption is well below 1%.

Solar energy is far and away the most abundant source of renewable energy. The rate at which the Earth intercepts solar radiation, commonly termed *insolation*, is enormous, amounting to approximately 165,000 terawatts (165,000 TW), or 11,000 times the average rate at which humans consumed energy from all sources in 2013. Through absorption by the Earth's land and oceans, solar radiation maintains temperatures conducive to plant and animal life. Through the process of photosynthesis, solar energy is converted to chemical energy in the form of biomass, which propagates through the food chain and can also be used as a biofuel.

Solar radiation is also responsible for temperature variations on land and sea. These variations sustain the Earth's hydrologic cycle and atmospheric winds, which can be tapped as sources of hydro, wind, and wave energy. Solar energy can also be converted directly to electricity by means of photovoltaic technologies and to space and process heat or indirectly to electricity by means of solar thermal technologies. Solar energy is the antithesis of fossil fuels. While fossil fuels are nonrenewable, concentrated (have a large energy density), and, to varying degrees, environmentally detrimental, solar energy is renewable, diffuse, and environmentally benign.

The distinction between renewable and nonrenewable forms of energy has an important bearing on the future of the human species. At some point, nonrenewable sources of energy will be depleted, and human innovation will have to achieve a sustainable energy future that relies exclusively on renewable sources.

1.3 Energy and economic growth

For centuries there has been a steady, seemingly inexorable increase in global energy consumption, and it is a sine qua non that economic growth is accompanied by increased energy consumption. Since the Industrial Revolution, living standards have been shaped by cheap and abundant energy. A huge global infrastructure has been created for the production and use of energy, enabled by capital markets, large corporations, and an abiding faith that it will have no end. Abundant and cheap energy has enabled globalization and has elevated expectations for higher living standards across the world.

In recent decades a nominal annual increase of 3% in gross world product (GWP) has been accompanied by an annual increase of about 2% in global energy consumption. The linkage between a nation's energy consumption and its economic activity is highlighted in Table 1.1 for representative nations at different stages of economic development. The first two columns of data provide energy consumption and gross domestic product (GDP) per capita, where the unit of energy (a gigajoule) is one billion joules and GDP is standardized on the basis of purchasing power parity (PPP).[2]

Several factors influence the relationship between a nation's economic output and its energy consumption. An economy relying heavily on manufacturing uses more energy than one based largely on services, while some nations simply use energy more efficiently than others. The third column of the table provides one measure of how effectively a nation uses

TABLE 1.1. *Circa 2011–12 primary energy consumption and GDP per capita and energy intensity for selected nations*

	Energy consumption (GJ/Person)[a]	GDP (PPP2012) (U.S.$/Person)[b]	Energy intensity (GJ/U.S.$1,000)
Group I			
Australia	271.5	43,300	6.27
Canada	418.4	43,400	9.64
Russia	242.6	18,000	13.5
Saudi Arabia	343.4	31,800	10.8
United States	330.0	50,700	6.17
Group II			
France	174.2	36,100	4.83
Germany	169.4	39,700	4.27
Japan	172.6	36,900	4.68
Switzerland	168.7	46,200	3.65
United Kingdom	143.3	37,500	3.82
Group III			
Brazil	62.2	12,100	5.14
Chile	84.7	18,700	4.53
China	86.5	9,300	9.30
India	21.0	3,900	5.38
Mexico	72.5	15,600	4.65
Group IV			
Bangladesh	6.8	2,100	3.24
Ethiopia	1.7	1,200	1.42
Haiti	3.2	1,300	2.46
Nigeria	4.7	2,800	1.68

[a] Data for 2011 from EIA (2013c).
[b] Data for 2012 from CIA (2013).

its energy. Termed energy intensity (EI), it provides the ratio of a nation's energy consumption to its GDP. A logical national goal would be to minimize the energy required to maintain a strong economy.

The selected nations are separated into four categories. Group I includes nations with large energy consumption and moderate to large GDP, while Group II embodies nations of comparable GDP and much lower energy consumption. With respect to economic output, Group II nations use energy more efficiently than those of Group I. Citizens of Group II enjoy living standards and a quality of life as good as or better than those of Group I while consuming much less energy per unit of economic output, and for Germany and Japan doing so with a large manufacturing base.

Group III consists of developing nations that have undergone rapid economic growth over the last two decades and, moving forward, are likely to grow more rapidly than the developed nations represented by Groups I and II. Two nations (Chile and Mexico) have energy intensities comparable to those of Group II; two nations (Brazil and India) lie between I and II; while China, a manufacturing juggernaut for which improving energy efficiency is a work in progress, is aligned with Group I. In contrast, Group IV represents some of the world's poorest nations for which both energy consumption and GDP are low.

Comprised largely of African, South American, and Asian nations, Groups III and IV are of special interest because they have the greatest potential for economic growth. But, as they grow, what trajectory of energy consumption will they follow? Will it be more closely aligned with Group I or II? Consider that from 2008 through 2013, primary energy consumption decreased by about 2.4% in the thirty-four developed nations of the Organisation for Economic Co-operation and Development (OECD),[3] while increasing by 24.2% in the largely developing non-OECD nations (BP, 2014a). In the context of climate change, why are these numbers important?

Economic activity is inextricably tied to energy consumption. But, when fossil fuels are burned, their carbon content is released to the atmosphere as carbon dioxide, where it becomes a major contributor to global warming. Although fossil fuels comprise only one of the five energy categories of Figure 1.1, they contribute disproportionately to meeting the world's needs for primary energy.

In 2011, fossil fuels provided about 82% of the world's total primary energy supply (TPES), with the remainder provided by nuclear (5%) and renewable (13%) energy (IEA, 2013a). Of the renewables, most of the energy was supplied by bio/hydro sources and about 1% from a combination of solar, wind, and geothermal energy. From 2011 through 2013, fossil fuels also accounted for about 82% of U.S. energy consumption, with nuclear and renewable energy each providing about 9% (EIA, 2014a). Fossil fuels also contribute significantly to generating the world's electricity, providing 68% of the primary energy used to produce 23,100 terawatt-hours (TWh) in 2013 (BP, 2014a). From 2011 through 2013, fossil fuels contributed 67% of the primary energy used to generate about 4,050 TWh per year in the United States (EIA, 2014a).

The bottom line is that the primary energy sustaining the global economy involves huge amounts of fossil fuels. Even with annual growth of 30% or more in the production of carbon-free solar and wind energy, growth is from a small base, and it will be many years before these sources can provide energy comparable to the scales associated with fossil fuels. By all estimates, global energy demand will continue to grow, and the demand for fossil fuels shows no sign of abating. Annual growth in GWP and global energy consumption – occurring mostly in developing, non-OECD nations – is projected to be 3.6% and 1.5%, respectively, through 2040, when demand for energy is expected to reach 820 quads with more than 75% supplied by fossil fuels (EIA, 2013b).

1.4 Energy, greenhouse gases, and the environment

Largely through their impact on air, water and/or land pollution, energy production and utilization are inextricably linked to the natural environment. If an energy source is to be used responsibly, harmful environmental consequences must be identified and reduced to acceptable levels.

Environmental concerns are not new, and in the second half of the twentieth century, several large movements were launched to curb environmental degradation. In the 1950s and 1960s, the focus was on moderating the use of harmful herbicides and pesticides in agriculture and on curbing water pollution. The clean air initiatives of the 1970s were directed at reducing emissions of pollutants such as oxides of nitrogen and sulfur from automobiles, aircraft and power plants. In the 1970s and 1980s, concerns for radioactive wastes produced by nuclear power plants, along with accidents at the Three-Mile Island (USA) and Chernobyl (USSR) plants, put a damper on further development of nuclear power in many nations. In the late 1980s and early 1990s, concerns for the depletion of stratospheric ozone resulted in the replacement of chlorofluorocarbons (CFCs) by more benign fluids in refrigeration and air-conditioning systems.

By the turn of the millennium, however, few environmental issues were drawing more attention than climate change, or more specifically, climate change due to human (anthropogenic) activity. Grade school children and their parents were learning about the greenhouse effect, greenhouse gases (GHGs) and global warming, while scientists and politicians across the globe were debating the gravity of the problem. But there's an important distinction to be made between anthropogenic climate change and other environmental issues. Because GHGs do not pose an immediate threat to human health and welfare, it is more difficult to make the case for mitigation.

Harmful effects of discharging GHGs into the atmosphere are manifested slowly, and absent measures to reduce emissions, serious consequences would increasingly be felt by future generations. This tendency to defer mitigation measures finds sustenance in two premises. Because the measures have associated costs, why spend today's dollars to deal with a problem that is not at hand? We can deal with the problem if and when we have to. And, if there is a problem, it's global in nature, since all nations share the same atmosphere. Why should one nation or a group of nations step up and bear the costs of reducing GHG emissions if all nations aren't willing to do so? Today, there are those who believe that global warming and climate change represent serious threats to future generations, while others are inclined to discount their significance. What is it about this issue that we can claim with certainty?

We know with absolute certainty that some atmospheric gases absorb radiant energy emitted by the Earth's surface, energy that would otherwise be transmitted directly to outer space. By trapping this energy, the gases act much like the glass cover of a greenhouse, which transmits solar radiation into the greenhouse but restricts outflow of radiation emitted by contents of the greenhouse.

Greenhouse gases exist naturally in the atmosphere, largely in the form of carbon dioxide (CO_2) and water vapor $(H_2O)_{vapor}$. Without them the Earth would be a colder and less hospitable planet. But what happens when human activities release greenhouse gases to the atmosphere at a rate that exceeds the ability of the Earth's ecosystems to remove them? Few would dispute the contention that, as the concentrations of greenhouse gases increase, more of the Earth's emitted radiation is absorbed by the atmosphere and the Earth's temperature must increase. But by how much, and is the effect significant or negligible relative to changes driven by natural agents? If anthropogenic agents are significant, what effect would global warming have on the Earth's environmental, economic and social systems?

Those who express concern for anthropogenic climate change point to the steady increase in atmospheric GHG concentrations since the middle of the eighteenth century, which marks the onset of the Industrial Revolution. Although greenhouse gases come in many forms, such as methane (CH_4), nitrous oxide (N_2O) and numerous industrial chemicals, carbon dioxide receives the greatest attention. From 1760 to 2014, atmospheric concentrations of CO_2 increased from approximately 280 to 400 parts per million (ppm) by volume, largely due to the burning of fossil fuels and secondarily to deforestation and other land-use changes.

There are two onerous implications of using fossil fuels: (1) the store of highly concentrated and valuable forms of energy is irreversibly reduced, and (2) the natural environment is degraded by their use. In both cases, there is a depletion of natural capital, in one case manifested by loss of the fuels themselves and in the other by degradation of the environment into which their waste products are discarded. According to Daly (1996, p. 49), "The evolution of the human economy has passed from an era in which man-made capital was the limiting factor in economic development to an era in which remaining natural capital is the limiting factor." It is debatable whether that transition has already occurred, and whether it is imminent will depend a good deal on mankind's ability to innovate. Nevertheless, the point is well taken. There is an upper limit to the use of natural capital that is determined by the "regenerative or absorptive capacity" of the environment, a limit or "anthropogenic optimum" for which the "marginal benefit to human beings of additional man-made capital is just equal to the marginal cost to human beings of sacrificed natural capital." In the context of global warming, what is the absorptive capacity of the Earth's atmosphere – an important constituent of natural capital – for greenhouse gases?

Although the use of fossil fuels was virtually nonexistent before 1760, it has since grown exponentially, becoming a cornerstone of human economic activity. In the preceding section we noted that fossil fuels account for more than 80% of global energy consumption. In 2013, CO_2 emissions from fossil fuels reached a new high of 35.1 billion metric tons (35.1 $Gt\text{-}CO_2$), with contributions of approximately 43%, 37% and 20% from the combustion of coal, oil and natural gas, respectively (BP, 2014a). With rapid economic and population growth in developing nations more than offsetting recessionary effects in developed nations, emissions increased at an average annual rate of 2.1% from 2008 to 2013. And with sustained economic and population growth in developing regions of the world, as well as continued high demand in developed nations, fossil fuel consumption and CO_2 emissions will continue their upward trajectory for the foreseeable future.

A critical question concerns the extent to which the concentration of atmospheric carbon dioxide (and other GHGs) can increase and still remain at levels that allow for human adaptation to potential climate change. One view is that a CO_2 concentration of 450 ppm represents a threshold for which the increase in the global mean surface temperature above preindustrial levels would be limited to two degrees Celsius (2°C) and adaptation to the effects of climate change would be manageable. To remain within this threshold, the world would have to act immediately and decisively to reduce emissions. Yet, emissions continue to rise, and with business-as-usual the atmospheric CO_2 concentration would increase by more than 2 ppm per year, exceeding the 450 ppm threshold by 2040 and putting the Earth's average surface temperature on a trajectory to rise well above 2°C.

1.5 Energy, economy, the environment, and sustainability

The term "sustainability," or more precisely "sustainable development," is inextricably linked to the use of natural resources and environmental conservation. It was first used by the Brundtland Commission and defined as development that "meets the needs of the present without compromising the ability of future generations to meet their own needs" (WCED, 1987). At face value the term implies management of resources in ways that ensure availability for future generations. But, there's more.

As noted by Allenby (2003), the Brundtland Commission recognized the need to address global inequities by encouraging economic growth in poor nations and by moderating consumption in wealthy nations. Specifically, its report states that "meeting essential needs requires not only a new era of economic growth for nations in which the majority are poor, but an assurance that those poor get their fair share of the resources required to sustain that growth." It goes on to say that "[s]ustainable global development requires that those who are more affluent adopt lifestyles within the planet's ecological means."

The foregoing statements are likely to chagrin libertarians and devotees of free markets. But their ethical and social implications are clear. The statements imply an ethics of space or geography, that is, one that transcends national boundaries and is aligned with principles of social justice. The definition itself also implies an ethics of time, one that transcends generations. The ethics of space challenges the global allocation of natural resources. The ethics of time questions the morality of maximizing resource utilization to meet current needs and wants, leaving future generations to make do with less.

Challenges to addressing the goal of sustainable development are economic and environmental, as well as technological, political and cultural. At opposite poles are the orthodox economist and the environmentalist (Wilson, 2002). Orthodox economic theory associates a nation's standard of living with its economic output and focuses on models of rational choice that enhance growth. The more we consume, the higher our standard of living. Bigger vehicles, larger homes heated and cooled to maintain "ideal" levels of comfort, large entertainment centers, and more are all intended to increase our standard of living. Success is measured in terms of GDP, and if problems associated with depletion of natural resources are acknowledged, it is often accompanied by the belief that human ingenuity (technology) will find ways of dealing with the problems.

In contrast, the environmentalist – citing examples of growth occurring at the expense of significant and irreparable environmental damage, social inequalities and disregard for future generations – looks at the long-term consequences and argues that a proper accounting of economic growth should include its impact on the environment and human development. Simply tracking GDP has limitations if, by extension, it is used as the sole measure of living standards or, more to the point, the quality of life. Along with protection of the environment, other factors need to be considered, such as health care, education, crime, personal freedom and income equality.

Daly (1996, p. 1) views sustainable development as a shift in how development and growth are viewed relative to the natural world. Defining development as qualitative improvement without exceeding the carrying capacity of the environment and growth as a quantitative increase in the value of goods and services, sustainability "involves replacing the economic norm of quantitative expansion (growth) with that of qualitative improvement (development) as the path of future progress." It is movement from an economics of *bigger* to one of *better*. The economics of growth focuses on quantitative measures such as increasing population and per capita consumption and reducing the cost of goods and services by commoditizing nonrenewable resources and externalizing costs associated with impairment of human health and habitat. The economics of development focuses on improving the quality of people's lives through health care, education, environmental protection, the efficiency of resource utilization and the preservation of natural capital. In the words of Hawken (2005, pp. 140, 167), the economics of development recognizes "the true, full costs of doing business and reassigning them to the marketplace, where they belong."

Daly (1996, p. 7) sees incongruity in sustainability and growth and labels the term "sustainable growth" an oxymoron. At some point economic growth must defer to sustainable development through stabilization of global population and per capita consumption consistent with the Earth's regenerative capacity. Allenby (2003) recognizes that integrating ethical matters with sustainability creates certain tensions, in one case between the concentration and distribution of wealth and in another between economic growth and environmental protection. Are there balances or compromises that would mitigate the tensions? If so, what are they and how could they be implemented? In an essay on sustainability, a student in one of my classes underscored the need for such balances by noting that "economic growth without sustainability is little more than profiteering, and sustainability without economic opportunity and development is a pathway to unfulfilled human potential and development."

Although GDP measures the value of a nation's goods and services, it cannot fully capture the quality of life of its citizens. In an effort to obtain a better measure of the quality of life, the United Nations established a Human Development Index (HDI), the value of which is based on a nation's average life expectancy, health care and educational systems, income equality and standard of living as determined by its GDP (UNDP, 2014). The index varies from zero to one, with circa-2013 results for four categories of human development ranging from very high (HDI > 0.8) to high (0.70 < HDI < 0.80), medium (0.55 < HDI < 0.70) and low (HDI < 0.55). To no surprise, the average per capita CO_2 emissions of nations in each of the four categories correlates with HDI, ranging from 11.2 tonnes for the very high category to 5.8, 1.8 and 0.4 for the high, medium and low groups, respectively. A higher quality of life aligns with higher GHG emissions, and correspondingly with greater energy consumption. All nations in Groups I and II of Table 1.1 are in the very high HDI category, and their per capita energy consumption is well above that of Group III and IV nations. All Group IV nations border on or are in the low HDI range. The data indicate that as developing nations achieve higher living standards, energy consumption will increase, in turn increasing GHG emissions through use of fossil fuels.

Climate change has become a significant subset of the sustainability issue. The atmosphere is a natural resource, shared by virtually all living things. If anthropogenic GHG emissions change the atmosphere's composition in ways that warm the Earth and alter climate, the emissions become a threat to sustainability.

1.6 A wicked problem

Anthropogenic climate change is a problem in need of a solution. But before we embark on the journey of seeking a solution, it behooves us to consider the scope of the problem we're trying to solve. Climate change is appropriately labeled a *wicked problem*, where "wicked" does not mean "evil" or "very," as in the sentence, "She's wicked smart." The term comes from an article written more than four decades ago (Rittel and Webber, 1973). It was brought to my attention by a colleague who teaches a course on design methodologies and adopts an eclectic view of the subject.

For Rittel and Webber, a wicked problem is one whose definition is elusive and for which a definitive solution may be lacking. Wicked problems are inherently societal problems that may have a scientific or technical component. But unlike the purely scientific or technical problem, which can be defined and for which solutions exist, societal problems are matters of public policy, and in a pluralistic society with diverse interests, traditions and values there is seldom consensus on the problem, much less its solution.

A wicked problem has many stakeholders, and any attempt at a solution will have multiple consequences as its implications ripple across the many affected parties. Whether the solution is right or wrong will not be judged by absolute or objective standards, but by the beliefs and values of the stakeholders. Are there common national or global values that provide the basis for an acceptable solution? In today's highly differentiated world, the answer is "probably not." As we'll find in Chapters 9 and 10, the social context in which climate change is framed becomes critical and even by invoking systems of ethical or religious beliefs, different conclusions can be drawn. In recognition of barriers to achieving a solution, Levin et al. (2010, 2012) have gone a step further in terming climate change a *super wicked* problem.

Confirmation of the problem is rooted in science; solutions are linked to energy technologies, economics and politics, as well as to human aspirations and behavior; and there are many stakeholders with competing interests. Political and economic factors loom large, influencing developed and developing nations in different ways and creating tensions within and between nations. Costs associated with addressing the problem at appropriate scales are large, challenging the ability of governments and capital markets to make appropriate investments. Human behavior and ideologies are also influential, as are ethical dimensions – philosophical and religious. With such an array of contributing factors, it's not surprising that divergent views have emerged, impeding efforts to deal with the problem. For wicked problems one solution does not fit all.

When my colleague asked his students to provide an example of a contemporary problem that could be classified as wicked, the responses overwhelmingly pointed to issues of sustainability and climate change. But before climate change can be considered a problem of any kind, wicked, serious or benign, there must be agreement that it is in fact a problem. In the United States, one of the most pluralistic nations in the world, there is no such agreement, and at this time prospects for reaching a consensus are slim. Without acceptance of the existence of a problem, there can be no agreement on a solution. Prospects become dimmer yet with recognition of the extended range of values and priorities that derive from the global nature of the issue.

While not using the word "wicked," Hulme (2009) and Giddens (2011) recognize related features of the climate change problem. Hulme acknowledges inherent complexities and uncertainties in underlying issues, particularly as they relate to economic, social, ecological and geopolitical factors, concluding that there are grounds for reasonable disagreement and debate. Giddens considers options for social action to curb warming and climate change but recognizes difficulties associated with reversing cultural norms that value the fruits of energy consumption but have little concern for the implications of consumption. In effect, if mitigation measures require behavioral and lifestyle changes, society is less inclined to adopt the measures than it is to let its progeny adapt to climate change. So, is the situation hopeless? No, but as we further our understanding of the nature of climate change and seek appropriate solutions, we may have to accept the fact that, although necessary, technological and policy options are not sufficient and may have to be accompanied by nothing less than a global sea change in cultural and behavioral norms.

1.7 Summary

Access to energy is essential to all forms of life and commerce, and the Earth is blessed with multiple sources of varying attributes. Some sources, namely fossil fuels, are nonrenewable and carbonaceous. Since the Industrial Revolution, fossil fuels have provided the lion's share of the world's primary energy supply, and for good reason. They are abundant, have large energy densities and can be supplied at comparatively low costs. Today fossil fuels supply about 80% of the world's primary energy consumption, and although they are nonrenewable, resources are more than sufficient to maintain this share well into, if not beyond, this century. But there is a downside. Emissions of pollutants such as oxides of sulfur and nitrogen, particulates and mercury are harmful to human health and the environment, and related costs are not borne by those who produce, market or consume the fuels, but by the larger society. Emissions of carbon dioxide contribute significantly to global warming, and related effects, although less immediate, are likely to be significant with costs borne by the larger society and future generations.

With business-as-usual, fossil fuels will remain the world's principal source of energy and atmospheric GHG emissions throughout this century. But, there is another way forward, one based on transitioning to renewable, nonpolluting and carbon-free sources of energy, adopting an energy efficiency/conservation mindset and embracing sustainability as a universal objective. If climate change is viewed as a serious environmental problem that merits immediate attention, society will have to alter its portfolio of energy production and utilization options, and it will have to look at a different economic model, one that favors sustainable development over quantitative growth.

Subsequent chapters consider the many issues that inform the climate change debate, with the goal of providing the reader a framework for drawing his/her own conclusions. Several questions are addressed in exploring the scientific basis for climate change. Are such changes entirely natural, or does a growing anthropogenic contribution provide a significant overlay to natural effects? What is the historical record of GHG emissions and global warming (or cooling)? What does the future hold? If plausible emission scenarios are threatening, what technology and policy options can be used to mitigate and adapt to the effects of warming? To what degree do economic, political/geopolitical and ethical considerations influence public debate on the matter? But first, some words of caution.

As previously noted, there are several features that distinguish climate change from other environmental issues. The knowledge base needed to reach definitive conclusions concerning the effects of global warming on the Earth's climate is far from complete. There is uncertainty in describing the complex and interrelated physical processes that drive the climate system and hence in establishing specific outcomes of warming on the climate. This uncertainty is exacerbated by the fact that anthropogenic and natural agents of climate change are superimposed. That is, they occur concurrently. Another difference has to do with *time scales*. Unlike other environmental issues, climate change is not generally regarded as an *immediate* threat, and if it is a threat, it may be decades before serious consequences materialize. With human instincts generally inclined to discount long-term effects, there is a natural impulse to defer dealing with the possible consequences of warming.

Add the fact that there are significant economic and political/geopolitical implications to dealing with climate change, and it is no surprise that the issue has become highly polarized and emotional. The reader is urged to rise above the acrimony and to keep an open mind in navigating a passage through the following chapters. An overriding objective is to provide a foundation from which rational conclusions can be drawn and future results can be assessed.

CHAPTER 2

The Earth's climate system

The Earth's climate system depends on many factors, and until the nineteenth century all of them were natural. But over the last two centuries exponential growth in human population and consumption has introduced an anthropogenic factor. How significant are anthropogenic effects? Can they alter the natural system, and if so, in what ways? To address these questions, we will consider the physical origins of climate change in Chapters 2 through 4. If you find some of the material challenging, please bear with me. It is fundamentally important to understanding the essence of the problem. Let's first clarify some terms.

2.1 Weather and climate

Although there is overlap at the margins, there is a difference between what is meant by *weather* and *climate*. Weather characterizes our immediate environment in terms of atmospheric conditions such as temperature, wind speed, cloud cover, and precipitation. As we commonly use it, the word describes conditions at the locale in which we reside or to which we are going and the time we are or will be there. At any locale, weather can vary significantly over small (hourly) time changes, as well as from one season to another. In contrast, climate represents long-term averages of atmospheric conditions for a particular region.

We can describe in general terms winter and summer climates of regions such as the American Southwest and Northeast, Siberia, and sub-Saharan Africa. The climate of any region can also experience statistically significant variations, but unlike weather, changes to climate have historically occurred over much longer time frames, commonly measured in millennia. It is in reference to such variations that we use the term *climate change*.

Another way to frame the distinction is to view weather as more variable over much smaller time scales.

It is tempting to draw conclusions about climate change from local weather patterns, particularly if they involve extreme weather events. The temptation should be resisted. Of course, climate change affects weather. But it can be misleading to draw conclusions about climate change from recent weather as, for example, last winter's conditions in South Bend, Indiana, or last summer's weather in Moscow, Russia. If climate change is inferred from changing weather patterns, the patterns should be examined globally and over an extended (decadal or longer) time frame.[1]

Climate change should also be distinguished from global warming. One follows from the other. Global warming refers to an increase in the Earth's average surface temperature, while climate change refers to long-term changes in atmospheric conditions that can result from global warming. Warming is the *cause*; climate change is the *effect*.

For the Earth as a whole, the climate system is determined by interactions between several large contributors. It is affected by the amount of incident solar radiation and the chemical composition and circulation of the *atmosphere*. It is also affected by processes associated with the hydrosphere and terrestrial (land-based) portions of the lithosphere. The hydrosphere encompasses all water bodies, but primarily the oceans, which comprise approximately 70% of the Earth's surface. Climate is strongly influenced by global ocean currents, as well as by the oceans' large capacity for storing thermal energy (and chemical species such as carbon dioxide). Not surprisingly, global circulations of the world's oceans provide for enormous flows of water and an accompanying redistribution of thermal energy from one region to another.

The biosphere and cryosphere also play prominent roles in the Earth's climate system. Through photosynthesis and respiration the biosphere exchanges carbon dioxide with its surroundings and plays a central role in the Earth's carbon cycle. It also influences climate through its role in the Earth's hydrologic cycle. The cryosphere consists of all terrestrial snow, ice, and permafrost, as well as ice residing on and below the surface of the oceans. An enormous amount of the Earth's water is stored as ice, much of it in *sheets* covering vast amounts of land and extending from coastal regions as shelves. Two major ice sheets remain after the last ice age, one on Greenland and the other on Antarctica. On land, water is also stored as ice in glaciers, and in the oceans as sea ice. Significant melting of the cryosphere would affect sea levels around the world and accelerate the pace of warming itself.

Because the Earth's climate is determined by a large and complex web of mutually dependent processes, any attempt to predict future climate change is a daunting challenge. But change due to natural agents is a well-documented aspect of the Earth's history.

2.2 Natural agents of climate change

The Holocene era, which spans 12,000 years since the last ice age, is the most recent manifestation of geologic time. It has been a period of enormous growth in human potential, as well as a time of unusual climate stability. But geologically speaking, climate stability is not the norm. During the Pleistocene era, which extends from 12,000 to 2.5 million years ago, the Earth experienced periodic transitions into and out of ice ages. But going back further in time from roughly 2.5 to 5.5 million years (the Pliocene era), conditions were again marked by relative climate stability, albeit at average Earth temperatures several degrees higher than in the Holocene. We therefore know that natural agents can tip the Earth's climate from one long-standing state to another.

Collectively, the Pleistocene and Holocene comprise the most recent (Quaternary) period of geologic time, and should patterns of climate change in this period be sustained, we can expect the Holocene to end some millennia hence and the Earth to enter a new ice age. Transitions into and out of an ice age are determined by changes in the rate at which the Earth intercepts solar radiation, which depends on the Earth's motion relative to the Sun.

As shown in Figure 2.1, the Earth annually executes a slightly elliptical orbit about the Sun, while it rotates about its polar axis every twenty-four hours. The polar axis is displaced from the axis of the orbital plane by a tilt angle, which causes the Northern Hemisphere to lean toward and away from the Sun during the summer and winter months, respectively. The angle is not fixed, but exhibits a small periodic variation over approximately 40,000 years (40 kyr). The polar axis also exhibits a slight precessional motion (exaggerated in the figure) that varies periodically over approximately 20 kyr. Yet a third periodicity of about 100 kyr is associated with a slight variation in the eccentricity of the orbital plane.

Each of the foregoing periodicities influences the amount of solar radiation intercepted by the Earth, and although changes in the radiation are small relative to the nominal value, they are sufficient to transition the Earth into and out of ice ages. Termed the Milankovitch theory, the record of glaciations (ice ages) and interglacials (warm periods) during the

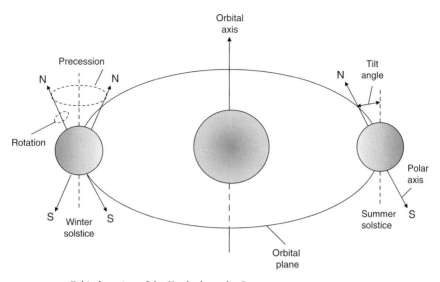

FIGURE 2.1. Orbital motion of the Earth about the Sun.

Pleistocene is well correlated by the periodicities (IPCC, 2007a). Over the last 700,000 years, glacial cycles were strongly influenced by changes in the eccentricity of the Earth's orbit over periods of approximately 100 kyr; before then, the dominant influence was the change in tilt angle over periods of 40 kyr.

The Earth's last interglacial period occurred about 125,000 years ago (125 ka), during which sea levels were approximately 5 meters higher than today. Since the last glacial maximum approximately 21 ka, the Earth has been warming, but at a rate that was approximately ten times slower than that of the twentieth century (IPCC, 2007a). In coming centuries global temperatures are unlikely to be affected by natural (orbitally driven) cooling, and transition into another ice age is unlikely for at least 30 kyr.

The existence of natural drivers – also called forcings – of climate change has prompted some to discount the influence of anthropogenic forcings such as GHG emissions. However, with few exceptions, natural forcings occur over much larger time scales (tens of millennia) and at much slower rates than those attributable to GHG emissions. One exception pertains to another solar cycle of much smaller periodicity. The amount of solar radiation emitted by the Sun increases with the number of sunspots appearing on its surface, which can range from near zero (a solar minimum) to as many as 250 (a solar maximum).

Although the extraterrestrial solar flux, also termed the Total Solar Irradiance (TSI), varies with sunspot activity, the variation between minimum and maximum is small,[2] as is the effect on the Earth's climate system over the typical eleven-year cycle. Nevertheless, since the thirteenth century there have been four abnormal and extended periods of low sunspot activity ranging from several decades to nearly a century. Termed grand minima, they can contribute to global cooling. The little ice age of the seventeenth century corresponded to a seventy-year period of abnormally weak sunspot activity and what is termed the Maunder Minimum. The fact that sunspot activity decreased during the first decade of the twenty-first century to near zero by 2010 raised the specter of another grand minimum, one whose cooling effect could temporarily reduce the impact of GHG emissions. However, the number of sunspots has since begun to increase (NASA, 2014), suggesting that the usual cycle is being followed.

Satellite solar radiation measurements have been made since 1979 and have yielded TSI values ranging from 1,360 to 1,375 W/m², which represent about a ±0.5% departure from the nominal value of 1,366 W/m² (NAS, 2012). Because the measurements were made by several satellites with different instruments, the variations are believed to be largely of instrumental and not of solar origin. Over the last fifty years, satellite TSI measurements have not revealed variability that could have contributed to global warming, and variations due to sunspot cycles have been within a 0.1% range (NAS, 2012).

Although it is not an agent of climate change, another natural phenomenon that enters the debate is the El Niño Southern Oscillation (ENSO). ENSO is a cyclical ocean-atmospheric event that transitions between two extremes – El Niño and La Niña – every five to seven years. El Niño results from diminished trade winds and eastward migration of warm surface waters from the Indian Ocean to equatorial regions of the Pacific Ocean. But at some point, warming of the Pacific is disrupted by an upwelling of cold water that creates subnormal surface temperatures (La Niña). Waters of the central Pacific Ocean become warmer and colder for El Niño and La Niña, respectively, and both events influence global weather patterns.[3]

Two questions arise in discussions of ENSO and global warming. Could global warming affect ENSO by eliminating La Niña and allowing for continuous warming of the Pacific Ocean? For now, the answer appears to be no (Watanabe et al., 2011). But, could warming amplify the severity of ENSO weather events? The intensity and frequency of such events have been increasing since the 1970s, and the 2010 La Niña was certainly one for the record books. But for now there's no definitive evidence linking increased intensity and frequency of the events to global warming.

2.3 Earth's global energy budget and the greenhouse effect

A key indicator of potential changes in the climate system is the average temperature at the Earth's surface, which is determined by radiation transfer into and out of the Earth-atmosphere system. This transfer is influenced by the fact that certain atmospheric constituents (greenhouse gases) are more effective in absorbing long-wave radiation emitted by the Earth (terrestrial radiation) than the shorter wavelengths of incident solar radiation. Termed the greenhouse effect, this influence is a natural phenomenon without which the Earth would be a far colder planet. Although the concentration of CO_2 in the Earth's atmospheric is only 0.04% by volume, it and other GHGs such as water vapor maintain the average surface temperature at a relatively balmy 15°C, well above the value of –20°C that would exist without GHGs.

The Earth's average temperature is determined by the extent to which it absorbs radiation incident from the Sun and the extent to which it emits radiation to outer space. If absorption and emission are equal, the temperature remains constant and a state of thermal equilibrium is said to exist. To understand the physical origins of climate change, it is useful to delineate the various components of radiation transfer.

Consider the conditions shown in Figure 2.2, where the extent of the atmosphere relative to the Earth's diameter D_e has been greatly exaggerated. At the outer edge (top) of the atmosphere (TOA), the Earth sees the Sun as a beam of parallel rays. Think of a hypothetical surface oriented normal to the rays. The rate at which radiant energy passes through a unit area of the surface is termed the solar constant S_c, which has a nominal value of $S_c \approx 1366$ W/m^2.[5] The product of the solar constant and the Earth's cross-sectional area, $S_c\left(\pi D_e^2 / 4\right)$, provides the rate at which the Earth-atmosphere system intercepts solar radiation. Dividing this rate by the surface area of the Earth, πD_e^2, provides the average rate at which solar radiation is incident per unit area of surface at the TOA, $G_{S,o} = S_c / 4 \approx 341.5$ W/m^2. The symbol G is termed the irradiation and is used to characterize the rate at which radiant energy is incident on a surface per unit area of the surface. The subscripts S and o refer to solar radiation and conditions at the TOA, respectively.

As solar radiation propagates through the Earth's atmosphere, it experiences reflection within the atmosphere and at the Earth's surface. As

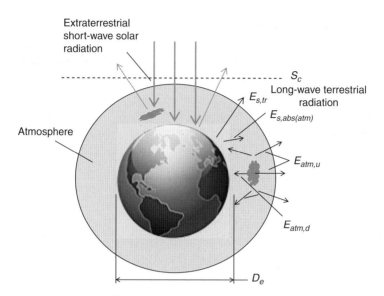

FIGURE 2.2. Terrestrial and extraterrestrial radiation fluxes.

indicated by the outwardly pointing arrows of Figure 2.2, some of the radiation is reflected back to space. All gaseous constituents of the atmosphere contribute to reflection, as do clouds and aerosols, which are suspensions of microscopic liquid droplets and solid particles. Aerosols occur naturally from events such as sandstorms, lightning-induced forest fires, and volcanic eruptions, as well as anthropogenically from the combustion of coal, petroleum products, and biomass. One such aerosol is produced when sulfur dioxide (SO_2), a by-product of coal and oil combustion, reacts with water vapor to form atmospheric sulfates. Although they ultimately descend to the surface as acid rain, they reflect solar radiation back to space while they remain in the atmosphere. Reflection by the Earth's surface depends on the nature of the surface. As much as 90% of the incident solar radiation is reflected by snow and ice and as little as 10% by open waters and leaf-covered canopies. Reflection is larger for barren lands than for foliated land.

Solar radiation that is not reflected back to space is absorbed in the atmosphere and by the Earth's land and oceans. Absorption by gases in the atmosphere is dominated by ozone (O_3) at lower wavelengths and water vapor at higher wavelengths. Absorption by aerosols at all wavelengths of the solar spectrum is dominated by soot (carbon particles) produced mostly by the incomplete combustion of coal, petroleum products, and biofuels such as wood, vegetable oil, and animal dung. Absorption at the surface is largest for open waters and leaf-covered canopies and smallest for ice/snow-covered surfaces and barren land. Absorption of most of the solar radiation making it to the Earth's surface occurs in the oceans.

From the foregoing considerations, propagation of solar irradiation through the atmosphere can be resolved into two components, one for reflection and the other for absorption,

$$G_{S,o} = G_{S,ref} + G_{S,abs} \qquad (2.1)$$

where each component includes contributions due to atmospheric (*atm*) and surface (*s*) effects.

$$G_{S,ref} = G_{S,ref(atm)} + G_{S,ref(s)} \qquad (2.2a)$$

$$G_{S,abs} = G_{S,abs(atm)} + G_{S,abs(s)} \qquad (2.2b)$$

An important parameter, termed the albedo α, is defined as the fraction of the incident solar radiation that is reflected back to space,

$$\alpha = \frac{G_{S,ref}}{G_{S,o}} = \frac{G_{S,ref(atm)} + G_{S,ref(s)}}{G_{S,o}} \qquad (2.3)$$

The globally averaged value of the albedo ($\alpha \approx 0.3$) provides a measure of the extent to which the planet is warmed by the Sun. A reduction in α corresponds to an increase in the amount of solar radiation absorbed by the Earth's surface-atmosphere system and therefore global warming. Conversely, an increase in α corresponds to reduced absorption and global cooling. Locally, reflection varies greatly with the nature of the surface and whether skies are clear or cloudy. The albedo is large (~ 0.9) for snow- and ice-covered surfaces, is larger for barren than for foliated land, and is small for open waters and leaf-covered canopies (~ 0.1).

Unlike solar radiation, which is emitted by a high-temperature source (the Sun) and therefore concentrated at low wavelengths from approximately 0.2 to 3 micrometers (μm), emission by the Earth's surface and atmosphere (terrestrial radiation) is associated with moderate temperatures and occupies the spectral region from approximately 5 to 50 μm. This difference in wavelengths has a profound effect on radiation transfer in the atmosphere. Because solar radiation is concentrated at low wavelengths, it is weakly absorbed by GHGs, while terrestrial radiation is more susceptible to absorption because of its much longer wavelengths. Therein lays the basis of the greenhouse effect. GHGs in the atmosphere are comparatively transparent to incoming solar radiation, allowing most of it to pass and to be absorbed by the oceans and land. In contrast, GHGs are more opaque to outgoing terrestrial radiation, absorbing much of the radiation and subsequently emitting downward to the Earth's surface and outward to space. With increasing opacity attributable to anthropogenic GHG emissions, terrestrial radiation experiences increasing atmospheric absorption, thereby reducing transmission of this radiation to outer space.

Referring again to Figure 2.2 and moving in the clockwise direction, the first arrow represents the portion of radiation emitted by the Earth's surface E_s that is transmitted to outer space, $E_{s,tr}$, and the second arrow represents the rest of the surface emission, $E_{s,abs(atm)}$, which does not make it to outer space but is instead absorbed by the atmosphere and clouds. In turn, the atmosphere (including clouds) emits long-wave radiation, some of which is transmitted upward to outer space, $E_{atm,u}$, and some in the downward direction for absorption by the Earth's surface, $E_{atm,d}$.

If a state of thermal equilibrium is to exist, the net transfer of solar radiation to the Earth (the amount absorbed by the Earth-atmosphere system), $G_{S,o} - G_{S,ref}$, must be balanced by the transfer of terrestrial radiation from the Earth to outer space, $E_{s,tr} + E_{atm,u} \equiv E_o$. Accordingly, equilibrium exists if

$$G_{S,o} - G_{S,ref} = E_o \qquad\qquad (2.4)$$

If the equality is not satisfied and the net transfer of solar radiation to the Earth exceeds the transfer of terrestrial radiation from the Earth, the Earth's temperature must increase (global warming) until the terrestrial radiation rises to a level that balances the solar input. If the inverse occurs (transfer of terrestrial radiation from the Earth exceeds the net transfer of solar radiation to the Earth), the path to equilibrium corresponds to a reduction in the Earth's temperature (global cooling) until the solar input and terrestrial output are equivalent

Efforts to quantify the Earth's energy budget date back to the early twentieth century (Dines, 1917), and through the years advancements have been made, particularly through the use of improved radiation models and the advent of satellite measurements. Kiehl and Trenberth (1997) reviewed the existing state of knowledge and provided estimates of the energy budget based on the latest measurements and models. Twelve years later Trenberth et al. (2009) updated the estimates on the basis of improved measurements, models, and procedures for partitioning energy among the different terms. We will briefly discuss their results, but let's first acknowledge some limitations.

Resolving components of the energy budget is a complex task, and because it requires analysis and adjustment of numerous data sources and models, for now, the results do not provide sufficient accuracy for assessing the extent to which global warming is, or is not, occurring. Although uncertainties decrease with implementation of improved measurements and models, they were estimated by Trenberth et al. (2009) to be $\pm 3\%$ for TOA terms and as large as $\pm 10\%$ for terms at the Earth's surface. Nevertheless, attempts to more accurately quantify terms in the energy balance contribute to a better understanding of how energy is partitioned in the Earth-atmosphere system.

Estimates for the period from March 2000 to May 2004 are shown in Figure 2.3. To three significant figures, conservation of energy (Equation 2.4) is satisfied at the TOA, where the net transfer of solar radiation to the Earth $\left(G_{S,o} - G_{S,ref} = 239 \text{ W/m}^2 \right)$ corresponds to the transfer of terrestrial radiation from Earth $\left(E_o = E_{s,tr} + E_{atm,u} = 239 \text{ W/m}^2 \right)$. However, to four significant figures, the left-hand side of the equation (239.4 W/m²) exceeds the right-hand side (238.5 W/m²), implying a net transfer of 0.9 W/m² to the Earth-atmosphere system.

$$\left(G_{S,o} - G_{S,ref} \right) - \left(E_{s,tr} + E_{atm,u} \right) > 0 \tag{2.5}$$

The imbalance corresponds to results determined from an atmospheric radiation model (Hansen et al., 2005) and implies that thermal equilibrium does not exist. That is, thermal energy storage in the Earth-atmosphere system, and hence the system temperature, must increase due to the imbalance.

From Figure 2.3, energy transfer to the Earth's surface includes the effects of absorbed solar radiation $\left(G_{S,abs(s)} = 161 \text{ W/m}^2 \right)$ and downward-propagating atmospheric (back) radiation $\left(E_{atm,d} = 333 \text{ W/m}^2 \right)$ for a total of 494 W/m². Transfer from the surface includes the effects of emission $\left(E_s = 396 \text{ W/m}^2 \right)$, as well as convection (17 W/m²) and evaporation (80 W/m²), for a total of 493 W/m². The imbalance of 1 W/m² (~ 0.9 W/m²) corresponds to the net rate at which energy is added per unit area of the Earth's surface.

The most recent attempt to improve the global radiation balance was made by Wild et al. (2013). Results were based on integrating satellite measurements for fluxes at the TOA and new measurements for fluxes at the Earth's surface with updated radiation models. The net transfer of solar radiation to the Earth, $G_{S,o} - G_{S,ref} = 240 \text{ W/m}^2$, exceeded the outgoing thermal radiation, $E_o = E_{s,tr} + E_{atm,u} = 239 \text{ W/m}^2$, suggesting an imbalance

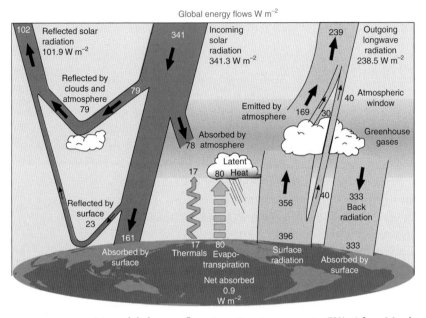

FIGURE 2.3. Mean global energy fluxes in watts per square meter (W/m²) from March 2000 to May 2004 for the Earth-atmosphere system. From Trenberth et al. (2009). For a color version of this figure, please see the color plate section.

of 1 W/m². However, to obtain agreement with estimates of the rate at which thermal energy was added to the oceans and to accommodate estimates of the other fluxes, the authors reduced their estimate of the imbalance to 0.6 W/m².

There is clearly a good deal of uncertainty in using the energy budget methodology to accurately determine net radiation transfer to the Earth-atmosphere system and hence the extent of global warming. The methodology also reveals little about the manner in which energy is distributed in the system. How much is going into the oceans and land masses, the atmosphere, and into melting land and sea ice? How is energy transferred within the oceans and between the oceans and the atmosphere? These questions were addressed by Trenberth and Fasullo (2010), who were unable to reconcile their 0.9 W/m² imbalance with measurable changes in the energy of the Earth's oceans and atmosphere and in the energy associated with melting of sea and land ice. Additional improvements in measurement and prediction methods are needed before the energy budget method can be used to draw definitive conclusions concerning net energy transfer to (or from) the Earth and the degree to which global warming (or cooling) is occurring. Nevertheless, it remains a useful model for describing factors that drive the climate system.

2.4 Summary

If the reader embraces only one takeaway from this chapter, it should be the fact that climate change is sensitive to the global energy budget and is driven by *energy imbalances*. If the Earth absorbs as much short-wave solar radiation as the long-wave radiation it emits to space, it is in a state of thermal equilibrium. There is no change in its thermal energy and hence average temperature. Nonequilibrium exists if absorption exceeds emission and the Earth's temperature increases over time. Or, if emission exceeds absorption, the Earth will cool. This behavior is an irrefutable consequence of the First Law of thermodynamics. It is also responsible for the many natural climate swings that have occurred before the emergence of modern humans, including transitions into and out of ice ages. For much of the last 12,000 years, these swings have been relatively mild, and the Earth has departed little from a state of thermal equilibrium. The question now before us is whether human activities, particularly in the form of GHG emissions, are disrupting this state of equilibrium and contributing to global warming.

CHAPTER 3

Greenhouse gases

The atmosphere is our constant companion. It envelops us 24/7, providing life-sustaining oxygen. In school we learned that, excluding water vapor, nitrogen and oxygen comprise about 98% of the atmosphere's dry air, with argon and carbon dioxide providing much of the remainder. In urban and industrial surroundings, we may also sense the intrusion of atmospheric pollutants such as ozone and sulfur dioxide. In addressing global warming, we turn our attention to atmospheric constituents that qualify as greenhouse gases (GHGs).

The atmospheric concentration of some GHGs is determined by both natural and anthropogenic effects. Hence, the gases would still exist if humans did not inhabit the planet. Water vapor (H_2O) and carbon dioxide (CO_2) are the biggest contributors to the Earth's natural greenhouse effect, with smaller roles played by methane (CH_4) and nitrous oxide (N_2O). But due to human activities, the atmospheric concentrations of these species are increasing, providing an anthropogenic component to the greenhouse effect. The effect is amplified by the existence of other, strictly anthropogenic GHGs that number in the hundreds (Ramaswamy et al., 2001; Stine and Sturges, 2007).

3.1 Distinguishing features

Not all GHGs contribute to global warming in the same way. For one thing, spectral absorption bands corresponding to discrete regions of the electromagnetic spectrum in which the gases absorb terrestrial radiation differ according to their strength and wavelengths. For example, methane molecules absorb terrestrial radiation at different wavelengths and more strongly than carbon dioxide molecules. Also, once released to the atmosphere, gases differ according to the amount of time they remain in the atmosphere.

Distinguishing features of five comparatively long-lived and well-mixed GHGs that account for about 96% of the anthropogenic greenhouse effect are provided in Table 3.1, along with the cumulative effect of fifteen minor, halogenated gases that account for much of the remaining 4%. Containing chlorine, fluorine, and/or bromine, halogens are man-made chemicals used in many industrial processes and commercial products such as refrigerants, aerosols, and foaming agents. Examples include the families of chlorofluorocarbons (CFCs) and hydrochlorofluorocarbons (HCFCs). Halogens containing chlorine or bromine contribute to another environmental problem, namely depletion of stratospheric ozone.[1] While the atmospheric concentrations of CO_2, CH_4, and N_2O are influenced by both natural and anthropogenic effects, the halogens are strictly anthropogenic.

Atmospheric concentrations

The second column of Table 3.1 provides circa-1750 atmospheric concentrations that are commonly taken as preindustrial baselines. Although direct concentration measurements only began in the nineteenth century (the instrument record), indirect measurements, as taken from ice core records, are used to infer concentrations well into geologic time. Since the beginning of the Holocene 11,000 years ago and until the Industrial Revolution, natural concentrations of CO_2, CH_4, and N_2O varied little from values of about 280 ppm (parts per million by volume), 720 ppb (parts per billion by volume), and 270 ppb, respectively. Preindustrial concentrations of anthropogenic gases such as CFC-11/12 and other long-lived halogens were essentially zero.

As shown in Figure 3.1a, concentrations of CO_2, CH_4, and N_2O began to increase by the middle of the eighteenth century and have since continued on a sharply ascendant trajectory. For each gas, the principal driver has been the use of fossil fuels, although other factors are important, such as deforestation (CO_2), livestock production (CH_4), and the use of synthetic fertilizers (N_2O). Because the data in Figure 3.1a correspond to annual averages, they mask variations that can occur over the course of a year. If the data for CO_2 are plotted using a finer time scale (Figure 3.1b), results reveal a sawtooth pattern for which the concentration decreases during the spring/summer months due to plant growth in the Northern Hemisphere

TABLE 3.1. *Characteristics of well-mixed greenhouse gases*

Gases	Natural concentration[a] circa 1750	Global annual mean concentration 2012	Average annual change in concentration 2005–2012	Atmospheric lifetime (yrs)	GWP (100yrs)	Radiative forcing[b] 1750–2013 (W/m²)
CO_2	280 ppm	393 ppm	2.0 ppm/yr	~5 to 200	1	1.88
CH_4	720 ppb	1,819 ppb	6.4 ppb/yr	12	25	0.50
N_2O	270 ppb	325 ppb	0.9 ppb/yr	114	298	0.18
CFC-12	0	534 ppt	−0.8 ppt/yr	100	10,900	0.17
CFC-11	0	241 ppt	−1.8 ppt/yr	45	4,750	0.06
15 Minor Halogens	0	–	–	~10 to 20,000	–	0.11

[a] Concentration:
  ppm (parts per million by volume)
  ppb (parts per billion by volume)
  ppt (parts per trillion by volume)

[b] Radiative forcings account for chemical reactions that may occur once a GHG is discharged to the atmosphere, thereby changing its concentration and that of species comprising products of the reactions.

Sources: From CDIAC (2013), NOAA (2014b) and Blasing (2014).

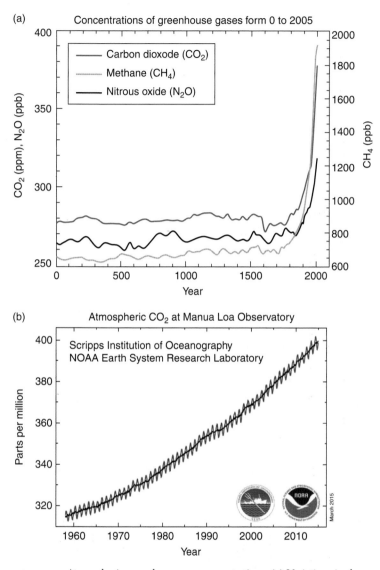

FIGURE 3.1. Atmospheric greenhouse gas concentrations: (a) Variations in the concentrations of CO_2, CH_4 and N_2O over the last two millennia (IPCC, 2007b). (b) Variation in the CO_2 concentration measured at the Mauna Loa Observatory from 1958 to December 2013 (NOAA, 2014a). For a color version of this figure, please see the color plate section.

and increases due to plant decay over the winter months. The overall seasonal variation is about 5 ppm.

As with other aspects of the global warming and climate change debate, reported carbon dioxide measurements have not gone unchallenged.

It therefore behooves us to step back and examine how the measurements are made. First, we should recognize that atmospheric CO_2 measurements depend on where and when the measurements are made. The data of Figure 3.1b were obtained at the Mauna Loa Observatory, which is at an elevation of 3,400 meters on the Island of Hawaii. Measurements have been made at the Observatory since 1958, and because they are obtained in a well-mixed, maritime atmosphere, they provide a good indication of the world's background CO_2 concentration.

Background CO_2 concentration measurements are uninfluenced by proximity of the measurement system to concentrated sources of emission to (or withdrawal from) the atmosphere. Concentrations can actually vary by more than 100 ppm over the surface of the Earth, particularly for continental land masses where levels can be exceptionally large in densely populated urban and industrialized areas or low for heavily forested regions during summer months. In urban regions where fossil fuels are burned in large amounts, ground-level concentrations are large and decrease with increasing elevation. Over forest canopies, ground-level values experience a diurnal cycle for which they are smaller during daylight hours. Over the Earth's oceans, concentrations at sea level can be up to 20 ppm lower than those at higher elevations due to transport of CO_2 from the atmosphere to the ocean. Large diurnal and seasonal variations also occur in urban and forested areas due to changes in human and photosynthetic activities. If a measure of the global atmospheric CO_2 concentration is to be determined over an extended period of time, measurements should be made at locations which are minimally affected by distributed sources and sinks. The Mauna Loa Observatory is one such location, and its measurements provide an accurate and important data base for tracking changes in the atmospheric CO_2 concentration.[2]

In 2012 the atmospheric CO_2 concentration increased by 2.1 ppm, slightly above the average rate of change of 2 ppm/yr from 2002 to 2011 and well above the average increase of approximately 1.6 ppm/yr from 1980 to 1999 (GCP, 2013; Blunden and Arndt, 2014). In 2013, the concentration increased by 2.8 ppm to a global average of 395.3 ppm for the year (Blunden and Arndt, 2014), and on May 9, 2013, the concentration recorded at the Mauna Loa Observatory exceeded 400 ppm for the first time. However, for those inclined to question the significance of the results, another issue must be addressed. How do we know that similar, if not more extreme trends haven't occurred in prehistoric times? The facts are that we do know, and they haven't!

For approximately 2 million years up to the Industrial Revolution, changes in the atmospheric CO_2 concentration have correlated with transitions between glacial and interglacial periods. Based on CO_2 measurements obtained from ice core records over the last 800,000 years and sea shell boron isotope records over the last 2.1 million years, the current atmospheric CO_2 concentration is larger than any previously experienced value (Siegenthaler et al., 2005; Hönisch et al., 2009). From 2 million to 900,000 years ago, concentrations for glacial and interglacial extremes ranged from approximately 213 to 283 ppm; from 800,000 to 450,000 years ago, they ranged from 181 to 252 ppm. Minimum and maximum values of 184 and 297 ppm are associated with the time frame from 418,000 years ago to onset of the current warming period.

From the last Glacial Maximum 24,000 years ago to the end of the glacial period 17,000 years ago, the CO_2 concentration was essentially unchanged at about 190 ppm (Brook, 2012; Schmitt et al., 2012). From that time to early stages of the Holocene 11,000 years ago, the concentration rose to 260 ppm, largely due to the release of CO_2 from oxidation of organic matter in the oceans, but mitigated in part by assimilation of CO_2 due to the terrestrial growth of biomass. Over the next 11,000 years of the interglacial, changes occurred very slowly, as the concentration increased to the preindustrial level of 280 ppm. Since then, it has spiked to 395 ppm in 2013 and continues to rise. Similar conditions characterize changes in the atmospheric concentrations of methane and nitrous oxide (Spahni et al., 2005).

For the five major GHGs, atmospheric concentrations for 2012 are provided in Table 3.1 and data from 1979 to 2013 are plotted in Figure 3.2. While the concentration of N_2O is increasing at a nearly constant rate of 0.9 ppb/yr, CO_2 is accumulating at a growing rate, from approximately 0.7 ppm/yr in the 1960s to 2 ppm/yr in the mid- to late 2000s. At this rate, the concentration would reach 450 ppm by 2040. Although the CH_4 concentration was largely unchanged from 1999 to 2006, it has since resumed an upward trajectory. The recent trend of decreasing CFC concentrations is due to restrictions on their use, prompted by concerns for their effect on stratospheric ozone. However, substitute chemicals such as HFC-134a and HCFC-22, while less destructive of stratospheric ozone, are potent GHGs whose concentrations are increasing.

Ever cautious in its pronouncements, the Intergovernmental Panel on Climate Change concluded – with very high confidence – that the atmospheric concentrations of CO_2, CH_4 and N_2O substantially exceed concentrations recorded over at least the last 800,000 years (IPCC, 2014A). The bottom line is that *the trend of increasing GHG concentrations and their anthropogenic origins are irrefutable*.

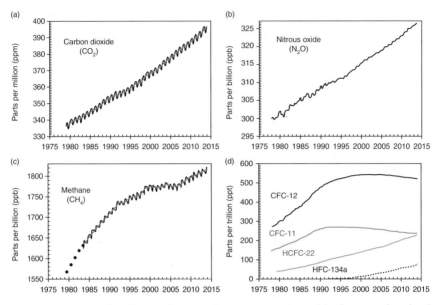

FIGURE 3.2. Variations in the atmospheric concentrations of the five major long-lived greenhouse gases from 1979 to August 2014 (NOAA, 2014b). For a color version of this figure, please see the color plate section.

Atmospheric lifetimes and global warming potentials

The next step in assessing the impact of GHGs on global warming deals with atmospheric lifetimes and global warming potentials (GWP). Following discharge, the lifetime of a GHG provides a measure of the time required for removal from the atmosphere by natural processes. Removal depends on the nature of the process, and rates vary significantly. Some species are removed by a single process. For CH_4 it is a chemical reaction involving the hydroxyl (OH) radical, and the lifetime is little more than a decade; for CFC-12 removal is by a stratospheric photochemical reaction, and the lifetime is about a century. Other species experience removal by multiple processes. Because CO_2 removal involves processes such as photosynthesis and absorption by the Earth's soils and oceans, it has a range of associated lifetimes from approximately 5 to 200 years (Table 3.1). The long atmospheric lifetimes associated with some species remind us that, even if we were to significantly reduce GHG emissions, *the contribution to global warming by existing gases would persist for decades to centuries.*

The concept of a GWP was introduced to compare the relative warming of different GHGs. It provides an approximate measure of the impact of a unit mass of a gas relative to a unit mass of CO_2, which is arbitrarily assigned a GWP of unity. Because the GWP of a GHG increases with the atmospheric lifetime of the species, its value is linked to a prescribed time interval. In Table 3.1, values of the GWP correspond to a period of 100 years. Relative to emissions of CO_2, the impact of any other GHG on warming is determined by multiplying the mass of its emissions by its GWP. The product is termed a carbon dioxide equivalent (CO_{2eq}).

From the results of Table 3.1, it follows that over 100 years the effect of discharging 1 kg of CH_4 into the atmosphere corresponds to discharging 25 kg of CO_2. The carbon dioxide equivalent, CO_{2eq}, of the methane discharge is therefore 25 kg.[3] In this manner, emissions of all GHGs can be standardized and their cumulative effect reported as a total mass of CO_{2eq}. It would therefore be misleading to link the impact of a GHG exclusively to the mass of its emissions. The 100-year GWPs of the other GHGs are larger yet, ranging from 298 for N_2O to values as high as 23,900 for the halogenated gases. Hence, although these gases are released to the atmosphere at rates well below that of CO_2, their contributions to warming cannot be summarily discounted. Accounting for all GHGs, the circa-2012 atmosphere had a CO_{2eq} concentration of 476 ppm, of which 393 ppm were due to CO_2 alone.

Greenhouse gas emissions can be reported in terms of the mass of carbon (C), as well as the mass of CO_2. To convert from one to another, it is simply a matter of multiplying (or dividing) by the ratio of the molecular weight of CO_2 to that of C ($44 \div 12 = 3.67$). Emissions amounting to 1 Mt-C (or 1 $Mt-C_{eq}$) therefore correspond to 3.67 $Mt-CO_2$ (or 3.67 $Mt-CO_{2eq}$). Regrettably, the two quantities are often confused, and it is important to recognize which quantity (C or CO_2) is associated with reported data.

Radiative forcings

The last column of Table 3.1 pertains to use of a radiative or climate forcing to estimate warming due to increasing GHG concentrations. A climate forcing is an agent of climate change that can be imposed by natural processes or human activities. A positive forcing warms the Earth; a negative forcing cools it.

Using the year 1750 as a baseline, a radiative forcing *RF* is the change in the net downward radiative flux at the tropopause (the boundary between the atmosphere's troposphere and stratosphere) that has occurred since that time due to changes in a parameter that affects the transfer of solar or terrestrial radiation to or from the Earth. Referring to Figure 2.3, the flux is a disturbance that alters the balance between long-wave radiation leaving the

Earth-atmosphere system (E_o) and the net transfer of solar radiation into the system ($G_{S,o} - G_{S,ref}$). Expressed in units of watts per square meter (W/m²), radiative forcings quantify the effect of GHG emissions on the Earth's radiation balance. Since GHGs disrupt this balance, the term is indeed a forcing since it drives changes in the Earth's temperature needed to restore equilibrium.

A positive forcing alters the global energy balance by reducing terrestrial radiation leaving the system or by increasing the net transfer of solar radiation into the system. It therefore increases the Earth's temperature (global warming) in order to restore equilibrium by increasing the emission of long-wave radiation. A negative forcing has the opposite effect. It increases long-wave radiation leaving the Earth-atmosphere system or reduces the net transfer of solar radiation into the system. In either case equilibrium is restored by a reduction in the emission of long-wave radiation, necessitating a reduction in the Earth's surface temperature (global cooling).

If thermal equilibrium of the Earth-atmosphere system corresponds to equivalent values of 239 W/m² for outgoing terrestrial radiation and net incoming solar radiation (Figure 2.3), what is the impact of GHG emissions on this balance? Table 3.1 provides the positive circa-2013 forcings of the most prominent well-mixed gases. In 2013, these gases added a combined positive forcing of 2.90 W/m² to the Earth's radiation budget, about a 32% increase from 1990 (NOAA, 2014b). The corresponding reduction in terrestrial radiation leaving the Earth's atmosphere creates an imbalance in the global energy budget for which incident solar radiation exceeds outgoing terrestrial radiation and warming occurs. Forcings associated with CO_2, CH_4, and N_2O are steadily climbing, as is the cumulative forcing associated with the fifteen minor halogenated gases. The decline in forcings for the two CFCs is due to reduced emissions imposed by the Montreal Protocol.[4]

Best estimates of circa-2011 forcings are illustrated by the bar chart of Figure 3.3, which is based on results reported in the Fifth Assessment Report of the IPCC (2013b).[5] Of the well-mixed GHGs (CO_2, CH_4, N_2O, and the halocarbons), changes in the concentration of carbon dioxide provide the largest forcing of 1.82±0.19 W/m², followed by methane (0.48±0.05 W/m²), the halocarbons (0.34±0.05 W/m²), and nitrous oxide (0.17±0.02 W/m²). Additional forcings are associated with formation of ozone in the troposphere (a positive forcing of 0.40±0.20 W/m²) and depletion in the stratosphere (a negative forcing of −0.05±0.10 W/m²), yielding a net forcing of 0.35±0.2 W/m². Tropospheric ozone is not discharged directly to the atmosphere, but is instead formed from chemical reactions triggered by GHGs such as CH_4 and N_2O. Ozone depletion in the stratosphere is due to reactions with anthropogenic halocarbons.

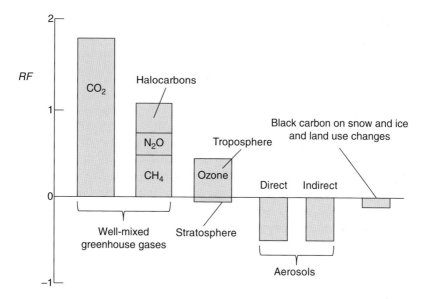

FIGURE 3.3. Nominal radiative forcings in watts per square meter due to changes from 1750 to 2011. Adapted from IPCC (2014a).

But GHGs do not provide the whole story. Anthropogenic forcings are also associated with atmospheric aerosols, which are liquid or solid particles ranging in size from nanometers to millimeters. Unlike GHGs, they do not become well mixed in the atmosphere and only remain suspended for a brief period before settling to the surface. They are of two primary forms – black carbon (soot) and sulfates – and they influence the global radiation balance directly and indirectly. The direct effect is due to interaction of radiation with the particles. Aerosols reduce the amount of solar radiation reaching the Earth's surface, in one case by absorption (soot) and in the other by reflection (sulfates). By absorbing solar radiation (as well as long-wave radiation from the Earth's surface), soot heats the atmosphere, thereby providing a positive forcing. By reflecting solar radiation back to space, sulfate particles have a negative forcing.

Sulfates and black carbon are products of burning coal and oil. Coal is widely used for power production, assorted industrial processes, and, in less developed regions of the world, residential heating and cooking. Oil in the form of diesel fuel is widely used in transportation vehicles, particularly large trucks, ships, and off-road (earthmoving and mining) vehicles, as well as for home heating and cooking. Soot is also produced by the incomplete combustion of biomass, principally from clearing forested lands and using agricultural remnants, wood, and animal dung for cooking and heating. In South Asia, persistent brown clouds attributable to burning biomass have been observed to coalesce over an area of the Indian Ocean as large as the United States (Venkataraman et al., 2005; Gustafsson et al., 2009; Rosenthal, 2009).

Absorbing (black carbon) and reflecting (sulfates) aerosols partially off-set each other, and in Figure 3.3 the net direct effect corresponds to a negative forcing of -0.45 ± 0.5 W/m².[6] Although the atmospheric lifetime of black carbon is brief (less than a month), it is the strongest absorber of solar radiation and its 100-year GWP exceeds 1,000. If its impact on warming is calculated over a shorter, 20-year interval, its GWP is more than 4,000. The same may be said for methane, whose GWP nearly triples if evaluated over 20 years rather than 100 years (Cross and Pierson, 2013). Compared to CO_2 and other GHGs, black carbon and methane – along with tropospheric ozone and certain HFCs – are short-lived contributors to global warming. Nevertheless, their contribution is significant and becomes more significant when considered over shorter time frames (Cross and Pierson, 2013). Although the circa-2013 combined *RF* of all HFCs is only 0.012 W/m², they are the fastest-growing component of GHG emissions and on a trajectory to double within a decade. Hence, while most attention is given to the long-term contribution of CO_2 to global warming, the significant, shorter-term effects of agents such as HFCs, soot, and CH_4 cannot be neglected.[7]

Aerosols also affect climate forcing indirectly through their influence on cloud formation. Clouds affect the Earth's radiation balance in two ways. By reflecting incident solar radiation back to space, they provide a negative forcing that acts to cool the planet. Conversely, by absorbing terrestrial radiation, they provide a positive forcing that warms the planet. Under normal conditions, reflection dominates and reduces the net flux of incident solar radiation by approximately 20 W/m² relative to a cloudless planet (Dessler, 2010).

Aerosols are considered coarse or fine according to whether their effective radius exceeds or is less than 0.5 μm, and they affect clouds in two ways, one microphysical and the other radiative. Typically, coarse aerosols occur naturally, while fine aerosols result from human activity such as the combustion of fossil fuels. The microphysical effect is attributed to the fact that fine (anthropogenic) aerosols provide convenient sites for the condensation of water vapor. They therefore contribute to cloud formation by increasing the number of smaller droplets that are less likely to coalesce and descend as precipitation (Rosenfeld, 2006). With a weakening of mechanisms for precipitation, clouds become larger, persist for longer periods, and reflect more sunlight to outer space. Absorption of solar radiation by the Earth's surface is reduced, and the climate forcing is negative. The net indirect effect of aerosols on climate forcing is -0.45 ± 0.5 W/m^2, and the combined (direct plus indirect) effect is -0.90 (-1.0, $+0.8$) W/m^2 (IPCC, 2014a). The large uncertainty bands are indicative of gaps in the knowledge base concerning the physics of aerosol-cloud interactions.

Climate forcing is also affected by a reduction in the Earth's albedo due to deposits of soot on snow and ice and an increase in the albedo due to changing land use. Although the global average forcing from soot deposited on snow and ice is only 0.04 (-0.02, $+0.05$) W/m^2, geographic variations are large and effects are felt most strongly in the Arctic (Kintisch, 2009a). A negative forcing of -0.15 ± 0.10 W/m^2 is associated with changing land use, largely from deforestation and urbanization, and a net negative forcing of about -0.1 ± 0.1 W/m^2 is attributed to the combined effect of soot deposits and changing land use (IPCC, 2014a).

The remaining climate forcings are relatively small (IPCC, 2014a) and are not shown in Figure 3.3. Positive forcings of 0.03 W/m^2 and 0.07 W/m^2, respectively, are associated with changes in the atmospheric concentrations of minor GHGs (HFCs, PFCs and SF$_6$) and from an increase in stratospheric water vapor due to the oxidation of CH$_4$. An additional forcing of 0.05 W/m^2 is attributed to linear contrails that form behind cruising aircraft and irregularly shaped cirrus clouds produced by the contrails (Burkhardt and Kärcher, 2011).

Changes in natural forcings are generally small and intermittent (IPCC, 2014a). During the industrial era, sunspot cycles have contributed to small variations in the insolation, and the long-term RF is estimated to be -0.04 W/m^2. On occasion, volcanic eruptions can produce large negative forcings, as evidenced by the value of -3.7 W/m^2 attributed to the eruption of Mount Pinatubo in 1991, but much smaller forcings of about -0.1 W/m^2 are more common. In each case the effect is short-lived, dissipating within a few years.

The best estimate of the net 2011 anthropogenic forcing relative to 1750 is 2.29 W/m^2, which is a 43% increase from the value of 1.60 W/m^2 reported for 2005 (IPCC, 2014a). Contrasting these results with values of 1.25 for 1980 and 0.57 W/m^2 for 1950, it's clear that the total RF is increasing and at an increasing rate. There is little to suggest that the trend will not continue. While lower estimates of the total for the remainder of the twenty-first century vary over a narrow band from 2.7 to 3.0 W/m^2, upper estimates increase from 3.3 W/m^2 in 2030 to 8.4 W/m^2 by 2100 (IPCC, 2014a).

3.2 Greenhouse gas emissions: recent trends

Global warming is driven by an increase in the radiative forcing, which in turn is driven by anthropogenic emissions of warming agents. What can be said about the nature and amount of these emissions and whether they are trending up or down? Let's begin by examining the contribution of one nation.

In 2012, the United States discharged 6,526 million metric tons (tonnes) of carbon dioxide equivalents (6,526 Mt-CO_{2eq}) of GHGs to the atmosphere (EPA, 2014a).[8] The symbol "t" designates one tonne (1 t = 10^3 kg = 10^6 g), and a million tonnes corresponds to one teragram (1 Mt = 10^{12} g = 1 Tg). Of the total emissions, CO_2 accounted for about 82.5% (5,383 Mt), the lion's share of which (94.2%) resulted from the combustion of fossil fuels. Of the emissions from combustion, 74.2% were associated with electricity genera-tion (2,023 Mt) and transportation (1,740 Mt). The data communicate three important messages: (1) carbon dioxide is the leading contributor to global warming; (2) fossil fuel combustion contributes disproportionately to CO_2 emissions; and (3) almost 75% of this contribution is linked to two eco-nomic sectors, power production and transportation. Important features of fossil fuels are discussed in Appendix B.

Although carbon dioxide is justifiably a focal point of attention, the sig-nificance of other GHGs cannot be discounted. In 2012, nitrous oxide and fluorinated gases contributed 410.1 Mt-CO_{2eq} (6.3%) and 165.0 Mt-CO_{2eq} (2.5%), respectively, to total U.S. emissions. Natural processes comprise approximately 70% of total N_2O emissions and are largely due to bacte-rial action on nitrogen in soils and oceans. The remaining 30% is due to human activity, primarily from tillage and fertilization of agricultural land and to a lesser extent from fossil fuel combustion and the production of nitric acid. Use of nitrogen fertilizers to increase agricultural production contributes to global warming in two ways, first from processes for making the fertilizer, which produce about 3 tonnes of CO_2 per tonne of fertilizer, and then from decomposition of the fertilizer in soil and release of N_2O to the atmosphere.[9] The fluorinated gases, which include perfluorinated chemicals (PFCs) and sulfur hexafluoride (SF_6), as well as the hydrofluo-rocarbons (HFCs), are exclusively anthropogenic and are used for a broad range of industrial processes and products. As agents of global warming, the significance of N_2O and fluorinated gases is underscored by their long atmospheric lifetimes and large GWPs.

Methane, which reportedly contributed 567.3 Mt-CO_{2eq} (8.7%) to U.S. 2012 GHG emissions, merits special attention. Anthropogenic sources – consisting of landfills, fermentation of enteric (intestinal) matter by livestock, natural gas and oil production and distribution systems, and coal mining – accounted for approximately 60% of the total.[10] Of these sources, natural gas production and distribution systems have come under intense scrutiny.

As we'll find in Chapter 6, there is significant potential to reduce total GHG emissions by substituting natural gas for coal in generating electricity. Because gas-fired power plants operate more efficiently than coal-fired plants and the carbon content of gas – which consists largely of methane – is less than that of coal, CO_2 emissions from a gas-fired plant are about half those of a coal-fired plant for equivalent electrical energy production. But leakage during production, processing, and transmission of the gas to the power plant can negate the effect of reducing emissions by transitioning from coal to gas (Tollefson, 2013). To reduce emissions by transitioning from coal to natural gas, leakage should be less than 3%, and to realize the potential for a significant reduction, it should be less than 1%. However, there is concern that leakage from production, storage, and distribution systems has been underestimated (Howarth et al., 2011; Zeller, 2011; Tollefson, 2012, 2013; Karion et al., 2013), in which case estimates of CH_4 emissions would be underreported. Leakage rates as high as 11.7% have been measured, well above industry and EPA estimates.

As we'll find in Section 4.3, there is also concern that global warming is increasing methane emissions from natural sources. Measurements made in surface waters and the atmosphere have revealed outgassing from permafrost in sea beds of the Arctic continental shelf and wetlands in temperate latitudes of the Northern Hemisphere (Shakhova et al., 2010; Heimann, 2010). Were warming to enhance these emissions and also to cause release of methane from clathrates (methane hydrates extensively found in sea beds around the world), CH_4 could well displace CO_2 as the dominant GHG. The underlying message is that methane emissions cannot be ignored and must be more carefully monitored. Sources of natural gas and methane emissions are discussed more fully in Appendix C.

Recent trends provide some mildly encouraging signs for U.S. GHG emissions. Although emissions increased at an average annual rate of 1.4% from 1990 to 2000, growth slowed to an average rate of 0.3% from 2000 to 2007 and was followed by a general decline from peak emissions of 7,263 $Gt\text{-}CO_{2eq}$ in 2007. From 2005 to 2012, total GHG emissions dropped by about 10%, with a year-over-year drop of more than 3% from 2011 to 2012, the significance of which was underscored by the fact that it coincided with a 0.7% increase in population and a 2.8% increase in GDP (EPA, 2014a). Much of the decline from 2005 to 2012 is attributable to a deep recession

TABLE 3.2. *Annual carbon dioxide emissions from the consumption of fossil fuels*

	CO2 emissions (Mt-CO2)						
	1970	1980	1990	2000	2005	2010	2013
U.S.	4,683	5,159	5,445	6,377	6,494	6,143	5,931
China	734	1,500	2,459	3,430	5,574	7,954	9,524
India	210	324	581	953	1,180	1,641	1,931
OECD	10,281	11,820	12,417	14,189	14,785	14,229	13,940
Non-OECD	4,712	7,502	10,196	11,193	14,668	18,647	21,155
World Total	14,993	19,322	22,613	25,382	29,453	32,876	35,094

Source: From BP (2014).

that had impaired economic activity and, correspondingly, GHG emissions. But even as the economy began to recover, emissions were suppressed by reduced demand for transportation fuels and a shift from coal- to gas-fired power generation. From 2005 to 2012, coal's share of U.S. power generation dropped from 50% to 37%, and in 2012, CO_2 emissions from energy consumption was the lowest since 1994. However, whether the United States can sustain a downward trend in its emissions remains to be seen. Driven in part by coal's rebound to a 39% share of the power production market, U.S. emissions increased by 2.9% from 2012 to 2013.

Like the United States, emissions of other OECD nations have been flat or declining slightly since 2005, although the decline has been dwarfed by growth in non-OECD nations, especially in Asia. Consider the results of Table 3.2, which track global emissions of CO_2 from fossil fuels, along with data for the United States, China, India, and the cumulative contributions of OECD (developed) and non-OECD (largely developing) nations. Since 1970, global emissions have increased without pause, achieving a high of nearly 35,100 Mt-CO_2 in 2013. A 5.8% increase from 2009 to 2010 was the largest year-over-year growth since 1969. In a departure from the norm, the increase even exceeded the 5.6% jump in energy consumption and was due to a 7.6% increase in global coal consumption, with non-OECD countries accounting for 69% of consumption and China alone accounting for 48% of the total. Since 2010, annual emissions have increased by about 2%. Combined with rapid economic growth, a large dependence on coal in non-OECD nations virtually ensures continued growth in global emissions.

Table 3.2 also reveals a significant rearrangement of the *elephants in the room*. In 1970, the United States accounted for 31% of global emissions. But, with emissions rising slowly since 2000 and peaking in 2007, its contribution to the total was down to 17% by 2013. In contrast, China surpassed the United States as the world's largest emitter in 2006 and in 2013 contributed 27% of the global total. Until recently, China could argue that its per capita emissions remained small relative to those of OECD nations, despite rapid growth in total emissions. But, in 2013, China's per capita emissions rose above those of the EU for the first time. China has become and will remain the world's largest contributor to GHG emissions. But, among developing nations of Asia, it is not alone.

India is the world's third-largest producer of GHGs, and its annual emissions are increasing by about 5.5%. Were this rate to remain unchanged – likely in view of the extent to which it must still develop its economy (Figure 1.1) and its abundant coal resources – and U.S. emissions were to remain unchanged, India would become the world's second-largest emitter by 2035.

A similar contrast exists between OECD and non-OECD nations. In 1970, OECD nations accounted for 69% of global emissions; in 2013, non-OECD nations accounted for 60% of the total. And while OECD emissions peaked in 2005, non-OECD emissions continued to rise. To the extent that global warming is a problem, developing nations have become significant contributors. And with hundreds of millions of citizens yet to be freed from poverty, non-OECD emissions will continue to grow and will be the principal source of rising global emissions.

While it is important to track annual GHG emissions, it is cumulative emissions that drive climate change. Using the period from 1861 to 1880 as a reference and accounting for non-CO_2 radiative forcings, it's estimated that to achieve a better than 66% probability that the global mean surface temperature will not rise by more than 2°C, cumulative CO_2 emissions must not exceed 2,900 Gt-CO_2. Through 2011, about 1,890 Gt-CO_2 had already been emitted (IPCC, 2014a). While developed nations are responsible for most of the cumulative emissions, it will not be long before developing economies achieve this distinction, with China well on its way to surpassing the United States. If and when annual global emissions begin to decrease, cumulative emissions would continue to rise along with atmospheric CO_2 concentrations.

3.3 A macro view of contributing factors

Greenhouse gas emissions drive global warming, but what drives emissions? At an elemental level, it's people. The exponential growth in the atmospheric concentrations of CO_2, N_2O, and CH_4 that began in the eighteenth century tracks well with exponential growth in the world's population (Falkowski and Tchernov, 2004). But it's about more than the number of people. Emissions also depend on how people live and how much they consume.

In Chapter 1, we saw that living standards are strongly linked to energy consumption. And, in this chapter we've seen that emissions are strongly tied to energy consumption. And we know that some forms of energy, namely fossil fuels, release GHGs, principally carbon dioxide, while others are carbon free. Another factor is therefore the carbon content of the energy source.

The foregoing factors can be combined in a deceptively simple but conceptually useful equation for the annual carbon dioxide emissions, CO_2A, of a nation (or the world).

$$CO_2A = P \times \frac{GDP}{P} \times EI \times (CO_2I)_{ave} \tag{3.1}$$

Factors that contribute to emissions include: (1) population (P), (2) standard of living, which by default is measured by gross domestic product (GDP) per capita, (3) the amount of energy used per unit of GDP, termed the energy intensity (EI), and (4) the average carbon content of the energy sources (fuel mix). With GDP quantified in dollars, EI has units of megajoules per dollar (MJ/$). A nation's fuel mix can be quantified in terms of the weighted average CO_2 intensity $(CO_2I)_{ave}$ of its fuel sources. Emissions increase with increasing population and per capita GDP, but decrease with decreasing energy intensity and replacement of carbon-based fuels by renewable or nuclear energy sources.

In October 2011, the world's human population reached 7 billion and remains on a trajectory of growth.[11] Moreover, human expectations for a higher standard of living are globalizing, with prospects for realization improving in many developing nations. Barring a prolonged global economic crisis, it is therefore likely that the first two terms on the right-hand side of Equation 3.1 will continue to increase. If GHG emissions are to be curbed, at least in the near term, emphasis will then have to be placed on reducing the third and fourth terms. Such reductions depend on the development of technologies that increase the efficiency of energy utilization, the adoption of conservation practices that transcend technology-based efficiency measures, the replacement of high CO_2I energy sources with those of lower or zero intensity, and/or sequestration of CO_2 emissions.

The contribution of each term on the right-hand side of Equation 3.1 to changes in emissions from 1990 to 2002 has been delineated for representative developing and developed nations (Kintisch and Buckheit,

2006). Although China was able to reduce its energy intensity by 96%, increases in population, GDP per capita, and the carbon content of its fuel mix increased emissions by 15%, 122%, and 8%, respectively, yielding a net increase of 49%. In India, another rapidly developing nation, a 31% reduction in energy intensity was overwhelmed by increases of 28%, 55%, and 19%, respectively, in population, GDP per capita and carbon content of the fuel mix, causing emissions to grow by 70%. The conclusion is clear. Even with significant progress made in reducing energy consumption per unit of economic output in developing nations, emissions continue to increase with economic and population growth, augmented by reliance on energy sources of high carbon content. Even in the United States, which realized reductions of 20% and 1% in energy intensity and the carbon content of its fuel mix, emissions increased by 18% due to increases of 16% in population and 23% in GDP per capita. Among developed nations, Germany was an exemplar, where reductions of 21% and 10% in *EI* and fuel mix reduced emissions by 13%, despite increases of 15% and 4% in *GDP* per capita and population.

3.4 Whither emissions?

So much for where we've been! Where are we going? Can we predict future emissions and their impact on atmospheric GHG concentrations? How much confidence can we have in the predictions?

Models developed to predict the extent of global warming and its ramifications require knowledge of changes in radiative forcing functions, which in turn depend on estimates of future GHG emissions. But these estimates are based on assumptions for a range of factors. At what pace will economic and population growth continue? How quickly will advancements in energy technologies proceed? Emissions are reduced by increasing the efficiency of processes and devices that produce and use energy. Emissions are also reduced by replacing fossil fuels with carbon-free sources of primary energy. These pathways depend critically on technology. How quickly can engineers develop systems and processes that reduce emissions at acceptable costs?

Predictions of future emissions are also linked to patterns of human behavior and cultural change. Will more of the world's population embrace consumption as global disparities in wealth diminish? Recent trends in China, Russia, and India suggest this to be a plausible scenario. Conversely, will citizens of wealthy nations moderate their consumption? Worldwide, will people embrace energy efficiency and conservation measures and demand greater access to modern and efficient public transportation systems? These questions are difficult to answer, and attempts to do so are subject to uncertainty. To hedge one's estimates of future emissions, the logical approach is to consider a range of scenarios based on different assumptions and designed to bracket future emissions.

An effort to predict emissions through the twenty-first century was made in 1992 by the Intergovernmental Panel on Climate Change (IPCC) and disseminated in its Second Assessment Report (IPCC, 1995). Predictions were linked to assumptions concerning global economic and population growth and the mix of energy sources (fossil versus non-fossil). Lower and upper limits corresponded to year-2100 human populations of 6.1 and 17.6 billion and to average annual economic growth rates of 1.2% and 3.5%. The scenario yielding the largest year-2100 emissions of approximately 75 billion tonnes of carbon dioxide (75 Gt-CO_2) assumed moderate population growth, rapid economic growth, and plentiful supplies of fossil fuels. The amount is huge, more than doubling circa-2014 emissions.

Emission scenarios for the period from 2000 to 2100 were refined in a Special Report on Emission Scenarios (SRES) (Nakicenovic et al., 2000) and the *Third Assessment Report* of the IPCC (2001a). It's instructional to examine three cases – labeled A1, A2, and B1 – which bracket the range of possibilities for GHG emissions.

Scenario A1: Strong global economic growth occurs throughout the century, but with global population peaking in mid-century at approximately 9 billion people and declining thereafter. Efficient energy technologies are developed and deployed throughout the world. Differences in per capita income across the globe diminish, with increasing economic and social interactions and hence continuation of what we now term *globalization*. Special cases within this category differ according to energy utilization options. The extremes are represented by the following two cases.

Scenario A1FI: Alternative energy sources (nuclear and renewable) are introduced slowly, and fossil fuels are used extensively.

Scenario A1T: Emphasis is placed on rapid deployment of new technologies and alternative sources of energy.

Scenario A2: Like A1, there is emphasis on developing efficient energy technologies, but unlike A1, the globalization movement falters. Significant regional differences in economic development and fertility persist, and global population increases throughout the century to approximately 10 billion people by 2100.

TABLE 3.3. *Projected year 2050 and 2100 atmospheric CO_2 emissions and concentrations*

Year	Annual emissions (Gt-C)	Atmospheric concentrations (ppm)
2000	8.0	370
2050		
A1FI	24.0	600
A1T	12.0	500
A2	17.4	540
B1	9.2	480
2100		
A1FI	28.2	970
A1T	4.3	560
A2	30.0	850
B1	5.2	545

Source: From IPCC (2001a).

Scenario B1: Global population declines rapidly following a mid-century peak, and there is a high level of international cooperation, driven by a commitment to economic, social, and environmental sustainability. Globally, there is also a transition from manufacturing (energy and material intensive) to service-based economies and aggressive deployment of alternative and energy-efficient technologies.

Models based on the foregoing scenarios were used to estimate twenty-first-century emissions and their effect on atmospheric CO_2 concentrations. As shown in Table 3.3, scenarios A1T and B1 are the most benign, while A1FI and A2 provide worst-case conditions. Beginning with year-2000 emissions of 8 Gt-C/yr, the worst-case scenarios predict emissions of 28 to 30 Gt-C by 2100, while the best-case scenarios yield a mid-century peak and a decline to emissions of 4 to 5 Gt-C by 2100. But even for the more benign scenarios, the atmospheric CO_2 concentration would increase throughout the century and the best-case (A1T, B1) year-2100 conditions would exceed what many believe to be acceptable upper limits.

Two limits are commonly used as approximate markers of the consequences of rising atmospheric CO_2 concentrations. If the concentration can be stabilized at no more than 450 ppm – a condition for which the average surface temperature of the Earth would not exceed the preindustrial value by more than 2°C – it is believed that adaptation to the consequences

TABLE 3.4. *Projected cumulative CO_2 emissions from 2012 to 2100 and circa-2100 atmospheric CO_2 concentrations for representative concentration pathways*

RCP (W/m²)	Cumulative CO_2 emissions (Gt-C) 2012–2100		Concentration (ppm) year–2100	
	Mean	*Range*	CO_2	CO_2eq
2.6	270	140–410	421	475
4.5	780	595–1,005	538	630
6.0	1,060	840–1,250	670	800
8.5	1,685	1,415–1,910	936	1,313

Source: From IPCC (2014a).

of climate change would be manageable. However, should the concentration stabilize at more than 550 ppm, adaptation would be problematic and potentially unmanageable. Neither of the thresholds is achieved for the foregoing scenarios. Even for scenarios B1 and A1T, stabilization does not occur by 2100, and is only achieved early in the twenty-second century at concentrations of approximately 560 and 600 ppm, respectively. For the other emission scenarios, concentrations remain on a sharply ascendant trajectory, stabilizing centuries later at values well in excess of year-2100 levels.

Recognizing the dependence of climate change on cumulative GHG emissions, the IPCC considered a different set of scenarios – called Representative Concentration Pathways (RCPs) – in its Fifth Assessment Report (IPCC, 2014a). Four pathways were linked to cumulative CO_2 emissions from 2012 to 2100 and to the corresponding value of the total year-2100 radiative forcing. The pathways account for all anthropogenic sources but exclude potential contributions from decaying permafrost. Referring to Table 3.4, RCP 2.6 involves a climate policy that aggressively implements emission reduction measures, maintains cumulative emissions within the designated range and yields a mean circa-2100 RF of 2.6 W/m² for nominal cumulative emissions of 270 Gt-C. At the other end of the spectrum, RCP 8.5 represents a policy of weak mitigation and extensive use of fossil fuels, and a circa-2100 RF of 8.5 W/m² corresponds to nominal cumulative emissions of 1685 Gt-C. For RCP 2.6 the RF peaks before year 2100 and subsequently declines; for RCP 8.5 it continues to increase well beyond 2100. Table 3.4 also provides the circa-2100 atmospheric CO_2 concentration for each scenario, as well as the combined effect of CO_2, CH_4 and N_2O emissions on CO_{2eq}. Given current trends, it is becoming a near certainty that atmospheric CO_2 concentrations will rise above the 450 ppm benchmark and ever more likely that 550 ppm will be exceeded.[12]

Although the SRES and RCP scenarios of the IPCC provide a range of possibilities for twenty-first-century GHG emissions, more specific projections – typically over shorter time frames – are left to other organizations. In its 2010 publication of World Energy Outlook (IEA, 2010), the International Energy Agency provided a twenty-five-year forecast of energy-related CO_2 emissions and, for a business-as-usual scenario, predicted a rise in annual emissions from approximately 30 Gt-CO_2 in 2010 to 43 Gt-CO_2 (from 8.2 to 11.7 Gt-C) in 2035. The estimate is consistent with more recent projections (BP, 2014b) that estimate a 29% increase from 2012 to a value of 45 Gt-CO_2 by 2035. However, with actual emissions of 35 Gt-CO_2 in 2013 and continuation of recent increases of 2% or more per year, circa-2035 emissions would be well above 45 Gt-CO_2 and the A1T/ B1 and RCP 2.6/ 4.5 scenarios of the IPCC would significantly understate the actual rise in emissions. *The world is awash in fossil fuels, and the incentives to exploit them are large.*

3.5 The carbon cycle

Once they enter the atmosphere, greenhouse gases do not remain in perpetuity. Various natural processes provide for removal or conversion to another gas. It is these processes that determine the atmospheric lifetimes of Table 3.1.

The fate of atmospheric carbon dioxide is governed by the Earth's carbon cycle (Prentice et al., 2001). The cycle involves transfer of CO_2 between four major repositories: the atmosphere, hydrosphere, lithosphere, and the Earth's pool of unused fossil fuels (Figure 3.4). A fifth repository (geological pools) consists of rock formations within which carbon is immobile. Storage in the hydrosphere is largely in the Earth's oceans and includes a biogenic (organic) component, $CaCO_3$ shells, and dissolved forms of CO_2 and $CaCO_3$. Storage in the lithosphere includes a terrestrial biotic (plant) component, as well as organic and inorganic constituents of soil (Lal, 2004).

Circa 2000, the carbon content of the active repositories was approximately 750 Gt and 2,500 Gt for the atmosphere and lithosphere, respectively, between 5,000 and 10,000 Gt for the pool of unused fossil fuels, and 93,000 Gt for the oceans (Peña and Grünbaum, 2001). The oceans clearly provide a huge repository for carbon. Selecting an intermediate value of 7,500 Gt for the carbon in unused fossil fuels, storage in the oceans, fossil fuels, and land is, respectively, 124, 10, and 3.33 times that of the atmosphere.

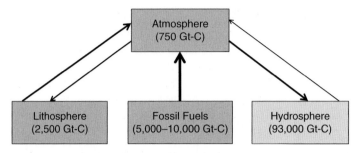

FIGURE 3.4. Carbon dioxide exchange in the global carbon cycle. Estimates of circa-2000 carbon stored in each repository are taken from Peña and Grünwald (2001).

If equilibrium exists, the carbon cycle calls for unchanging carbon content in each of the repositories. Equilibrium does not preclude transfer of carbon between repositories, but it does require a balance between inflow and outflow for each repository. For example, if the atmospheric CO_2 concentration is to remain unchanged, the rate at which CO_2 is discharged to the atmosphere must equal the rate at which it is transferred from the atmosphere to the lithosphere and hydrosphere. Under nonequilibrium conditions, the atmospheric concentration would increase if the rate at which CO_2 is discharged to the atmosphere exceeds the rate at which it is transferred from the atmosphere. Such a condition was initiated by onset of the Industrial Revolution.

With growing use of fossil fuels, natural emissions of CO_2 are augmented by anthropogenic emissions, causing the atmospheric CO_2 concentration to increase sharply from its preindustrial value of 280 ppm (Figure 3.1a). The increase in concentration would be larger were it not for the fact that approximately 50% of the emissions are transferred from the atmosphere to upper layers of the ocean. Although net transfer of CO_2 from the lithosphere to the atmosphere has varied since the IR, it has been largely positive in recent years. While photosynthesis and reforestation withdraw CO_2 from the atmosphere, the combustion of biomass (including forest fires), deforestation, and the respiration of organic matter transfer CO_2 from the lithosphere to the atmosphere.[13] Differences between the widths of opposing arrows in Figure 3.4 reflect the relative magnitudes of competing processes and the fact that net transfer is to the atmosphere.

An important feature of the carbon cycle is the large differences in time scales associated with the transport processes, ranging from comparatively fast (less than a year) for dispersion of CO_2 in the atmosphere to very slow (millennia) for dispersion in the oceans. Although CO_2 is soluble in ocean waters, the rate at which it is transferred from the atmosphere is limited by the rate at which it is transferred from surface to deeper layers of the ocean (McKinley et al., 2011). Warmer, thermally stratified upper layers suppress mixing that would enhance transfer to deeper layers, and transfer of CO_2 from the atmosphere to the oceans significantly lags accumulation in the atmosphere. Time scales associated with terrestrial plant growth are relatively short (days to decades), while those associated with the many transformation processes occurring within the soil vary from days to millennia.

Time scales and the inherent inertia of large contributors to the Earth's climate system strongly influence the pace of climate change. Especially important are the atmospheric lifetimes of GHGs (Table 3.1) and factors governing changes in the Earth's oceans and cryosphere, of which the Greenland and Antarctic ice sheets play prominent roles. As described in Appendix D, the effects of today's emissions are not felt immediately but unfold over time frames that can encompass millennia. Even if GHG emissions were to be sharply curtailed, their long atmospheric lifetimes ensure continuation of a large RF well into the future.

The atmospheric CO_2 concentration is known to increase by 1 ppm for each 2.13 Gt-C of net CO_2 transfer to the atmosphere (Schmitz, 2002). The concentration increased by an average of 1.53 ppm/yr from 1980 to 1999, far exceeding the average increase of 0.40 ppm/yr from 1800 to 1994. And, the *rate* of growth continues to increase, with an average value of 1.93 ppm/yr recorded for the period from 2000 to 2006 (Canadell et al., 2007) and with annual growth of more than 2 ppm in recent years. The increased rate is due to an accelerated pace of global economic development, an increase in the carbon intensity associated with energy utilization, and the inability of assimilation by the Earth's oceans and land to keep pace with the emissions.

3.6 Summary

Rapidly rising atmospheric GHG concentrations are a reality and are driven by human activity. Since the Industrial Revolution, both the concentration of CO_2 and the rate at which it is increasing exceed levels experienced at any time in the previous 2 million years. The carbon cycle has become unbalanced. Simply put, the rate at which natural processes remove CO_2 from the atmosphere has lagged the rate at which CO_2 is discharged to the atmosphere, causing CO_2 to accumulate in the atmosphere. In addition, time scales and inertia associated with the Earth's climate system ensure that the effects of past and current emissions will be manifested over centuries, if not millennia, only to be augmented by future emissions.

If an increasing atmospheric CO_2 concentration is a fait accompli, what would be an acceptable level of accumulation? Is there a tipping point beyond which we dare not go? We know that the increase in radiative forcing with increasing concentration is a major factor in global warming, and we'd like the CO_2 concentration to remain below a value for which the effects of warming are small or at least manageable. An atmospheric concentration of 450 ppm is often cited as a threshold for which the Earth's average surface temperature would exceed the preindustrial level by about 2°C and beyond which adaptation to the effects of climate change would be difficult. It can be said with near certainty that this threshold will be exceeded. What about the second threshold of 550 ppm, beyond which consequences of climate change may become unmanageable? On our current emissions trajectory, we would get there well before the end of the century.

Beginning with a circa-2013 concentration of approximately 397 ppm (846 Gt of atmospheric carbon), an increase to 550 ppm would add another 326 Gt-C to the atmosphere. If we continue to increase the concentration at an average rate of 2 ppm/yr, we'll be there by 2090. Since approximately 50% of CO_2 emissions enter the world's oceans and forests, the atmospheric concentration would increase by 1 ppm for every 4.26 Gt of emitted carbon, which would correspond to emission of approximately 652 (2×326) Gt. But keep in mind that the full effect of the new emissions would not be felt for centuries, nor would emissions cease after 2090. They are not controlled by an on-off switch, and the CO_2 concentration would continue to rise. To stabilize at 450 ppm, emissions would have to peak within several years and drop rapidly thereafter; to stabilize at 550 ppm, emissions would have to peak within a few decades and drop well below the 2013 level by the end of the century.

GHG emissions have been a topic of concern and intense debate for more than twenty-five years. Although efforts have been made to slow the growth in emissions, results to date have been incremental at best. Emissions continue to rise on a seemingly inexorable trajectory, and the world is at a crossroad. Substantive measures to reduce emissions can no longer be deferred. If emissions don't peak within the next few years and begin a steady and significant decline, the goal of keeping global warming within 2°C will not be met.

CHAPTER 4

Global warming

We know that atmospheric GHG concentrations have been increasing for more than two centuries in concert with increasing GHG emissions. We also know that increasing GHG concentrations create a positive radiative forcing that reduces the outflow of thermal radiation from the Earth to outer space. Hence, with little change in solar radiation, the global radiation balance is disrupted by a net transfer of radiation to the Earth, whose temperature must then increase until a balance between the radiative inflows and outflows is restored.

Our confidence in the validity of the foregoing assertions is *rock solid*. But to what extent has the Earth's temperature increased, and how much will it increase in years to come? The first question is answered by temperature records, the second by the use of climate models.

4.1 The Earth's temperature history

Proxy and instrument records

Since the early nineteenth century, temperatures at the Earth's surface and lower atmosphere have been obtained from direct measurements, beginning with the use of thermometers and progressing to more sophisticated electronic and radiometric sensors. Such measurements comprise what's known as the *instrument record*. For earlier periods attempts to reconstruct the Earth's temperature history rely on *proxy data* obtained from paleoclimatic artifacts such as tree rings, ice cores, marine sediments, and sea coral. Although the data track past climate conditions and natural variability of the Earth's surface temperature, it is tied to discrete places and times and is by no means complete or without uncertainty. To a lesser extent, the same can be said of the instrument record.

Although the instrument record provides data from thousands of weather stations and seagoing vessels around the world, gaps exist in both space and time. Conditions around a weather station can also change over time due to land development practices, or the site itself can change, as can the instruments used for the measurements. Such variations must be reconciled before the measurements can be used to determine changes in the global or hemispheric average temperature over time. Different approaches can be taken to the reconciliation process, commonly termed *homogenization*, and to interpretation of the results.

Changes in the Earth's globally and annually averaged surface temperature are shown in Figure 4.1 for the period from 1880 to 2011. The results are based on direct measurements and, by referencing them to the average temperature from 1961 to 1990, are presented as a temperature anomaly. Blue and orange bars correspond to La Niña and El Niño years, respectively. With below-average sea temperatures in upper layers of the Pacific, a La Niña involves energy transfer from (cooling of) the atmosphere; conversely, with above-average sea surface temperatures, an El Niño provides energy transfer to (heating of) the atmosphere. ENSO events add a layer of natural variability to anthropogenic effects, augmenting warming during an El Niño and attenuating it during a La Niña. The last year (1997–8) marked by a strong El Niño (Null, 2014) coincided with an exceptionally high anomaly of more than 0.5°C, as did the last year (2009–10) of a moderate El Niño. Conversely, recent years of strong (2010–11, 1999–2000) and moderate (2007–8) La Niña coincided with depressed temperature anomalies. Nevertheless, overall the data reveal an increase in the anomaly of approximately 0.6°C from 1880 to 2011, with the most significant changes occurring from 1910 to 1940 and 1980 to 2000. Thirteen of the warmest years occurred in the fifteen years from 1997 to 2011, with the two warmest years in 1998 and 2010 corresponding to El Niño events and 2011 corresponding to the warmest La Niña year on record (CSIRO, 2012). The subsequent period from 2012 through 2014 was marked by absence of a significant ENSO event.

The fact that the overall trend of increasing temperature is interrupted by departures due to short-term natural drivers of climate underscores the importance of assessing the impact of anthropogenic effects on longer – decadal or larger – time scales. The data also mask regional differences in the temperature anomaly, which tends to be larger at higher latitudes and over land than at lower latitudes and over oceans.[1]

Figure 4.2 takes us further back in time by providing the temperature anomaly for the Northern Hemisphere over the last millennium. Annual

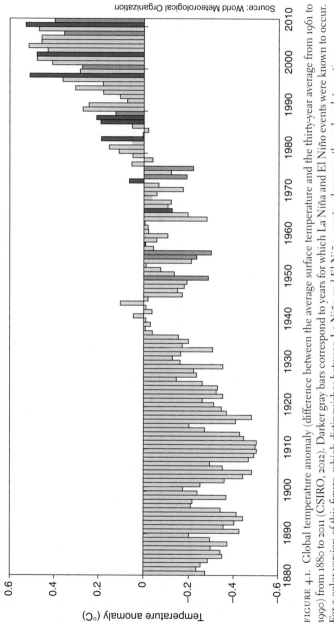

FIGURE 4.1. Global temperature anomaly (difference between the average surface temperature and the thirty-year average from 1961 to 1990) from 1880 to 2011 (CSIRO, 2012). Darker gray bars correspond to years for which La Niña and El Niño events were known to occur. For a color version of this figure, which distinguishes between La Niña and El Niño events, please see the color plate section.

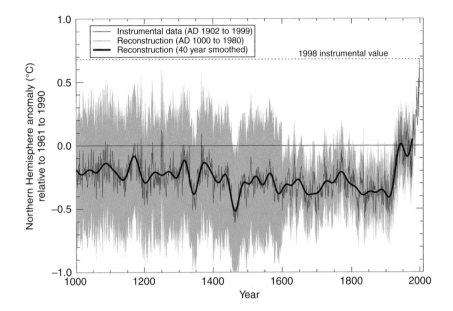

FIGURE 4.2. Temperature anomaly (difference between the average surface temperature and the thirty-year average from 1961 to 1990) for the Northern Hemisphere from
1000 to 2000 (IPCC, 2001b). For a color version of this figure, please see the color
plate section.

variations in the reconstructed proxy data (in blue) are represented by fluctuations about the bold line, which provides a running forty-year average of
the data. The shaded (gray) region indicates the range of uncertainty in the
data, which is larger for the period from AD 1000 to 1600. The proxy data
are calibrated against and merged with the instrument data (in red) for the
last century of the millennium. The twentieth-century trend of increasing
temperature is again evident, contrasting sharply with the ±0.20°C oscillation about a nominal value of minus 0.30°C for the period from 1000 to
1900. The results suggest that the 1990s were the warmest decade of the
last millennium and that each of the last three decades was successively
warmer than any previous decade.

   In view of the large uncertainties in the proxy data, it is reasonable to
question whether they provide a reliable window to the past and whether it is
appropriate to merge the data with the more recent and accurate instrument
record. The second question was laid to rest by a recent study (Anderson et al.,
2013) in which proxy temperatures were compiled over the past 130 years and
compared with the instrument record. The proxy record confirmed the overall warming trend revealed by the instrument data, including accelerated

warming from 1920 to 1940 and 1980 to 2010, and affirmed the validity of using proxies to reconstruct the Earth's thermal history.

Not without controversy

With no basis for attributing the warming to natural agents and with ever increasing GHG emissions, the foregoing results point to a human finger-print. But this conclusion is not without controversy.

In its entirety the solid line of Figure 4.2 became known as the *hockey stick*, but a more appropriate metaphor would be a lightning rod. The reconstructed data suggest comparatively small variations in the mean temperature of the Northern Hemisphere over the first 900 years of the millennium (the handle of the hockey stick), followed by a dramatic increase over the last 100 years (the blade of the stick). However, because the data mask a Medieval Warm Period that occurred in the eleventh cen-tury and a Little Ice Age in the sixteenth century, questions were raised about the methodology used by Mann et al. (1998, 1999) to reconstruct the proxy data.

Soon and Baliunas (2003) drew attention to absence of the Medieval warming period, while McIntyre and McKitrick (2005) concluded that the statistical procedures used by Mann and colleagues masked the extent to which temperature variations were induced by natural agents. Similar conclusions were reached by Moberg et al. (2005), who used a different procedure to reconstruct the data. Their results yielded Medieval warm-ing comparable to that experienced in much of the twentieth century, as well as pronounced cooling in the sixteenth century. Accordingly, they rea-soned that the "large natural variability in the past suggests an important role of natural multi centennial variability that is likely to continue." In other words, natural effects, such as changes in solar irradiation and aerosol concentrations, were significant agents of global warming (or cooling), and the handle of the hockey stick is not as flat as some may wish to believe. That said, anthropogenic effects could not be dismissed.

Moberg et al. (2005) went on to state that their findings do "not imply that the global warming of the last few decades has been caused by natural forcing factors alone," and they "find no evidence for any earlier periods in the last two millennia with warmer conditions than the post-1990 period." Their conclusions were affirmed by Osborn and Briffa (2006), who exam-ined millennial proxy data for the Northern Hemisphere and discerned Medieval warming from approximately 890 to 1170 and a Little Ice Age from approximately 1580 to 1850. However, they also found the most significant and widespread (geographically) warming to have occurred from the mid-to late twentieth century.

In the United States, the hockey stick drew intense political inter-est, with one congressman requesting input from the National Research Council (NRC), an arm of the U.S. National Academies, and another unequivocally rejecting its validity. From the NRC study (NRC, 2006), it was concluded with high confidence that, over the last few decades, the global mean surface temperature exceeded that for any comparable period during the last four centuries. Although it attached less confi-dence in reconstruction of pre-1600 proxy data, the study found it "plau-sible that the Northern Hemisphere was warmer during the last few decades of the 20th century than during any comparable period over the preceding millennium."

Nevertheless, the political drumbeat continued, and on July 19, 2006, a subcommittee of the U.S. House Committee on Energy held a six-hour meeting for the purpose of discrediting the Mann et al. work (Monastersky, 2006). Despite expert testimony given by the chair of the NRC study, the chair of the House Committee remained unconvinced that global warm-ing has anthropogenic origins or that it even exists. Additionally, a major U.S. newspaper (WSJ, 2006) continued to debunk the hockey stick on the basis of deficiencies in statistical methodologies. Meanwhile the scientific effort continued, with subsequent studies augmenting the case for anthro-pogenic warming.

Mann et al. (2008) subsequently reconstructed hemispheric and global surface temperatures over the past 2,000 years using a much larger set of proxy data, as well as updated instrumental results. Their key conclusion remained the same: twentieth-century warming is unprecedented. From reconstruction of seventy-three globally distributed records based on eight different temperature proxies, Marcott et al. (2013) obtained similar results. But by reconstructing data over the last 11,300 years – essentially the entire Holocene – their study went much further. Following a warm period from approximately 10,000 to 5,000 years ago, the Earth cooled by about 0.7 °C, reaching its lowest temperature during the Little Ice Age, only to rise rapidly in modern times, reaching the highest levels seen in more than 4,000 years. Although current temperatures do not exceed levels reached during the early Holocene, they are projected to do so by the end of the century. Collectively, the foregoing results suggest little reason to doubt anthropogenic origins of the twentieth-century warming.

Proxy data and the hockey stick aren't the only contentious issues, how-ever. It's generally accepted that the Earth's temperature is linked to the atmospheric CO_2 concentration. But, what's the driver? In preindustrial times, did increased concentrations precede or follow increased temperatures? Our knowledge of atmospheric physics suggests that changes in temperature should lag changes in concentration. But could other processes create the opposite effect? Could increased CO_2 concentrations be a consequence of increased temperatures? If so, as some data suggest, is there reason to doubt the association of current warming with increased GHG emissions? And, looking again at Figure 4.1, what are we to make of the fact that there has been little change in temperature since 1998, despite a significant increase in GHG emissions over that period? Could the temperature have reached a plateau? We'll examine both issues, beginning with the connection between temperature and the atmospheric CO_2 concentration.

The CO_2–temperature linkage

One way to resolve this *chicken-and-egg* question is to examine how changes in atmospheric CO_2 concentration and temperature tracked one another during previous periods of climate change. Since ice cores provide a well-preserved and accurate 800,000-year record of atmospheric CO_2 concentrations and Antarctic temperatures, they are well suited for this purpose. Monnin et al. (2001) used the ice cores to obtain a record of concentrations and temperatures from the last glacial maximum to the start of the Holocene (about 22,000 to 11,000 years ago). Over that period the CO_2 concentration increased by about 40% and seemingly lagged the increase in temperature by 800±600 years. That is, rising temperatures led rising CO_2 concentrations. The results were not lost on those dismissive of anthropogenic agents of warming, and the study was widely used to block action that would reduce GHG emissions and mitigate warming. In 2007, the paper was used to rebuke former Vice President Al Gore during testimony on the scientific basis of warming before a Congressional committee. In the words of Representative Joe Barton (Gillis, 2013), "The temperature appears to drive CO_2, not vice versa. On this point Mr. Vice President, you're not just off a little. You're totally wrong." At the time, Representative Barton's comments were not unfounded, but neither was the vice president wrong.

Carbon dioxide concentrations measured from air bubbles trapped in ice cores can be accurately dated, but it's more difficult to date the ice itself, which is used to determine temperatures from the amount of heavy and light isotopes in the water molecules. Although it's well known that the trapped air is younger than the surrounding ice, uncertainties associated with establishing the age difference are large (Brook, 2013). Parrenin et al. (2013) addressed this problem by developing improved methods of determining the difference between the age of the ice and the CO_2. To within an uncertainty of less than 200 years, they concluded that Antarctic temperatures did not begin to increase centuries before the increase in atmospheric CO_2 and that the changes were essentially synchronous. Working with Antarctic ice cores and improved methodologies for dating the ice, Pedro et al. (2012) also concluded that coupling between increasing temperatures and atmospheric CO_2 concentrations was closer than previous estimates, and Shakun et al. (2012), using data from eighty locations to obtain a global record following the last ice age, concluded that warming followed increasing CO_2 concentrations at most locations.

With the current state of knowledge, what can be said with confidence is that large changes in temperature and concentration were strongly linked during the last deglaciation. What cannot be said with confidence is whether one or the other led by a century or so. But the point is moot. As discussed in Section 2.2, deglaciations are driven by an increase in solar radiation due to changes in the Earth's orbit. As oceans warm, they release CO_2 to the atmosphere, augmenting the warming due to increased insolation. But during the Holocene there has been little change in the insolation. What has changed is the use of fossil fuels and an attendant increase in the atmospheric CO_2 concentration over the last 200 years. The essential point is that changes in temperature and CO_2 concentration go hand-in-glove.

Temperature stasis

The plateau in temperature shown in Figure 4.1 for the first decade of this century is also revealed in Figure 4.3 as annual (top) and decadal (bottom) averages. The plateau has persisted, and through 2014, it had spanned fifteen years, during which there had been no statistically significant warming. The trend has been encouraging for skeptics of global warming but troubling for scientists seeking to understand cause and effect. Especially troubling was the fact that, while measured temperatures lagged predictions, falling well below upper estimates and moving ever closer to lower estimates (Stott et al., 2013), CO_2 emissions were still increasing and at a pace corresponding to the worst case (A1FI) of the IPCC SRES scenarios (CCRC, 2009). What's happening? If there is an imbalance in the Earth's radiation budget with net transfer to the Earth, where is the energy going? What do differences between predictions and observations say about the models and our understanding of climate science?

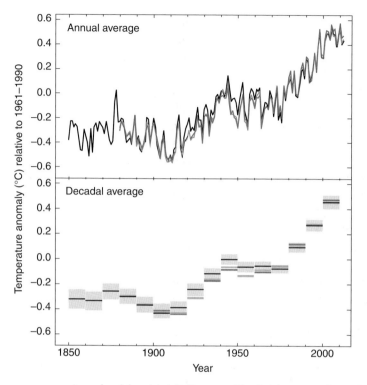

FIGURE 4.3. Annual and decadal globally averaged land and ocean surface temperature anomaly from 1850 to 2012 (IPCC, 2014a). For a color version of this figure, please see the color plate section.

We know that surface temperature is affected by both natural and anthropogenic agents. While anthropogenic GHG emissions contribute to surface warming, natural agents or other anthropogenic influences may act to enhance or impede the warming. We've seen that manifestations of the ENSO cycle can affect surface temperatures, increasing them during an El Niño and cooling them during a La Niña. The sharp rise in temperature experienced in 1998 was due to an exceptionally strong El Niño. Changes in solar input as well as departures from normal volcanic activity can also augment or curtail warming. Characterization of warming during the last half of the twentieth century as the blade of a hockey stick is therefore misleading, since it creates expectations for a monotonically increasing temperature due to GHG emissions, missing the point that other factors can cause departures – up or down – from expectations. If climate models are unable to accurately represent the other factors, they are unable to capture the departures.

Because the first decade of the twenty-first century also coincided with the declining phase of a sunspot cycle and concluded with a pronounced minimum of nearly zero spots, it might be argued that the temperature plateau was in part caused by an attendant reduction in solar radiation. However, the reduction was short-lived, and even as the ascending phase of the sunspot cycle returned, the plateau persisted. Accurate measurements of incident solar radiation by satellite-borne instrumentation indicate negligible changes since 1978 (National Academies, 2008).

Another explanation deals with large uncertainties in the Arctic temperature record due to inadequate spatial coverage of the measurements (Curry, 2014). Because the region has been warming so rapidly, the uncertainties can have a significant bearing on determination of the global average temperature. Using alternative approaches to reduce uncertainties associated with the Arctic temperature record, Cowtan and Way (2014) found that adjustments to the record were material and increased the global mean temperature by 0.11 to 0.12°C per decade over the period from 1997 to 2012.

Yet, another explanation for the stasis involves the effect of aerosols, particularly sulfates resulting from coal combustion (Kaufmann et al., 2011). Globally, coal consumption increased by 68% from 2003 to 2010. In China, consumption doubled from 2003 to 2007 and increased another 30% from 2007 to 2010. Although atmospheric aerosols produced by sulfur emissions – whether from coal or volcanic eruptions – are short-lived, they still have a near-term cooling effect due to reflection of solar radiation (a negative forcing), which counters warming due to increased CO_2 concentrations (a positive forcing). A period for which there is no change in the net forcing would be consistent with little or no change in temperature. Similar conditions may have marked the 1950s and early 1960s (Figure 4.1), when a large increase in U.S. and European coal consumption coincided with a temporary leveling of global temperatures.

Increasingly, the most plausible explanation involves thermal energy transport and storage in the oceans. More than 90% of global warming is manifested by heating of the oceans, and until recently, most of the heating was believed to be concentrated above a depth of 700 meters (the upper ocean), where the thermal energy is estimated to have increased by 17×10^{22} joules from 1971 to 2010 (IPCC, 2014a). However, an analysis of thermal energy stored in the world's oceans (Balmaseda et al., 2013) concluded that storage increased sharply in the Pacific Ocean after 1998 and that 30% of the warming since 2004 occurred in deeper waters below 700 meters. The trend was attributed to changing atmospheric circulation patterns and extended La Niña events, which increased mixing between near-surface and deeper ocean waters. The implication is that energy transport from upper to lower layers of the ocean has been underestimated and that global warming has continued unabated, despite the decade-long plateau in surface temperatures. That is, the plateau is not indicative of a cessation in warming but of continued warming manifested by a larger increase in the thermal energy of deeper ocean waters.

The foregoing results are consistent with those of Meehl et al. (2011, 2013), who used climate model simulations to show that for a TOA energy imbalance of 1 W/m², there is an increase in thermal energy transfer to ocean layers below 750 meters during decades for which there is little change in the globally averaged surface temperature. Conversely, for decades in which there is a significant increase in surface warming, there is an increase in transfer to upper ocean layers. In another simulation Kosaka and Xie (2013) linked the current stasis to a decadal La Niña cooling effect in the eastern equatorial Pacific. Collectively, the results suggest that the stasis is due to variability in ocean currents that transfers thermal energy to deeper layers of the Pacific Ocean.

The notion that the missing energy resides in deeper ocean layers was strengthened by the results of Chen and Tung (2014), albeit with one important caveat. Their work is especially significant because it uses data from a multinational program called Argo, which involves a global array of about 3,500 floats measuring ocean temperatures at depths up to 2,000 meters. The authors concluded that between 1999 and 2010, the energy stored at depths from 300 to 1,500 meters increased by about 69×10^{21} joules – a huge amount sufficient to explain the hiatus in surface temperature occurring over the same time frame. The caveat is that the increase occurred almost exclusively in the Atlantic and Southern Oceans and not in the Pacific, which other studies have designated as the source of the hiatus.

Chen and Tung also concluded that the surface temperature is influenced by a multidecadal trend involving two time scales. The longer time scale of forty or more years corresponds to a periodic variation in which the temperature transitions between accelerated warming and stasis. The warming that occurred from about 1975 to 2000 and the hiatus that began in 2000 and could continue for another decade or more is the most recent manifestation of the cycle, with the previous version characterized by warming from about 1915 to 1945 and stasis from 1945 to 1975. Within the longer time frame, the surface temperature is also influenced by superposition of ENSO events occurring over a shorter interval of two to seven years. But if the hiatus is to be explained by a significant accumulation of energy

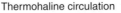

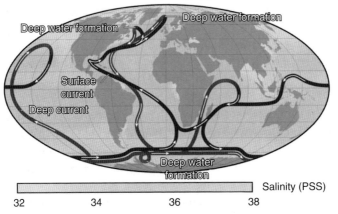

FIGURE 4.4. Thermohaline circulation associated with the Great Ocean Conveyor Belt. Courtesy of the National Aeronautics and Space Administration. For a color version of this figure, please see the color plate section.

stored in deeper layers of the Atlantic and Southern Oceans, what is the process driving this accumulation?

Chen and Tung attribute the long-term trend to periodic variations in the intensity of the *Atlantic Meridional Overturning Circulation* (AMOC), also known as the *thermohaline circulation* (THC) or the *great ocean conveyor belt*. Like ENSO, the AMOC has a significant effect on climate, but differs in its physical origins. ENSO is a cyclical ocean-atmospheric event that originates from diminished trade winds and eastward migration of warm surface waters from the Indian Ocean to equatorial regions of the Pacific Ocean. In contrast, the THC is driven by vertical distributions of temperature and salinity in waters of the North Atlantic.

The density of cold, salty water exceeds that of water that is warm and/or of low salinity. Hence, if cold ocean waters of high salinity overlie warmer water of lower salinity, conditions are unstable and buoyancy forces drive the cold, salt-rich water to deeper regions of the ocean. Such is the case in the North Atlantic. As shown in Figure 4.4, surface layers of water heading north from the tropics experience an increase in salinity due to evaporation and a decrease in temperature due to heat transfer to the atmosphere. Reaching the North Atlantic, buoyancy forces drive the water downward to deeper layers of the ocean, initiating a southward flow for thousands of miles to the Southern Ocean before bifurcating and flowing northward and eastward into the Indian and Pacific Oceans, respectively. As the temperature and salinity of the water increase and decrease, respectively, during its journey, upwelling eventually occurs in the Indian and Pacific oceans, and the process repeats itself with resumption of the northerly flow. Use of the term "the great ocean conveyor belt" follows from participation of the Atlantic, Southern, Indian, and Pacific Oceans in the circulation pattern.

Chen and Tung suggest that cyclical variations in the strength of the THC are responsible for the long-term cycle involving stasis and accelerated warming of the atmosphere. When the strength exceeds the long-term average, less heat is transferred to the atmosphere during its northward journey in the Atlantic and more is retained for transfer to deeper waters. As the current weakens and drops below its long-term average, more heat is transferred to the atmosphere at the expense of energy stored in deeper waters. If this interpretation is correct, the current stasis could last until 2030 before accelerated warming resumes. Over extended cycles, the trajectory of atmospheric warming would then be characterized by a stair-step pattern for which a half-period of increasing temperature is followed by a half-period of temperature stasis.

Although there is an ongoing debate over whether energy storage in deeper waters is principally in the Pacific or Atlantic/Southern Oceans (Kintisch, 2014), the foregoing studies provide a plausible explanation for the current surface temperature stasis. The Earth is continuing to warm during the hiatus but with the missing energy likely residing in the world's oceans. This explanation has been affirmed by other studies, in one case by Durack et al. (2014), who found that warming in the top 700 meters of ocean in the Southern Hemisphere has been underestimated by 24–58%, and in another by Llovel et al. (2014), who reported significant warming to a depth of 2 km. A growing body of evidence suggests that the oceans are warming faster than previously thought, and the rate at which the Earth is warming may, in fact, be at the high end of earlier predictions. That said, efforts to fully understand energy transfer in the land-ocean-atmosphere system remain a work in progress.

The instrument record redux

In the early 2010s, a group of climate skeptics – with financial support from individuals and organizations that rejected the notion of anthropogenic warming – independently and critically reviewed the temperature record for the last two centuries. Termed the *Berkeley Earth Surface Temperature Project*, the study included an exhaustive review of existing temperature data, as well as algorithms used for homogenization, averaging, and error analysis (Berkeley, 2011). To address sources of skepticism, the group

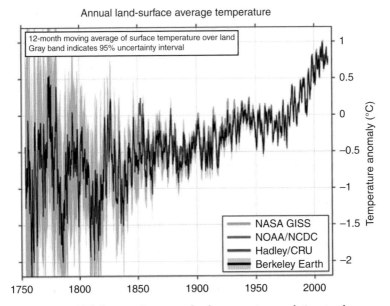

Annual land-surface average temperature

12-month moving average of surface temperature over land
Gray band indicates 95% uncertainty interval

NASA GISS
NOAA/NCDC
Hadley/CRU
Berkeley Earth

Temperature anomaly (°C)

FIGURE 4.5. Global *annual* average land temperatures relative to the 1950–79 mean. Results reported by the Hadley Centre of the UK Meteorological Office and the Climate Research Unit of the University of East Anglia (HadCRU), the Goddard Institute of Space Studies (GISS) of the National Aeronautics and Space Administration, the National Oceanic and Atmospheric Administration (NOAA), and the Berkeley Earth Surface Temperature Project (http://berkeleyearth.org/resources .php). Accessed January 15, 2015. For a color version of this figure, please see the color plate section.

considered data from almost 40,000 land and sea measurement stations and developed new algorithms for processing the data. When the investigators obtained what they believed to be their most reliable representation of the data, they compared their results with temperature records determined by three other groups: the National Aeronautics and Space Administration's Goddard Institute of Space Sciences (GISS), the National Oceanic and Atmospheric Administration (NOAA), and the United Kingdom's Hadley Centre and Climate Research Unit (HadCRU).

Annual variations in the temperature anomaly (temperature relative to the 1950–79 mean) are shown in Figure 4.5. In view of the use of different data sets and algorithms, the general agreement between the results suggests that measurement data are being appropriately vetted and processed by multiple groups. Over time, uncertainties decreased due to the addition of new measurement stations, particularly after 1950. Correspondence of the sharp fluctuations between data sets suggests that they represent

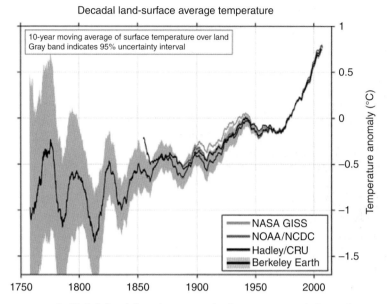

FIGURE 4.6. Global *decadal* moving average land temperatures relative to the 1950–79 mean. Results reported by the Hadley Centre of the UK Meteorological Office and the Climate Research Unit of the University of East Anglia (HadCRU), the Goddard Institute of Space Studies (GISS) of the National Aeronautics and Space Administration, the National Oceanic and Atmospheric Administration (NOAA), and the Berkeley Earth Surface Temperature Project. (http://berkeleyearth.org/resources.php). Accessed January, 15, 2015. For a color version of this figure, please see the color plate section.

actual temperature variations and are not algorithmic aberrations. Each of the plots indicates a significant twentieth-century rise in temperature – slightly more so for the Berkeley and NOAA results – as well as a leveling of temperature for the first decade of the twenty-first century. So far so good! But, when it comes to global warming, few things happen without controversy.

Results of the Berkeley study were also plotted in the format of Figure 4.6. In contrast to Figure 4.5, which provides a rolling annual average of the data, Figure 4.6 represents a rolling decadal average, one that averages out the temperature stasis of the last decade. This representation informed a press release (Muller, 2011), which included comments to the effect that "global warming is real" and that the Berkeley study "finds reliable evidence of a rise in the world average land surface temperature of approximately 1°C since the mid-1950s." By not mentioning the stasis of the last decade, the statements prompted an immediate disclaimer by one of the study's participants and, not surprisingly, a firestorm in the blogosphere (Rose, 2011).

As previously noted, apart from GHG emissions, other factors such as ENSO and AMOC cycles can influence the average surface temperature, in some cases significantly as for the 1997–8 El Niño and the 2010–11 La Niña. Because such factors are variable, they contribute to fluctuations or noise in the data that can obscure signals provided by longer-term trends. Year-to-year data will always exhibit natural variability, and to mitigate such effects in assessing the influence of GHG emissions, it's better to work with a moving decadal average, as in Figure 4.6, than moving one-year averages, as in Figure 4.5. Stated differently, judgments on anthropogenic global warming should not be based on year-over-year changes but on changes over a longer term.[2]

In a subsequent op-ed piece, one of the project's participants (Muller, 2012) acknowledged that "we (may) currently (be) no warmer than we were a thousand years ago during the Medieval Warm Period," while also writing, "Our results show that the average temperature of the earth's land has risen by two and a half degrees (Fahrenheit) over the past 250 years, including an increase of one and a half degrees over the most recent 50 years. Moreover, it appears likely that essentially all of this increase results from the human emissions of greenhouse gases." He goes on:

> We carefully studied issues raised by the skeptics: biases from urban heating (we duplicated our results using rural data alone), from data selection (prior groups selected fewer than 20 percent of the available temperature stations; we used virtually 100 percent), from poor station quality (we separately analyzed good stations and poor ones), and from human intervention and data adjustment (our work is completely automated and hands-off).... [N]one of these potentially troublesome effects unduly biased our conclusions.

The Berkeley study represents a comprehensive and thorough assessment of industrial-era temperature data. It affirms the premise that warming is real and, in concert with other studies, links the warming to human activities.

Since the advent of satellite measurements in 1979, average surface temperatures have increased by about 0.15°C per decade. As shown in Figure 4.3, even with the post-2000 stasis, the decadal average temperature for 2000–9 exceeded that of 1990–9 by approximately 0.2°C. In addition to multidecadal tropospheric warming, measurements also reveal stratospheric cooling, interrupted only by short-term volcanic emissions (Santer et al., 2013). Existence of warming and cooling in the lower and upper atmosphere, respectively, is precisely what one would expect from increased absorption of terrestrial radiation by tropospheric GHGs and reduced absorption of solar radiation due to depletion of stratospheric ozone (National Academies, 2008).

4.2 Climate models and future warming

Although proxy and instrument data can be used to infer past climate conditions, there is no recourse but to use mathematical models for projecting future conditions. But, straight away, let's be clear about one thing. There's no getting around uncertainties in the models, beginning with projections of future GHG emissions. As noted in Chapter 3, uncertainties in the projections are linked to assumptions concerning global economic and population growth, public policy and geopolitical considerations, and the advancement of energy technologies. But that's just the tip of the iceberg. Additional uncertainties accrue from the need for climate models to simulate a wide range of physical, chemical, and biological processes (Baede et al., 2001; Cubasch et al., 2001; Stocker et al., 2001).

As discussed in Section 2.1, major components of the climate system include the atmosphere, hydrosphere, biosphere, lithosphere, and cryosphere. These components are not independent, but are strongly coupled through processes involving exchanges of energy and matter. Models must account for energy, moisture, and carbon exchange between the oceans and the atmosphere, which, in turn, are affected by coupled oceanic and atmospheric circulation patterns. Termed Atmosphere-Ocean General Circulation Models (AOGCMs), they must also consider the dynamics of ice and snow accumulation and depletion in the cryosphere. In addition to predicting short-term meteorological conditions such as seasonal temperature and precipitation patterns and cyclical events such as ENSO, the models are intended to track changes extending from decades to millennia. They can be adjusted and validated by comparing predictions with past climate records and then used to project future trends. There are no guarantees, however, that the projections will be reliable, even over a few years much less for decades or centuries.

Development of reliable climate models is a daunting task. Today's most detailed models map the atmosphere/land/ocean system by a three-dimensional grid for which complex equations governing coupled atmospheric and oceanic circulations and the transport of energy and chemical species are solved as a function of space and time. Even using today's most powerful computers to solve the equations, limitations abound. Spatial grids are coarse, with cell dimensions of approximately 200 km making it difficult to resolve effects occurring on regional scales (Kerr, 2011), much less smaller but densely populated urban scales (Fernando and Klaić, 2011). There are also uncertainties in the coefficients used to quantify atmospheric chemical reactions and radiative processes, as well as species and energy exchange in the atmosphere-ocean system. That said, model improvements continue to be made; knowledge of the many physical parameters required by the models is advancing; ever more powerful computers are increasing the spatial and temporal resolution of the predictions; and there is an ever expanding empirical data base for validating the models.

There will always be room for improving climate models, and their predictions will never be spot on. Nevertheless, they can be used to clarify important climate relationships, including the combined influence of anthropogenic and natural agents. They are our only means of projecting future trends. If the ultimate goal of providing reliable forecasts of events such as flooding and drought on yearly to centennial time scales and on local to global length scales can be achieved, the models would become indispensable tools for farmers, urban planners, insurance companies, and other organizations seeking guidance on the course of climate change and related risks (NRC, 2012a).

For now, the models should be viewed as approximate forecasting tools, with statements of future outcomes couched in terms of the probability of their occurrence. For example, in its fourth assessment report (AR4), the Intergovernmental Panel on Climate Change used "likely" to characterize outcomes of 66% or more probability and "virtually certain" for outcomes of more than 99% probability (IPCC, 2007c). In AR5 it added "very likely" to represent probabilities of 90% or more. The bottom line is that, although future conditions cannot be predicted with precision, a range of possible outcomes can be determined with probabilities assigned to the outcomes. So, what do today's climate models tell us about the future of global warming?

Using the Earth's year-1990 average surface temperature as a reference, nominal AR3 predictions of the year-2100 temperature rise range from a low of 2°C to a high of 4.5°C for scenarios B1 and A1FI, respectively (IPCC, 2001a). Accounting for the effects of model variations and uncertainties, the range broadens from 1.4°C to 5.7°C. All models predict warming in excess of that experienced during later stages of the twentieth century, and increases for higher latitudes of the Northern Hemisphere are predicted to exceed global averages.

Predictions of future warming reported in the more recent AR5 study (IPCC, 2014a) are summarized in Table 4.1. Results represent the difference

TABLE 4.1. *Difference between the average global mean surface temperature estimated for the period from 2081 to 2100 and the temporal average recorded from 1986 to 2005 for representative concentration pathways*

RCP (W/m²)	ΔT (°C)	
	Mean	Range
2.6	1.0	0.3–1.7
4.5	1.8	1.1–2.6
6.0	2.2	1.4–3.1
8.5	3.7	2.6–4.8

Source: From (IPCC, 2014a).

between the temporal mean temperature predicted over the period from 2081 to 2100 and the temporal mean of measurements made from 1986 to 2005. Likely ranges and nominal values of the temperature difference are reported for each Representative Concentration Pathway (RCP). Adding the measured temperature rise of 0.61(±0.06)°C obtained from temporal mean values for the periods from 1850 to 1900 and 1986 to 2005 to the nominal values of Table 4.1, it was concluded that the circa 2081–2100 mean temperature is likely (probability of 66 to 100%) to exceed the 1850–1900 mean by 1.5°C for all but RCP 2.6 and by 2.0°C for RCP 6.0 and 8.5 (IPCC, 2014a).

The foregoing results reveal a cautious, somewhat ambiguous approach to reporting results, one that can be rationalized on the basis of model uncertainties but can also provide fodder for those wishing to dismiss the significance of global warming. Before embracing dismissal, it would be well to revisit Section 3.4 and note that CO_2 emissions are currently on a trajectory that exceeds the rosier SRES A1T/B1 and RCP 2.6/4.5 scenarios. Even if emissions were to remain at a 2012 level of 35 Gt-CO_2/yr (9.5 Gt-C/yr), cumulative 2012–2100 emissions would exceed the nominal RCP 4.5 value of 780 Gt-C (Table 3.4). Were emissions to increase 1% per year to 84 Gt-CO_2 (23 Gt-C) by 2100, cumulative emissions would approximate those of RCP 8.5, raising the specter of a 2.6°C to 4.8°C rise in temperature above the 1986–2005 mean. Even this result may be too conservative. In a study that considered 400 scenarios of global economic activity and climate processes, a probability of 90% was attributed to prospects of the year-2100 temperature exceeding the preindustrial value by 3.5° to 7.4°C (Chandler, 2009).

Another way to assess future outcomes begins with recognition that the global mean surface temperature increases almost linearly with cumulative

CO_2 emissions (IPCC, 2014a). To assign a probability of 66% or more for limiting the temperature rise to less than 2°C relative to the period from 1861 to 1880, cumulative emissions must remain below 1,000 Gt-C. Since approximately 500 Gt were emitted from 1870 to 2010, that leaves a cushion of less than 500 Gt. Referring to Table 3.4, only the RCP 2.6 scenario remains within the cushion. In the final report summarizing conclusions drawn from its Fifth Assessment, the IPCC (2014c) does not equivocate. To remain below 2°C, circa-2010 global emissions would have to be reduced by 40% to 70% before 2050 and dropped to zero by 2100.

Because climate models neglect several contributions to warming, such as the release of carbon dioxide and methane due to melting of Arctic permafrost, the foregoing estimates are not indicative of worst-case conditions. And year-2100 results are by no means the final word. As discussed in Chapter 3 and Appendix D, it could be many years beyond 2100 before the Earth achieved a new thermal equilibrium in response to stabilized atmospheric CO_2 concentrations. For stabilization at any level above 450 ppm, the Earth's temperature would continue to rise before achieving equilibrium decades, if not centuries, later. The need to draw the GHG emissions trajectory downward – sooner rather later – cannot be ignored.[3]

4.3 Feedback mechanisms

There's more to global warming than GHG emissions. Through *feedback mechanisms*, warming due to emissions can beget more warming (or cooling).

A feedback mechanism is a process that is affected by changing conditions and, in turn, can augment (a *positive feedback*) or diminish (a *negative feedback*) the changes. In the context of global warming, the term refers to processes that are affected by warming and, in turn, increase or reduce the warming. Climate feedbacks either amplify or dampen the response to a given forcing. The mechanisms involve complex interactions between the Earth's atmosphere, hydrosphere, cryosphere, land, and biosphere. Consider the following examples.

Effect of atmospheric water vapor

Water vapor is a prominent GHG and is central to multifaceted feedback effects. As the temperature of the Earth-atmosphere system increases due to emissions of CO_2 and other GHGs, the atmospheric concentration of water vapor is affected by increased evaporation from the Earth's water bodies as well as by an increase in the atmosphere's capacity for storing vapor. Not only is more vapor discharged to the atmosphere, but the atmosphere is able to retain more of what it receives. Both effects contribute to increasing the amount of tropospheric water vapor (Soden et al., 2005). The additional vapor increases absorption of terrestrial radiation by the atmosphere, thereby enhancing the warming effect (a positive feedback).[4] But there may be a competing effect.

Clouds consist of a multiphase mixture of water vapor, droplets, and ice particles, as well as solid and liquid aerosols. They are a critical component of the hydrological cycle and strongly influence the global radiation balance. Additional vapor can increase cloud cover, which increases the reflection of solar radiation back to space (a negative feedback), while also increasing the absorption of radiation emitted by the Earth's surface (a positive feedback), which in turn increases evaporation of droplets and thinning of the clouds. Although there are conflicting results on the response of clouds to global warming,[5] the net feedback is almost certainly positive, with climate forcings comparable to the contribution of other warming agents (Dessler and Sherwood, 2009).

A common misperception is that, because the atmospheric concentration of water vapor exceeds that of CO_2 by well over an order of magnitude, GHG emissions have a negligible effect on the global energy balance. To the contrary, even without anthropogenic effects, CO_2 contributes significantly to the absorption of terrestrial radiation, and its contribution increases as its concentration increases in response to carbon emissions. Of course, absorption by water vapor is also important and increases with emissions due to the positive feedback.

Effect of ice cover on the Earth's albedo

Any effect of global warming that increases ice (or snow) cover on the Earth's land and seas increases the Earth's albedo (the fraction of incident solar radiation reflected back to space) and mitigates warming (a negative feedback); any effect that reduces this cover reduces the albedo and enhances warming (a positive feedback). Nowhere has this effect been more pronounced than at northern latitudes and especially in Arctic regions.

By increasing the atmospheric concentration of water vapor, global warming increases the amount of moisture available for precipitation. One outcome is an increase in snowfall responsible for sustaining land-based ice sheets. For such an outcome, there is no change in the Earth's albedo and hence no climate feedback. However, if snow falls on land that would not otherwise be snow-covered, more solar radiation is reflected back to space, increasing the albedo and creating a negative feedback. Such is the case in Siberia, where increased precipitation has increased the spatial extent and duration of snow cover and accordingly the albedo. The effect has lowered winter temperatures in Siberia and has contributed to colder winters in East Asia, the Eastern United States and Northern Europe (Cohen, 2010). However, in other Arctic regions such as Alaska, there has been a reduction in snow and ice cover due to global warming, reducing the albedo and creating a positive feedback that enhances warming (Chapin III et al., 2005).

Whether the extent of land-based ice increases, decreases, or is unchanged, the albedo is also affected by the accumulation of airborne soot on the ice. By increasing the absorption of solar radiation and reducing the albedo, the accumulation of soot on both land- and sea-based ice accelerates Arctic warming (Kintisch, 2009a). Another amplifier of Arctic warming is the reduction in summer sea ice cover (Foley, 2005). Replacement of sea ice by open water reduces the albedo, providing a positive feedback that enhances warming. Using twenty-eight years of satellite data, Riihelä et al. (2013) determined that the albedo of Arctic sea ice decreased by about 0.03 per decade.

Because winter air temperatures in the Arctic are well below those of sea waters and ice insulates the atmosphere from the water, a decline in ice cover also increases heat transfer from the water to the atmosphere and, correspondingly, winter air temperatures. Increasing air temperatures reduce the extent and duration of land snow cover at higher latitudes, increasing the spatial extent of reduced albedo. Screen and Simmonds (2010) have shown that, although changes in water vapor and cloud cover are contributing factors, enhanced warming of the Arctic is due primarily to the effect of decreasing ice and snow cover on the albedo. All things considered, the net effect of losses in land and sea ice is to reduce the albedo, creating a positive feedback that increases global warming.

Decomposition of organic matter

There is growing evidence that increasing temperatures are increasing respiration rates of soil-based organic carbon, with an attendant increase in the release of CO_2 to the atmosphere. In situ measurements performed for subsurface peat deposits in subartic regions reveal a 50–60% increase in respiration for a 1°C rise in temperature (Dorrepaal et al., 2009). Since peatlands contain approximately one-third of the world's soil-based organic carbon – more than half of the atmosphere's current carbon content – the phenomenon could provide a significant positive feedback for global warming. Were the temperature of northern peatland soils to increase by 1°C, emissions could increase by up to 100 Mt-C/yr.

Large pools of carbon also reside in land-based and subsea Arctic permafrost, a frozen combination of soil, rock, and organic matter that has yet to decompose. But as warming occurs from the top down, bacteria break down the organic matter, converting its carbon content to CO_2. A case in point is the 7,000 km-long coast of the East Siberian Arctic Shelf, where thawing of the permafrost is accelerating conversion of carbon to carbon dioxide and accelerating Arctic and global warming (Vonk et al., 2012).

More insidious is the production of methane by the anaerobic decomposition of organic matter in wetlands and permafrost. Decomposition increases with temperature, creating a positive feedback mechanism. Outgassing of methane has been observed from thawing permafrost in regions of Alaska and Siberia (Zimov et al., 2006), as well as from permafrost in sea beds of the Arctic continental shelf and wetlands in temperate latitudes of the Northern Hemisphere (Shakhova et al., 2010; Heimann, 2010). Although it is too early to predict that rapid Arctic warming will trigger the release of massive amounts of methane, accelerated exudation could cause CO_2 to be displaced, if not dwarfed, by CH_4 as the principal agent of global warming.

About 25% of the land surface in the Northern Hemisphere is covered by permafrost, as is much of the Arctic's shallow continental shelf, and together they are estimated to contain up to 1,700 Gt-C, nearly twice the circa-2012 content of the atmosphere (UNEP, 2012). Depending on the extent to which Arctic temperatures continue to increase, the permafrost carbon feedback could emit from 43 to 135 Gt-CO_{2eq} by 2100 and from 246 to 415 Gt-CO_{2eq} by 2200. Add the release of some of the 800 Gt-C entrapped in methane hydrates below the subsea permafrost, and it would not be hyperbole to say that the consequences could be catastrophic. Moreover, should the Arctic begin to release significant amounts of CO_2 and methane, early detection would be impaired by absence of a widely distributed measurement system.

4.4 Summary

The issue of global warming due to human GHG emissions has been contentious, both scientifically and politically. Over the last two decades, much of the contention centered on the initial failure of reconstructed proxy measurements to reveal significant natural warming and cooling trends during early and middle stages of the last millennium. Representing the first 900 years of the millennium as a period of comparative temperature stability made it easier to discount the effect of natural causes on the sharp rise in temperature recorded for the twentieth century. But even when the natural warming and cooling trends are captured by the reconstructed proxy measurements, conclusions concerning recent trends remain the same. The unprecedented extent of twentieth-century warming and the rate at which it occurred are attributable to human activity.

Global warming is determined by (1) GHG emissions and their radiative forcings, (2) feedback effects, and (3) the manner in which net transfer of energy to the Earth is apportioned within its atmosphere, oceans, and land. There is little uncertainty about mechanisms associated with GHG emissions. It is well understood that the gases absorb radiation at wavelengths associated with emission from the Earth's surface. By discharging them to the atmosphere, human activity adds a positive forcing that alters the Earth's radiation balance and contributes to global warming. Apart from uncertainties associated with the effect of warming on clouds, feedbacks are well understood and largely augment warming due to GHG emissions. Although there is still much to be learned about the role of oceanic and atmospheric currents on distributing energy between the atmosphere and oceans, it can be said with confidence that anthropogenic disruption of the global energy balance involves a net transfer of energy to the Earth.

Despite uncertainties, the science underlying anthropogenic contributions to global warming is strong and getting stronger. Denial doesn't make it any less so. Yes, there are natural warming and cooling agents that cycle the Earth into and out of ice ages. But periodicities measure in the tens of thousands of years, and for several millennia the Earth has enjoyed the relative tranquility of interglacial conditions. In contrast, anthropogenic effects, which are superimposed on natural agents, are felt over smaller time scales of decades to centuries. The rate at which warming occurred during the last century exceeded that of any period during at least the last millennium, with the 1990s and 1998 likely to have been the warmest decade and year of the millennium (IPCC, 2007a). Each of the last three decades of the twentieth century and the first decade of the twenty-first century were warmer than its predecessor, with increasing temperatures recorded in ocean waters and the troposphere. There is no disputing that GHG emissions have warmed the planet at a rate that far exceeds the normal pace of warming for an interglacial. The issue then becomes one of future warming.

We rely on global climate models to project future temperatures, and we know that inherent uncertainties yield a wide range of possible temperatures, requiring us to speak of probabilities associated with temperatures in a particular range. Nevertheless, the models all point to increasing temperatures. If we're inclined to minimize the threat of global warming, we might favor the lower end of the predictions, which indicate a temperature rise of about 1°C during the twenty-first century. Or, if we view global warming as a serious threat, we might be inclined to focus on an upper estimate of 4–5°C. But it's sobering to recognize that continuing on the current trajectory of GHG emissions points to realization of the upper estimates, with prospects enhanced by positive feedback effects.

CHAPTER 5

Consequences of global warming

The foregoing results provide strong evidence that global warming is real and attributable in no small measure to human activities, a conclusion strengthened by other indicators that are updated annually (Blunden and Arndt, 2014; EPA, 2014b). But why should we be concerned? What is the impact of warming? How will it affect the natural environment? The built environment? Food production? Human health and security? And, although we've been careful to differentiate between climate and weather, could warming affect the frequency and intensity of extreme weather events?

The ramifications of global climate change (GCC) and their likelihood have been discussed extensively (IPCC, 2007d, 2014a; Cullen, 2010). Cullen extrapolates past and current warming trends into a prediction of circa-2050 climate. Droughts are projected to be more severe in some regions and floods in others. Depletion of ground water and food sources is predicted, as is deterioration of infrastructure in low-lying regions vulnerable to rising sea levels and storm surges of growing frequency and intensity. And the world would also have to contend with growing migrations of climate refugees.

The potential effects on ecological and socioeconomic systems are numerous and may be positive or negative, with significant differences from one region of the globe to another. Let's begin with the region in which GCC and its ramifications are most evident.

5.1 The Arctic: canary in a mine shaft

The Arctic occupies the region north of the Arctic Circle and includes northern portions of eight nations – Canada, Denmark (Greenland), Finland, Iceland, Norway, Russia, Sweden, and the United States

(Alaska) – as well as the Arctic Ocean. Because warming is occurring much faster in the Arctic than in any other region of the world, it has become the proverbial *canary in the mine shaft*, a testimonial to the reality of global warming.

Since the 1950s, Arctic temperatures have increased at a rate more than twice the global average. Using Alaska as an example, its temperature increased by about 2°C, more than twice that of the lower forty-eight states. Were the trend to continue, an estimate of 2°C for the twenty-first-century rise in the global average temperature would translate to a 4°C increase for the Arctic. But that estimate may be too conservative. In a study sponsored by the eight Arctic nations (Hassol, S.J., 2004), it was concluded that Arctic temperatures could rise as much as 7°C by the end of the century. Similar results were obtained from simulations for RCP 8.5 of AR5 (IPCC, 2014a), which predicted a difference of 7°C to 11°C between the average temperature for the period from 2081 to 2100 and the average for 1986–2005. The phenomenon, termed *Arctic amplification*, is attributed to several of the positive feedbacks described in Section 4.3. An increase in the concentration of atmospheric water vapor, as well as effects on the albedo of declining sea ice and airborne soot, each act to augment warming due to GHG emissions.

Arctic amplification has had a significant effect on the amount of perennial sea ice – a key measure of Arctic climate stability. Sometime in September the portion of the Arctic Ocean covered by ice reaches its annual minimum (perennial ice) as conditions transition from summer melting to winter freezing. From the time the ice was first monitored by nuclear submarines in the 1950s to continuous satellite observation since 1979, the area covered by summer sea ice has been in steady decline. From 1979 to 2001, the area decreased by an average of 6.5% per decade, and since then it has been decreasing twice as fast. It achieved a record low in 2007 (NOAA, 2010), its second lowest minimum in 2011 (WMO, 2011), and a record-smashing minimum in 2012 – 18% below the 2007 value and 49% below the 1979 minimum (Kerr, 2012). The reduction in area is accompanied by thinning of the ice and a reduction in volume – about 36% for the average from 2010 to 2012 relative to the average from 2003 to 2008 (JPL, 2013). It is very likely that additional shrinkage will occur, with circa-2100 estimates ranging from 43% to 94% for RCP 2.6 and 8.5, respectively (IPCC, 2014a).

The rapid pace of declining summer sea ice begs the question: When will it completely vanish? Climate models have been used to answer this question, and results obtained from twenty-three different models range from complete loss as early as 2020 to only a slight loss by 2100 (Kerr, 2009). Although a September ice-free condition by mid-century was deemed likely for the worst-case RCP 8.5 scenario of AR5, it could not be stated with confidence that such a condition would be achieved in this century for the other scenarios (IPCC, 2014a). Declining sea ice is not the only manifestation of warming. Since 1979, springtime terrestrial snow cover in the Arctic and elsewhere in the Northern Hemisphere has been declining at a rate of about 22% per decade (Derksen and Brown, 2012), and the trend is expected to continue (IPCC, 2014a).

There are some advantages to a warming Arctic. Declining summer sea ice opens the door to commercial opportunities such as greater access to large fishing grounds, expedited transport of goods, and offshore exploration for oil and gas (Krauss et al., 2005; National Academies, 2010). For several months each summer, seagoing passages could reduce shipping distances between Asian and Western markets by thousands of miles. In terms of undiscovered but recoverable resources, the region north of the Arctic Circle is estimated to have about 70 billion barrels of oil, 47 trillion cubic meters of natural gas, and 44 billion barrels of natural gas liquids (USGS, 2008). Add minerals such as gold, copper, nickel, uranium, and rare earths to these resources, and it's easy to see why nations and corporations are eager to explore investment opportunities. That said, challenges to extracting the resources would still be daunting. Weather conditions would be harsh, the threat of icebergs would remain, and remediation of oil spills would be difficult if not impossible. Year-to-year uncertainties in predicting summer sea ice and the need for icebreakers and ice-capable container vessels would also limit exploitation of new shipping routes (Economist, 2012a).

Ecological implications of a warming Arctic are mixed. The positive feedback mechanism associated with microbial conversion of organic matter to carbon dioxide and methane in thawing permafrost enhances warming. But the process also releases nitrates that stimulate plant growth and uptake of atmospheric CO_2. Overall, some ecosystems would benefit, as others are threatened, and some species would prosper while others struggle. But the fact remains that some species, if not entire ecosystems, will disappear.

Nations bordering the Arctic Sea have begun to think seriously about exploiting economic opportunities and have thus far indicated a willingness to cooperate. Aided by the Law of the Sea, which provides nations with title to the sea bed 200 nautical miles from their shores, or more if it is an extension of their continental shelves, it should be possible to resolve disputes. Nevertheless, there is the potential for competition and conflict, with several nations having indicated a willingness to protect their interests militarily (C2ES, 2012).

Melting of Arctic sea ice does not alter sea level, since much of the ice is submerged. The same cannot be said for the Earth's glaciers and ice sheets.

5.2 Changing sea levels

Like the Earth's average surface temperature, changes in sea level provide an important measure of global warming. And as for temperature, agents of change are both natural and anthropogenic.

Natural variations are due to the periodic warming and cooling that accompany changes in the Earth's orbit about the Sun (Section 2.2). Termed orbital cycles in sea level, periodicities are measured in tens of thousands of years. During a glacial maximum, the sea level is at a minimum, with much of the Earth's water stored in continental (land-based) ice sheets and glaciers. Were all that ice to melt, sea levels would increase by more than 200 meters, or approximately 5% of the average ocean depth. Suborbital cycles of much smaller (millennial) periodicity can also occur, with attendant sea level variations of approximately 15 meters (Alley et al., 2005; Thompson and Goldstein, 2005). But as a superposition to natural effects, anthropogenic changes in atmospheric GHG concentrations and global warming are becoming significant agents of changing sea levels (Church et al., 2001).

Global warming affects sea level in two ways. First, as ocean waters warm, the level increases as the volume of the water increases due to thermal expansion. The effect is termed *steric* change. Second, as land-based ice warms, melts, and in some cases fractures at outlets to the sea, levels increase as water is added to the oceans. The effect is termed *eustatic* change. Although most of the eustatic changes to date have been due to melting glaciers, contributions from the Greenland and Antarctica ice sheets are rising and are likely to become the largest source of future change (IPCC, 2014a).

The effect of glaciers

Because they exist in climates where melting temperatures are more readily achieved, mountain glaciers and ice caps are particularly sensitive to warming and glacial recession is occurring in the Himalayas, Alps, Andes, and Rockies. In Glacier National Park the number of glaciers has decreased from approximately 150 at the time the park was established in 1910 to 30 in 2004, with complete melting expected by 2040 (Appenzeller and Dimick, 2004). Many of the 35,000 glaciers that comprise the Himalayan mountain range are also receding (Service, 2012), although complete melting is unlikely before the end of the century. But can the losses be attributed to anthropogenic global warming?

Because glaciers began to recede in the nineteenth century – before anthropogenic effects became significant – natural climate variability cannot be ignored. In fact, only about 25% of the mass lost between 1850 and 2010 can be attributed to human activities (Marzeion et al., 2014). However, by the early 1970s, a transition from dominance of natural to anthropogenic forcing had occurred, and during the period from 1991 to 2010, anthropogenic effects accounted for about 70% of the lost mass. Today, anthropogenic warming is the dominant forcing and will remain so as temperatures continue to rise. The rate of ice loss from glaciers around the world has been increasing and *very likely* amounted to an annual average of approximately 275±135 Gt/yr over the period from 1993 to 2009 (IPCC, 2014a).

Consequences of retreating glaciers go well beyond their contribution to rising seas. Consider the Himalayas, for which seasonal water runoff is critical to numerous ecosystems and to agriculture and aquaculture for a billion or more Asians. Eventual demise of the glaciers would not be simply a matter of lost grandeur; it would have serious consequences for the natural environment and the livelihood of many of the world's poorest people.

The effect of ice sheets

About 125,000 years ago, during the last interglacial period, sea levels were much higher than they are now, and reductions in the size of the world's major ice sheets were a significant contributing factor (Dutton and Lambeck, 2012). An ice sheet gains mass from snowfall and loses it in two ways, one from surface melting and runoff into the sea and the other from breakup (calving) of ice shelves. Also known as outlet or marine-terminating glaciers, ice shelves extend into the sea and are fed by ice flowing from interior regions of the sheet. They can be several kilometers thick and extend tens of kilometers into the sea before breaking up. If the mass loss due to surface melting and calving exceeds accumulation due to snowfall on the sheet, there is a net gain in the amount of seawater and an increase in sea level.

Outlet glaciers associated with an ice sheet are dynamic entities that move at some velocity and push ice shelves into the sea. The velocity depends on several mechanisms. By enhancing melting at the bottom of ice shelves, warming seawater reduces their resistance to the movement of outlet glaciers. In addition, if atmospheric temperatures are large enough to induce surface melting, water can seep into cracks, enlarging them and increasing prospects for shelf fracture and breakup. Meltwater can also make its way through large cracks – termed moulins – to the bottom of a glacier and, by lubricating its interface with the bedrock, increase its velocity.[1]

Measuring mass loss (or gain) for an ice sheet is not a trivial matter. Consider the Greenland Ice Sheet (GIS), which spans more than two million square miles. Beginning in the early 1990s, changes in its mass were inferred from satellite measurements that track changes in elevation and subsequently from measurements of changes in the local gravitational force imposed by the ice sheet. Measurements for the period from 1992 to 2003 (Johannessen et al., 2005) indicated an average net loss of 224km³/yr (cubic kilometers per year), while other results suggested much smaller losses ranging from no net loss between 1992 and 2002 and a net loss of 82 km³/yr for 2002 to 2004 (Alley et al., 2005; Kerr, 2006b). From 2002 to 2005, measurements ranged from an average loss of 110 km³/yr (Lutchke et al., 2006) to 239 km³/yr (Chen et al., 2006).

Although the range of the foregoing results is indicative of uncertainties in the measurement methods, the trend is one of an increasing rate of mass loss. The trend was affirmed by van den Broecke et al. (2009), who reported an average loss of 180 km³/yr from 2002 to 2008 but 294 km³/yr when averaged over 2006–2008. Although the increasing rate of mass loss has been attributed to a more than twofold increase in the velocity of outlet glaciers (Rignot and Kanagaratnam, 2006), the effects of surface melting cannot be discounted. In 2012, summer melting covered 98.6% of the GIS surface (Nghiem et al., 2012). If completely melted, the GIS would raise the average sea level by 7 meters (Dowdeswell, 2006). As reported in AR5, the average rate of mass loss from the GIS is very likely to have increased more than sixfold from 30(±40) Gt/yr for the period from 1992 to 2001 to 215(±58) Gt/yr for the 2002–2011 period (IPCC, 2014a). And it will continue to increase.

Projecting to the year 2100 and accounting for surface losses as well as outlet glaciers, Pfeffer et al. (2008) estimated a cumulative mass loss of 59,000 to 194,000 Gt of ice. With a one-millimeter rise in the global mean sea level corresponding to a 360 Gt (392 km³) loss of land ice (Alley et al., 2005), the mass loss would correspond to an increase in sea level of 165–538 mm. Although the lower estimate was deemed more probable and consistent with assessments of the evolving dynamics of more than 200 of Greenland's outlet glaciers (Moon et al., 2012), resolving the contributions of outlet glaciers and surface melting remains a work in progress, and the models may in fact be underestimating future contributions of the GIS to rising sea levels. Altimeter measurements reveal that increasing air and ocean temperatures are contributing to mass loss from a 600-km-long ice stream that covers 16% of the GIS and had been neglected in previous models (Khan et al., 2014). Consistent with the pace of Arctic warming, summer melting of the GIS is also increasing (Freedman, 2012), encompassing nearly 100% of the surface. In addition to enhancing mass loss due to runoff, melting contributes yet another positive feedback mechanism, since the surface water absorbs more solar radiation than would otherwise be absorbed by ice.

Despite its size and the large amount of water stored as ice, the GIS pales by comparison to that of Antarctica. The Antarctic Ice Sheet (AIS) sits on a continent comparable in area to the United States, has a nominal thickness of more than two kilometers, and stores approximately 90% of the world's ice. It is surrounded by the Southern Ocean, and its perimeter consists largely of floating ice shelves sustained by the flow of ice from inland glaciers. The shelves gain mass from snowfall and lose mass from calving and melting from below.

Portions of the West Antarctic Ice Sheet (WAIS) projecting into the Amundsen Sea as ice shelves are thinning, and movement of the sheet toward the sea is accelerating (Kerr, 2004b, 2006b). A key feature of the WAIS is that most of it lies on bedrock below sea level, making it especially susceptible to melting by warming seas. The first indication of imminent problems occurred with the collapse of a large 3,200 km^2 (square kilometer) section of the Larsen Ice Shelf in 2002, and subsequent breakdowns over the next decade yielded a cumulative loss of about 25,000 km^2. Were the WAIS to completely melt, it would raise the mean level of the Earth's oceans by 5 meters (Oppenheimer and Alley, 2004), sufficient to submerge much of Bangladesh, Louisiana, Florida, and other coastal regions of the world.

Locations at which outlet glaciers of the WAIS float off the sea bed – termed grounding lines, which demarcate portions resting on bedrock from portions floating in the sea – are moving inland, further reducing the resistance to flow and enhancing prospects for breakup. A large section of the WAIS is now known to be irreversibly retreating. Satellite radar measurements used to track movement of grounding lines from 1992 to 2011 reveal retreat of more than 30 km for each of two glaciers and up to 14 km for two others (Rignot et al., 2014). Recession caused by melting at the lower surface of a glacier due to contact with warming seawater reduces resistance to motion of the glacier, creating a positive feedback mechanism that points to irreversibility of the retreat. Irreversibility of the process was confirmed by a dynamic model of one of the glaciers, with complete loss (disintegration) predicted to occur over several centuries (Joughlin et al., 2014). When fully melted, the glaciers considered in the two studies would raise sea level by more than one meter.

Although the East Antarctic Ice Sheet (EAIS) has long been viewed as stable and not likely to contribute to rising sea levels, concerns are beginning to emerge (Fox, 2010). Small amounts of ice loss were detected in 2009, but of greater concern are findings that several large basins may lie well below sea level. If much of the EAIS is not above sea level, as previously believed, contact with the ocean would expose the ice sheet to warmer waters and to greater basal melting.

Efforts to reconcile results obtained from different measurement techniques for the period from 1992 to 2011 yielded average annual mass losses of 142 (± 49), 65 (± 26), and 20 (± 14) Gt/yr for the GIS, WAIS, and Antarctic Peninsula, respectively, and a slight gain of 14 (± 43 Gt/yr) for the EAIS (Shepherd et al., 2012). But, the cumulative average GIS/AIS loss of 213 ± 72 Gt/yr over the twenty-year period does not tell the whole story. The loss rate increased over time, with the average rate from 2005 to 2010 exceeding that from 1992 to 2000 by more than a factor of three. Although calving had been viewed as the dominant loss mechanism for the AIS, recent findings indicate that bottom (basal) melting accounts for more than 50% of the total, underscoring the importance of temperatures in the Southern Ocean to the overall AIS mass balance (Rignot et al., 2013). Overall, ice loss from the AIS has *likely* increased from an average annual rate of 30±67 Gt/yr for the period from 1992 to 2001 to 147±75 Gt/yr from 2002 to 2011, with losses mainly from West Antarctica and the northern Antarctic Peninsula (IPCC, 2014a).

Rising seas

Sea levels have been steadily rising, and changes in the global-average relative to circa-1880 conditions are shown in Figure 5.1. The green and orange lines correspond to tidal gauge and satellite altimetry measurements, respectively, and the shaded region indicates the range of measurement uncertainties. From 1880 to 2012, the sea level increased by approximately 225 mm and at an average rate of 1.70 mm per year during the twentieth century. However, the rate is clearly increasing over time, and from 1993 to 2012, the average rate was approximately 3.2 mm/yr. Comparable results were

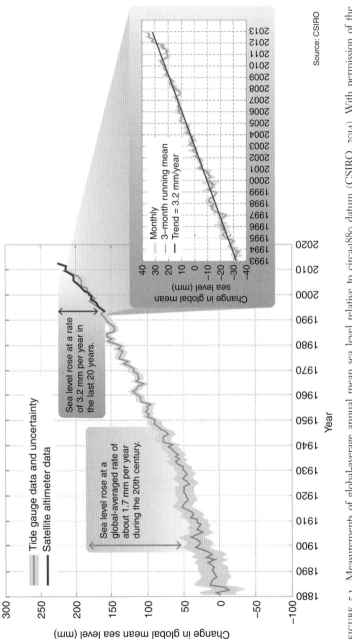

FIGURE 5.1. Measurements of global-average annual mean sea level relative to circa-1880 datum (CSIRO, 2014). With permission of the Commonwealth Scientific and Industrial Research Organization of Australia. For a color version of this figure, please see the color plate section.

TABLE 5.1. *Difference between the average global mean sea level rise estimated for the period from 2081–2100 and the temporal average recorded from 1986–2005 for representative concentration pathways*

RCP (W/m²)	ΔH (m)	
	Mean	Range
2.6	0.40	0.26–0.55
4.5	0.47	0.32–0.63
6.0	0.48	0.33–0.63
8.5	0.63	0.45–0.82

Source: From IPCC (2014a).

reported in AR5 (IPCC, 2014a), with contributions of 1.10±0.30 mm/yr from thermal expansion, 0.76±0.37 mm/yr from glacial melting, 0.33±0.08 and 0.27±0.11 mm/yr from the GIS and AIS, respectively, and 0.38±0.12 mm/yr from changes in land water storage. The increases are at the upper limit of projections made in AR3 and AR4 (IPCC, 2001c, 2007c) and trend 50% to 60% higher than the nominal IPCC estimates (Rahmstorf et al., 2012).[2]

What can we expect for the remainder of this century? Results of the AR5 projections are summarized in Table 5.1. Estimates for the rise in the global mean sea level from the average for 1986–2005 to the average for 2081–2100 range from a low of 260 mm to a high of 820 mm. For RCP 8.5 the circa-2100 increase is expected to range from 580 to 980 mm, with the average rate of increase for 2081–2100 ranging from 8 mm/yr to 16 mm/yr. However, IPCC estimates have been criticized for being too conservative. Through 2011, the models have consistently underpredicted measured results (Rahmstorf, 2012), and estimates reported in AR4 are believed to underestimate the year-2100 rise in sea level by as much as 1 meter (Kintisch, 2009b).

But, however much the seas rise by 2100, that's not the end of it. As discussed in Appendix D, time scales over which thermal equilibrium is achieved between the atmosphere and the oceans are measured in centuries. The large time constant is due to the slow rate at which thermal energy is transferred to deeper regions of the oceans and to the fact that the full impact of global warming would not be felt until well after GHG concentrations are stabilized. Although recent observations suggest that energy transfer to deeper ocean waters is being enhanced by changing atmosphere and ocean circulation patterns, the process still remains a limiting factor to achieving thermal equilibrium. Even if GHG concentrations were to remain at current levels, warming would continue and the steric effect of thermal expansion on sea level would persist into the next millennium. And glaciers and ice sheets would continue to melt.

The adverse effects of increasing sea level are many and dire. Large amounts of coastal land in many countries would be submerged. Salinization of neighboring aquifers, flood plains and deltas would be severe, impairing the quality and availability of fresh water. Rice-producing regions of Asia would be especially susceptible, as would the Mississippi Gulf states. If the sea level were to rise by 1 meter, which could happen in this century, Bangladesh, a nation of approximately 150 million people, would lose much of its capacity for rice production. Similar problems would beset the Nile delta and valley, which comprise only 3% of Egypt's land but are densely populated and the nation's major source of food production.

Globally, more than 100 million people live no more than a meter above sea level, and huge cities from Tokyo, Shanghai, and Cairo to London, New York, and New Orleans would be at risk (Appenzeller and Dimick, 2004; Gillis, 2010). Cities constructed on soft sediment and subject to land subsidence, such as New Orleans, would be affected most severely. In principle, levees and dikes could be constructed to protect the cities, but construction and annual maintenance costs, exacerbated by more frequent and intense storm surges, would soar. All signs are that sea levels will continue to rise and at an increasing rate.

5.3 Extreme weather events

In Section 2.1 we were careful to distinguish between weather and climate and to caution against using one to draw conclusions about the other. However, the increased frequency of what have been considered 100-year or more storms begs the question: Is there a connection between global warming and *extreme weather events* (Cullen, 2010; Carey, 2011a; Morello, 2011)? In the words of White House science advisor, John Holdren (Gillis, 2014), "Scientifically, no single episode of extreme weather, no storm, no flood, no drought can be said to have been caused by global climate change." EWEs are commonly attributed to natural variability in the climate caused by rare movements of air masses and interactions of high- and low-pressure systems. But could anthropogenic warming be increasing their frequency and intensity?

A linkage has been suggested between global warming and the occurrence of category 4 and 5 hurricanes/typhoons (NCAT45) having sustained wind speeds of more than 130 mph. Webster et al. (2005) reported a significant increase in the number and proportion of NCAT45 hurricanes and typhoons from 1970 to 2004, principally in the North and Southwest Pacific and Indian Oceans, while Trenbeth (2005) reported unprecedented activity in the North Atlantic and Pacific during the 2004 hurricane season. And for tropical storms originating in the Atlantic Ocean, the 2005 season was unmatched with a total of fifteen hurricanes and twenty-seven named storms. Hurricanes Katrina and Rita became category 5 storms as they moved over warm Gulf waters, and Wilma set records with speeds of approximately 185 mph at sea. Hoyos et al. (2006) went a step further by correlating NCAT45 storms from 1970 to 2004 against four causal agents and linking an increase in the frequency and energy of Atlantic storms to increasing sea surface temperatures (SST).

Despite the foregoing results, more data are needed before it can be said with confidence that a linkage exists between global warming and hurricane/typhoon activity. In addition to SST, other factors such as atmospheric humidity and wind shear affect storm intensity, making it difficult to definitively link increased activity to global warming.[3] Yet, as evidenced by Super Storm Sandy in 2012 and Super Typhoon Haiyan in 2013 – the largest storm on record – storms of unprecedented destructive power appear to be the new norm.

EWEs also involve torrential rains and ensuing flooding. Evaporation and the atmosphere's capacity to hold moisture increase with warming, and with more moisture in the atmosphere, there is more precipitation. Warming therefore amplifies the global water cycle by increasing the rate at which water evaporates and returns as rain. Research is increasingly pointing to a human fingerprint.

During the twentieth century, precipitation increased by more than 5% for mid- to high latitudes of the Northern Hemisphere and decreased by approximately 3% for subtropical regions (IPCC, 2014a). These trends are consistent with global climate models, which call for increased rainfall where it is already abundant and reduced precipitation where it is low. Land-based observations of precipitation patterns for the Northern Hemisphere from 1951 to 2005 were predicted by a model that included both natural and anthropogenic forcings but could not be explained with only natural forcings (Zhang et al., 2013). Extreme precipitation events that followed a twenty-year cycle in the 1950s were recurring at fifteen-year intervals by the 2000s. From ocean salinity measurements, Durack et al. (2012) found that the hydrological cycle is being amplified at a rate of about 8% per degree of warming. In a world that is 3°C warmer, a 24% increase in the global water cycle would significantly enhance torrential rains and flooding in some regions, while intensifying drought in others.

Increased flooding increases soil erosion, while prolonged drought reduces the availability and quality of fresh water and increases the risk of forest fires. Working with a large database of wildfires that occurred in the Western United States from 1970 to 2005, Westerling et al. (2006) observed that the frequency and duration of large fires increased dramatically in the mid-1980s and could be correlated with increased spring and summer temperatures and earlier spring snowmelt rather than with changing land-use practices. Barnett et al. (2008) found that up to 60% of the climate-related trends experienced in the Western United States during the last half of the twentieth century could not be attributed to natural causes and could only be explained by accounting for anthropogenic GHG emissions.

Extremes in flooding, drought, and tornadoes are occurring more frequently. Flooding experienced in the American Midwest during the summers of 1993, 2008 and 2011 was, in each case, thought to be a once-in-100-year (if not 500-year) event. Of comparable proportions were the record 2010 floods in Tennessee and Pakistan and the 2011 flooding in Australia, Columbia, and China (Carey, 2011a). To these events could be added the large number of severe tornadoes that swept across the United States in 2011 and the chronic drought afflicting many regions of the world, from the U.S. West to Australia, Southern Europe, and portions of Africa, all in the face of growing demands for water (WMO, 2011). Collectively, the foregoing events contributed to tens, if not hundreds, of billions of dollars of economic loss. In the United States alone twelve EWEs in 2011 each caused more than \$1 billion in damages for a total of \$52 billion (NOAA, 2011). In 2012, there were eleven such events, highlighted by Super Storm Sandy and long-term drought, which contributed \$65 and \$30 billion, respectively, to total losses of \$110 billion, second only to the \$160 billion incurred in 2005 from four major hurricanes (NOAA, 2013c). In 2013, Super Typhoon Haiyan wreaked havoc in the Philippines, causing extensive economic impairment as well as widespread destruction and human suffering.

"Climate attribution" is a term that links an increased frequency of EWEs to evidence of a human fingerprint (Carey, 2011b). Because the evidence relies heavily on computer models, which have limitations and can provide conflicting results, conclusions must be drawn cautiously. However, confidence in the results increases when they are confirmed by multiple groups using different methods. Such was the case when five separate research groups analyzed factors contributing to the record temperatures experienced by Australia in 2013 (AMS, 2014). Each group concluded that anthropogenic factors increased the likelihood and severity of Australia's unprecedented temperatures, with two of the groups expressing essentially 100% certainty in their conclusions. Heat waves comprised five of the sixteen EWEs assessed in the AMS report, and in each case anthropogenic factors were found to increase their severity and likelihood. Results were less conclusive for the other eleven events – such as torrential rains and droughts – where natural factors played significant roles.

Although the science is by no means complete, the trends are unmistakable and the evidence is mounting: the world appears to be on a path of experiencing ever more intense EWEs, ever more frequently and with ever increasing costs in human suffering and damage to the built and natural environments. To be convinced, one need look no further than the insurance industry, which more than a decade ago began to factor climate change into its models for assessing and pricing future risks (Mills, 2005; Carey, 2006b).

Increasingly, corporations and individual property owners are becoming more sensitive to the financial risks posed by climate change. Stimulated by huge claims for asset impairment due to storms, as well as concern for the potential effects of drought and disease on water resources, agricultural production, and human health, insurers are becoming more conscious of material risks and, in concert with other segments of the financial community, are encouraging individuals, corporations, and governments to assess and reduce their exposure (Carey, 2011a). Weather-related insurance losses have increased exponentially from approximately $1 billion per year in the 1970s to a decadal average of $17 billion per year from the mid-1990s to a record of $71 billion in hurricane-ravaged 2005 (Breslau, 2007).

Although EWEs have increased across the United States, it is its Eastern seaboard that is receiving the greatest scrutiny. Attribution of increasing temperatures in the Atlantic Ocean to global warming and the related threat of more intense storms from Florida to New England have prompted insurance companies to sharply increase premiums and deductibles for homeowners along the coast or to simply cancel policies to reduce their exposure to large claims. And the problem is, of course, global. Concerns for climate risk around the world have prompted insurers and reinsurers to encourage an "economics of climate adaptation." Translation: infrastructure should be developed or modified to provide greater resilience to major weather events and to thereby allow for future insurability (Greeley, 2011).

5.4 The built environment

As if infrastructure could tolerate another threat to its durability and func-
tionality! In the United States, roads, bridges, buildings, water management
systems, and the like are deteriorating, and the growing costs of maintenance and refurbishment are becoming ever more burdensome to local,
regional, and national governments. To this burden the costs of preparing
for and/or dealing with the impact of climate change must now be added.

The potential effects of climate change can be viewed through the lens
of a particularly vulnerable and commercially significant region extending
along the U.S. Gulf Coast and roughly 80 kilometers inland from Houston,
Texas to Mobile, Alabama (NRC, 2008; Savonis et al., 2008; Kintisch, 2008;
Schwartz, 2010). The region is home to approximately 10 million people
and transportation infrastructure critical to the nation and the region. In
addition to a large network of roads and rail lines, it has numerous airports,
the nation's largest concentration of seaports, and its largest and third-largest
inland waterways. Its pipelines transport two-thirds of the nation's oil
imports, almost half of its natural gas consumption, and more than 90% of
the oil and natural gas extracted from the Outer Continental Shelf. Much
of the region is at elevations below 30 meters and is prone to flooding, soil
erosion, loss of wetlands, and deterioration of barrier islands. Estimates of
the relative rise in sea level over the next fifty years range from 0.3 meters to
2 meters, with the largest increases projected for regions of Louisiana and
East Texas prone to subsidence. Without adequate protection by levees and
dikes, an increase of 1.22 meters (4 feet) would permanently inundate 27%
of the region's roads, 9% of its rail lines, and 72% of its ports.

With increasing susceptibility to greater wind speeds, storm surges, and
flooding, new tools must be developed for assessing risks and guiding risk
mitigation decisions concerning land use, retrofitting existing infrastructure, and building new infrastructure. But by themselves, assessment tools
are not enough. They must be accompanied by specific measures for hardening infrastructure and by public-private partnerships for meeting the
large costs of doing so.

Much of Holland lies below sea level, and for years it has provided
the technical expertise, financial resources, and political/social will to
deal with the threat of flooding. It now has plans to meet the challenges
posed by rising sea levels. Large cities such as New York and London
have also begun the planning process, although related costs comprise
a formidable barrier to implementation. For New York City it's sobering to note that of the ten most severe flooding events since 1900, three

have occurred since 2009, punctuated in October 2012 by the devastation wreaked by Super Storm Sandy.

Whether for a large city or a small nation, costs associated with protecting against the threat of rising sea levels and storm surges are large, and larger yet for extended coastlines along the Gulf of Mexico and the Atlantic seaboard. With damages inflicted by storms such as Katrina and Sandy amounting to tens of billions of dollars – the U.S. Congress authorized $60 billion for rebuilding in the aftermath of Sandy – how should the ever growing risks imposed by climate change be managed?

The growing consensus among insurers is that, whatever is done to mitigate its effects, climate change is inevitable and the need to adapt is unavoidable. In joining the Pew Center's Business Environmental Leadership Council, Marsh, Inc., the world's leading risk and insurance services firm, acknowledged climate change as a significant emerging business risk (Hoffman, 2006). In the words of its CEO, "Climate change is a complex global issue at the intersection of science, risk, and public policy. It is a challenge that our clients – and the world – will face for a very long time." In the words of Franklin Nutter, president of the Reinsurance Association of America (Sanders, 2012), "From our industry's perspective, the footprints of climate change are around us and the trend of increasing damage to property and threat to lives is clear." All regions of the world are affected, but none more than North America, where weather-related loss events increased fivefold from 1980 to 2010 (MunichRE, 2012).

A forte of the insurance industry is its ability to analyze risk. It should therefore come as no surprise that insurance companies are becoming ever more cautious in writing flood and wind damage policies for coastal regions. Policies are no longer being written in some areas and are becoming unaffordable in others. By default, risks are increasingly being borne by state insurance pools and the National Flood Insurance Program (NFIP), placing unsustainable financial burdens on state and federal taxpayers. But that's not all.

In 1988, Congress passed the Robert T. Stafford Disaster Relief and Emergency Act, which was designed to assist communities impacted by natural disasters such as tornados, hurricanes, and wildfires (FEMA, 1993). The law, which obligates the federal government to cover 75% of the cost of repairing infrastructure such as roads, sewers, bridges, and utilities, is administered by FEMA and is activated when a federal disaster is declared by the president. Together, the NFIP and Stafford Act have encouraged development in regions that are at high risk of destruction by natural disasters. And if American taxpayers are to continue to bear most of the rebuilding costs, what's to discourage the continuation of the practice?[4] In the wake of Super Storm Sandy, Kildow and Scorse (2012) provided voice to a growing sentiment: "Homeowners and businesses should be responsible for purchasing their own flood insurance on the private market, if they can find it. If they can't, then the market is telling them that where they live is too dangerous. If they choose to live in harm's way, they should bear the cost of that risk – not the taxpayers." More succinctly: *"Taxpayers should not pay for people to live in risky places."*

It's time to discourage property development along flood-prone coastlines and rivers and in regions susceptible to wildfires. We can begin by eliminating the NFIP and its costly practice of government subsidies for property insurance. Americans should assume full financial risks for their property investments, even (or especially) if insurance is unavailable from the private sector or the cost is deemed prohibitive. After all, that's how free markets are supposed to operate. If Americans are inclined to dismiss or ignore the threats of climate change, shouldn't they accept personal responsibility for associated risks? Also, the Stafford Act should be revisited, making its provisions less generous and giving municipal and state governments more incentive to curb development in high-risk areas.

5.5 The natural environment

The natural environment consists of many ecosystems, each comprised of different species of plants and animals that interact with each other and the land, water, and air around them. Living as we do in anthropogenic communities, it's easy to forget how dependent we are on these ecosystems. They maintain air and water quality, as well as soil fertility. They influence climate and mitigate the impact of extreme weather events. And they provide important sources of food and raw materials.

The *vulnerability* of an ecosystem to climate change depends on its *sensitivity* and *adaptability* to change. In certain regions of the world, some systems may be naturally robust and relatively insensitive to climate change, or they may be able to successfully adapt to change. Other ecosystems may be far less resilient. Such factors make assessments of the effects of climate change as uncertain as the changes themselves. Nevertheless, humankind would be remiss to ignore them. Consider two of the world's most important ecosystems: its forests and oceans.

Forests play an important role in the Earth's carbon cycle, and whether they provide a net *uptake* or *source* of atmospheric CO_2 depends on whether losses are less than or exceed gains due to new growth. An important case in point is old-growth forests of the Amazon, which store approximately 120 Gt-C and annually process approximately 18 Gt-C, more than twice annual emissions from burning fossil fuels. We know that losses are associated with overdevelopment and deforestation. But they can also occur as a consequence of climate change manifested by intense drought (Phillips et al., 2009). Water deficits would reduce growth, and for prolonged droughts forests would be displaced by savannas (tropical/subtropical grasslands). The amount of CO_2 discharged to the atmosphere would increase at the expense of carbon stored as biomass in the forests, providing yet another positive feedback that enhances global warming.

The world's oceans, which house some of the world's most important ecosystems, are being severely stressed by acidification and increasing temperatures due to CO_2 emissions, as well as by hypoxia (deoxygenation), pollution, and overfishing. Oceanic phytoplankton account for almost half of global photosynthesis, and billions of years ago it was phytoplankton that created our atmosphere. Warming, acidification, and hypoxia have contributed to past mass extinctions, and concerns are growing that the oceans may be "entering a phase of extinction of marine species unprecedented in human history" (ScienceDaily, 2011).

The oceans have absorbed about 30% of the CO_2 emitted since preindustrial times (IPCC, 2014a), and the current rate of absorption exceeds levels associated with the last mass extinction of marine species, more than 50 million years ago. The upside to the increased uptake of CO_2 is that accumulation in the atmosphere is slowed while production of phytoplankton is enhanced. The downside is that the oceans become more acidic (Kerr, 2010). From its preindustrial level, the average ocean pH has decreased by about 0.1 to a circa-2010 value of 8.06 (IPCC, 2014a). At first glance, such a change may appear inconsequential. But pH is measured on a logarithmic scale, and a reduction of 0.1 corresponds to a 26% increase in acidity. Additional reductions in the pH projected for 2100 range from a low of 0.06 for RCP 2.6 to 0.32 for RCP 8.5 (IPCC, 2014a). A reduction from 8.06 to 7.74 would more than double the ocean's acidity.

A reduction in the pH of ocean waters corresponds to a larger concentration of hydrogen ions (H^+), which react with carbonate ions (CO_3^{2-}) to form bicarbonate (hydrogen carbonate, HCO_3^-). With increasing acidity (increasing H^+), the carbonate concentration is reduced, as more is converted to bicarbonate. Since carbonate ions are essential to formation of calcium carbonate ($CaCO_3$) skeletons and shells, the growth of coral and other calcifying species such as shellfish and microscopic organisms is stunted, threatening fisheries that feed off the organisms. Acidification is proceeding at an unprecedented rate, and economic losses in this century could total $130 billion for the shellfish industry and up to $1 trillion from degradation of the world's coral reefs (IGBP, 2013).

Coral reefs have evolved over hundreds of millions of years into the largest living systems on Earth. They are found in both cold and temperate waters and host enormously diverse ecosystems with elaborate food chains and an array of important biochemical compounds and sea life. Although they cover less than 1% of the ocean floor, they house about 25% of its biodiversity and are a major source of protein for hundreds of millions of people. However, coral reefs are deteriorating due to increasing water temperature and acidification (Hoegh-Guldberg et al., 2007), and nearly 75% of the world's reefs are currently at risk. From a review of the effects of warming and acidification on marine ecosystems, Hoegh-Guldberg and Bruno (2010) conclude that "reducing greenhouse gas emissions (is a) priority, not only because it will reduce the huge costs of adaptation but also because it will reduce the growing risk of pushing our planet into an unknown and highly dangerous state." Global warming also reduces absorption of atmospheric oxygen by the oceans, increasing hypoxic zones and exacerbating detrimental effects on marine life (Deutsch et al., 2011).

5.6 Food production

Greenhouse gas emissions and global warming affect food production in diverse ways, good and bad. Factors that influence photosynthesis and plant growth include the atmospheric temperature and CO_2 concentration, the amount of solar radiation (insolation), and the availability of moisture and soil nutrients. Changes in any of these factors can enhance or inhibit growth, and each factor can limit growth. That is, regardless of changes in the other factors, growth cannot be increased without changing the limiting factor. Soil nutrients and water are the most common limiting conditions. Irrespective of existing temperature, CO_2 concentration, insolation, and nutrients, crops will not grow if there is insufficient water. Or, with sufficient water, growth is stunted by the absence of sufficient nutrients. Crop irrigation and fertilization are human interventions designed to preclude conditions for which water and nutrients become limiting factors.

A positive outcome of increasing atmospheric CO_2 concentration is its effect on enhancing the growth of cereal crops, which are the foundation of agricultural production. Carbon dioxide is an essential reactant for photosynthesis, and as long as it is a limiting factor in the process, increased plant growth will accompany increased concentrations. At some point, however,

the CO_2 concentration can reach a value for which it is no longer a limiting factor and sustained increases in CO_2 will have no additional effect on crop yields. → plateau

To a certain degree, global warming can also increase agricultural production by increasing temperatures at higher latitudes, lengthening the growing season and reducing the probability of frost damage. However, there are threshold temperatures beyond which yields would no longer increase. Each crop variety has a threshold temperature, or tipping point, above which crop yields decline with increasing temperature. In a study of corn, soybeans, and cotton, crops that have tipping points of 29°C, 30°C and 32°C, respectively, Schlenker and Roberts (2009) concluded that in this century yields would decline by 30–46% for a scenario of slow global warming and by 63–82% for rapid warming. Yields of wheat and corn are projected to drop by 5–15% per degree centigrade rise in temperature above the preindustrial level (NRC, 2012b). will GMO be able to solve this?

Lobell et al. (2008) used climate projections for 2030 and statistical crop models to assess climate risks to production in twelve food-insecure regions, largely in Asia, Africa, and Latin America. In 2008, three of the regions (South Asia, Southeast Asia, and China) accounted for more than 60% of the world's malnourished people. The potential impact of climate change varied across the twelve regions, with the highest risks predicted for crops such as rapeseed, wheat, maize, and rice in South Asia, Southeast Asia, and Southern Africa. The actual impact would be determined by the adaptive capacity of a region, which would depend on its social/political systems, its financial means, and access to technology that enables alternative crops. Poorer nations are less likely to have such resources, and large sectors of the developing world would be vulnerable to reductions in local food production. The impact would be worsened by a reduction in global production and an increase in food prices on the global market. While Africa and South/Southeast Asia are most vulnerable to warming, the two nations bearing the largest responsibility for warming (China and the United States) would experience fewer adverse effects and could see increased agricultural output in certain regions. environmental injustice

In another study, Battisti and Naylor (2009) used global climate models with historical examples of agricultural impairment by extreme temperatures to project a significant reduction in global food production by 2100. The direst effects were projected for tropical and subtropical climes, where to better than a 90% probability, temperatures during the growing season were predicted to exceed the most extreme temperatures recorded from 1900 to 2006. But, the effects were not confined to the Southern

Hemisphere. Citing France as an example, the extreme heat wave of 2003, which caused the average summer temperature to exceed the long-term mean by 3.6°C and devastated agricultural production, was predicted to become the norm by 2100. → extreme case now norm

The foregoing studies dealt largely with adverse effects of increasing temperatures on food production. But in many regions of the world, agricultural production depends critically on the existence of adequate levels of precipitation; in others, where dry-land farming is not an option, crops must be irrigated, requiring access to surface and/or ground water. Climate change is altering hydrological conditions in ways that benefit some regions but significantly disadvantage others through prolonged drought, severe flooding, reduced surface water flows, and/or saliniza-tion of fresh water supplies. In many regions of the world, the effect of groundwater depletion and contamination can be added to these threats. As populations grow and developing nations advance their economies, it will become more problematic to meet demands on water for agricul-tural production.

Absent the development of crop varieties that can tolerate higher tem-peratures and water stress, a global decline in grain (and livestock) pro-duction will accompany global warming. With declining production from existing farmland, a hungry world would likely respond by converting more wilderness to agriculture – in much the same way that the rush to biofuels did in the 2000s – creating another positive feedback by enhancing CO_2 emissions from deforestation.

Problems are exacerbated by the effect of declining agricultural produc-tion on food prices. Based on gradually increasing temperatures and chang-ing rainfall patterns, prices are projected to increase by 50% from 2010 to 2030 (Carty, 2012). But that's not the whole story. Prices are also affected by EWEs, and in one year events such as extreme drought or flooding in agricultural regions can cause commodity spikes comparable to a 50% rise over twenty years. Such spikes can be managed in wealthier nations, where people spend a relatively small portion of their income on food, but they are devastating in developing countries where many people spend most of their income on food. The bottom line is that the poor of the world will be hit hardest by the adverse effects of climate change on agricultural produc-tion and prices.

In its Fifth Assessment Report on the effects of climate change, the IPCC (2014a) attached a high probability to breakdown in food produc-tion systems due to warming and the effect of changing precipitation pat-terns, including extreme drought and flooding. The effects will be most pronounced for the world's poorest nations, many of which have already experienced the impact of supply disruptions on food prices and security.

5.7 Human health and security

In 2010, an agency of the U.S. federal government published a report that referred to climate change and energy as "two key issues that play a significant role in shaping the future security environment." The report went on to say that "climate change could have significant geopolitical impacts around the world, contributing to poverty, environmental degradation, and the further weakening of fragile governments," and that it "will contribute to food and water scarcity, will increase the spread of disease and may spur or exacerbate mass migration." It also noted that, while "climate change alone does not cause conflict, it may act as an accelerant of instability or conflict, placing a burden to respond on civilian institutions and militaries around the world." *Disease, weakening of fragile governments, political instability, conflict, and mass migrations* – the words are ominous. But did they appear in some obscure government report and perhaps lack credibility? No! The quotes come from a major quadrennial review conducted by the U.S. Department of Defense (DOD, 2010). Four years later the DOD (2014a) doubled down on its assessment by referring to the effects of climate change as "threat multipliers that will aggravate stressors abroad such as poverty, environmental degradation, political instability, and social tensions – conditions that can enable terrorist activity and other forms of violence." Shortly thereafter, in a subsequent report (DOD, 2014b), the threat was upgraded from imminent to present.

The notion of climate change as an "accelerant of instability and conflict" was affirmed by a study that correlated the history of human conflict with extreme climate events (Burke et al., 2013; Hsiang et al., 2013). A large knowledge base, encompassing thousands of years, was combed and statistical procedures were used to establish causal relationships. Violent crime and civil conflict increased with higher temperatures; ethnic conflict and invasions increased with drought and extreme rainfall; and civilizations collapsed in the face of climate change. Extreme weather heightens aggressive tendencies. It can also threaten survivability by creating scarcities of water, food, and shelter, particularly in poorer nations, and planting seeds of social strife and warfare. Quoting from Burke and colleagues, "Decision makers must show an understanding that climate can fundamentally shape social interactions ... (and) that our children and grandchildren could face an increasingly hot and angry planet."

The foregoing concerns were expressed in a separate and independent study conducted by the National Research Council (Broder, 2012). Quoting from the report, "It is prudent to expect that over the course of a decade some climate events … will produce consequences that exceed the capacity of the affected societies or global system to manage and that have global security implications serious enough to compel international response." Implications are that consequences will be sufficiently dire to overwhelm the capacity to deal with them, causing political turmoil, failed governments, violent conflict, and mass migrations. This message was also conveyed in AR5 of the IPCC (2014a), which projected a slowdown in economic growth and poverty reduction measures, along with growing food insecurity, both of which are well-known precursors to conflict. In turn, conflict diminishes the resource base required for adaptation to climate change, creating a positive feedback that worsens the problem.

A range of human health issues would also be exacerbated by global warming, beginning with increased illness and death due to heat stress, particularly for the poor and the aged. With northern latitudes of North America, Europe, and Central and Northern Asia expected to see the largest temperature changes, heat waves would become more frequent, more intense, and longer lasting (Meehl and Tebaldi, 2004). Northern latitudes of the United States would, for example, see a growing number of days per year with temperatures above 90–100°F. In addition, human health would become more susceptible to climate-based illnesses and disease vectors whose transmission depends on factors such as temperature, humidity and surface water. With growing stress on supplies of food and potable water, illnesses and death due to malnutrition and waterborne pathogens would rise. Major global health threats such as malaria, dengue, and diarrheal diseases would worsen, exacerbated by cycles in which torrential rains follow prolonged drought, with the loss of natural predators during drought amplifying the spread of pathogens following rain.

Finally, climate change in general and an increase in EWEs in particular will affect personal income and economic security (Scheffran et al., 2012) and will be felt disproportionately, amplifying existing inequalities within nations and between nations. The effects would exacerbate political instability and increase prospects for conflict within and between nations.

5.8 Abrupt climate change

The foregoing considerations deal largely with climate change that occurs gradually and with consequences manifested over a comparatively long time frame. However, there are conditions that can contribute to *abrupt climate change* (ACC). The term has several implications. It refers to a significant change in climate that occurs over a comparatively brief period, such as decades rather than centuries or millennia. It also implies climate change that occurs on a continental, if not global, scale and can persist for a long period of time. Adjusting to the change would be difficult and the ecological and economic consequences significant. It could be manifested by a large and persistent increase in the strength and frequency of severe storms, floods or droughts, a disruption of plant and animal reproduction cycles with attendant loss of biodiversity, and/or a significant decline in agricultural production.

ACC can occur in response to a rapidly varying input, such as discharge of large amounts of freshwater due to accelerated melting of ice sheets or a large increase in methane emissions due to melting of permafrost or clathrates. It would involve changes that *cross a threshold* and induce an *abrupt transition* to a new condition. Such transitions are well known to engineers and scientists. Complex natural and engineered systems are often characterized by quasi-equilibrium states for which changing inputs are absorbed with little effect on system behavior. Such systems may seem *robust* and able to accommodate large variations in external inputs without significant changes. However, imperceptible changes in the system may, in fact, be occurring that render it increasingly vulnerable to abrupt change.

For engineered systems such as an aging aircraft or the space shuttle, an undetected crack in a wing or heat shield can suddenly propagate, leading to catastrophic outcomes for their passengers. In such cases, time scales associated with departure from the behavioral norm are small, perhaps measured in seconds or microseconds. For the climate, time scales associated with abrupt change are larger, but they are not geological and are certainly much smaller than the norms of human experience.

The Earth has experienced abrupt climate change throughout its history, and during the Holocene human societies have adapted with varying degrees of success. Periods of ACC can be inferred from paleoclimate proxy data (NRC, 2002), and the most recent event – known as the Younger Dryas period – began as a disruption of the warming that followed the last major ice age (Alley, 2000). From proxy data obtained for central Greenland, the period is believed to have started 12,800 years ago with abrupt cooling of approximately 8°C and to have been followed 1,200 years later by an equally abrupt warming of approximately 12°C. Subsequent but far more gradual warming occurred for approximately 3,000 years, only to be interrupted 8,000 years ago by a brief period – termed the 8ka event – of cooling by approximately 6°C. Pursuant to this event and excluding post-1950 conditions, temperatures and climate remained relatively stable consistent with stability of the Earth's atmosphere and ice masses.

Subject to uncertainty, a plausible explanation for the 8ka cooling event is that the earlier warming trend had reduced the intensity of the Atlantic Meridional Overturning Circulation (AMOC) – also known as the great ocean conveyor belt or thermohaline circulation (THC). The AMOC was introduced in Section 4.1 as a possible explanation for significant accumulation of energy stored in deeper layers of the Atlantic and Southern Oceans and, correspondingly, the recent stasis in the Earth's average surface temperature. Energy transfer by the AMOC from the Southern to the Northern Hemisphere is enormous, with an estimated contribution of approximately 8 W/m^2 to the energy budget of the Northern Hemisphere (NRC, 2002). If global warming caused the THC (Figure 4.4) to shut down, average temperatures in the North Atlantic could decrease by 4–6°C, while warming would continue throughout the rest of the world. Although there are other contributing factors, such as the jet stream that delivers cold Arctic air to northeastern regions of North America, the THC is commonly used to explain temperature differences between these regions and North Central Europe.

During the warming that preceded the 8ka cooling, there would likely have been significant melting of North American ice masses and a discharge of large amounts of fresh water into the North Atlantic. Such a discharge would reduce the salinity and density of surface waters, thereby reducing down-welling at upper latitudes and hence the overall driving potential for the THC. If the 8ka event was initiated in this way, could global warming precipitate a future disruption of the THC?

Although computer models have shown the ability to trigger ACC by means of fresh water discharges into the North Atlantic (Liu et al., 2009; Timmermann and Menviel, 2009), and some data have suggested ongoing changes to the THC, the results are by no means definitive (Kerr, 2005, 2006c, 2008). What can be said at this time is that it's possible but by no means likely that ACC will be triggered by disruption of the THC, at least in this century. But more powerful agents of ACC may be lurking.

In Section 4.3 we discussed a positive feedback associated with the decomposition of organic matter. The carbon stored in permafrost soils as frozen organic matter exceeds the carbon content of the atmosphere by nearly a factor of two. As the soils thaw due to global warming, microbial decomposition of the biomass can contribute significantly to GHG emissions and increase warming (Schurr et al., 2008, 2009). But the impact of decaying biomass in melting permafrost is not the whole story. Permafrost and frozen sea beds are known to contain enormous amounts of methane in the form of clathrates (Appendix C).

Methane is a potent GHG, and methane molecules can be encapsulated by ice at low temperatures and high pressures. Huge clathrate deposits exist beneath the sea floor, and warming of a few degrees centigrade could trigger a massive discharge of methane. A doomsday scenario would be one for which Arctic thawing released hundreds of gigatonnes of methane over a few decades, increasing the radiative forcing by more than an order of magnitude. Although most clathrates are found in sediments hundreds of meters below the surface and are in no immediate danger of melting, the issue bears watching, particularly with growing interest in exploiting clathrates to augment production of natural gas (Bohannon, 2008). *we will make it worse*

In view of the uncertainties associated with today's most comprehensive models and the limited amount of definitive data, it is not possible to attach probabilities to the effect of global warming on future ACC events, nor can it be said that such events are imminent. However, they are within the realm of possibility and warrant measures to reduce potential risks.

5.9 Summary

The link between GHG emissions and warming of the climate system is unequivocal, making it difficult to deny an anthropogenic effect on climate change. Data dealing with conditions in the Earth's atmosphere, hydrosphere, and cryosphere are published annually for a large number of climate change indicators (Blunden and Arndt, 2014; EPA, 2014b) and provide growing evidence of a human impact on warming and climate change. The evidence can no longer be dismissed.

It is also becoming more difficult to dismiss the connection between extreme weather events and global warming. Whether manifested by drought, heat waves, wildfires, floods, tornados, or hurricanes, EWEs are occurring with increased frequency and intensity. Can any one event be attributed to global warming? No! Natural variability remains a factor. Are the events being exacerbated by global warming? Yes! Superimposed on natural causes, the effects of warming, such as rising sea levels and increasing atmospheric moisture, are amplifying the effects of EWEs. Are the effects harbingers of more to come? Yes!

Rising seas and increasing storm intensities provide a growing threat to the built environment, from roads, bridges, tunnels, rail lines, utility transmission lines, pipelines, and sewers to homes, factories, and commercial

good for nonenv. to read 3 see they will be affected too.

buildings. Capital investments in related infrastructure are enormous, as are the costs of repair, replacement, and lost economic activity posed by the threat. Rising seas also threaten low-lying coastal regions through loss of arable land and agricultural production from flooding and salinization of freshwater supplies. An increase of one meter in the average sea level by the end of the century is well within the realm of possibility, and from what is known about environmental time scales and inertia, sea levels would continue to rise for centuries thereafter, even if substantive measures were taken to curb GHG emissions. Food production would be further impaired by drought and the loss of runoff water from the world's mountain glaciers. The well-being of more than a billion people depends on the ecological and food production systems sustained by rivers receiving spring runoff from the Himalayan Mountains. And in the extreme, there is the threat of having to deal with the political instability, mass migrations, and conflict that could accompany loss of sustainability.

The adverse effects of global warming are real and increasingly menacing. Quoting Garvey (2008, p. 59),

> What's clear is that climate change involves harm … the planet warms, weather systems change. The harm won't be evenly spread: some places will become more habitable, but many more will face new extremes of weather. Sea levels will rise, flooding homes and destroying crops. Elsewhere water shortages will threaten. Disease will spread to new areas. There will be conflict. A lot of people will die or be uprooted or suffer in other ways. Species will disappear. Whole ecosystems might well be destroyed.

If we wait until these effects are in full bloom, only to be exacerbated by inertia in the system, it may be too late for decisive action.

For Americans who may feel immune to the effects of climate change, a report entitled "The Economic Risks of Climate Change in the United States" by Houser et al. (2014) should provide a sobering wake-up call. Using the tools of risk management analysis, the study has three foci: the vulnerability of coastal property and infrastructure to rising seas and storms; the effects of elevated temperatures on human health, labor productivity, and energy systems; and the effect of elevated temperatures on agricultural patterns and crop yields. In a business-as-usual scenario, related losses are large and increase throughout the century, putting significant pressure on the nation's economic system.

At this juncture we've concluded our treatment of the science underpinning global warming and climate change. Moving forward, we can be confident about the following: (1) GHG emissions and atmospheric concentrations will continue to increase; (2) we'll learn more about distribution and redistribution of energy between the Earth's oceans, atmosphere, and land, as well as feedback effects and reasons for the current temperature stasis; and (3) manifestations of warming will remain unchanged in kind but will affirm current trends associated with increasing levels, intensities, and/or frequencies.

With regard to scientific matters, the debate is all but over. The Earth will continue to warm due to human forcings, and manifestations of warming will become more pronounced. Other aspects of the debate are very much in play, however, and revolve about appropriate responses. How should technology and capital be deployed to mitigate or adapt to the problem? What is the role of public policy? How should decisions be guided by social and ethical considerations? These elements will be discussed in subsequent chapters, beginning with strengths and weaknesses of technological options.

CHAPTER 6

Mitigation, adaptation, and geoengineering

If anthropogenic global warming is acknowledged and related threats are taken seriously, what can be done about it? In broad terms, there are three paths that can be taken. The first two are by no means mutually exclusive and, in fact, should be pursued concurrently. One path involves *mitigation* measures. What can be done to reduce GHG emissions and stabilize atmospheric concentrations at levels of low risk? Globally, can CO_2 emissions be reduced to 20% of current levels by 2050 and atmospheric CO_2 stabilized at 450 ppm? And what if mitigation measures are insufficient to reduce atmospheric GHG concentrations to acceptable levels, a scenario that is becoming ever more likely?

The second path involves *adaptation*. Recognizing diminishing prospects for stabilizing GHG concentrations at acceptably low levels, what can be done to increase the resilience of human and natural systems to the effects of climate change? In this chapter we'll examine a range of mitigation and adaptation options. But should mitigation and adaptation measures both fall short of desired outcomes, what then? The chapter concludes with a discussion of measures falling under the rubric of *geoengineering* – a term used to characterize large-scale engineering endeavors designed to counteract the warming effects of GHG emissions.

We'll begin with mitigation. As shown in Figure 6.1, options for reducing GHG emissions fall into three categories. One category deals with increasing the efficiency of producing and using energy; the second deals with transitioning the world's energy portfolio from carbon-rich to carbon-free sources of energy; and the third deals with collecting and sequestering GHG gases before they are released to the atmosphere. In each case, the *scale* of implementation needed to achieve meaningful reductions is enormous, requiring large capital investments and global cooperation.

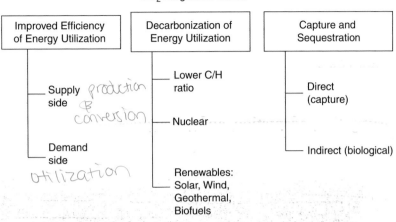

FIGURE 6.1. Options for reducing carbon dioxide emissions.

6.1 Energy efficiency and conservation

Beginning with a fossil fuel, we can reduce emissions by increasing efficiencies at all stages of the fuel's life cycle – by increasing the efficiency with which the fuel is produced and converted to other forms of energy and by increasing the efficiency associated with end use of the energy. Life cycle assessment (LCA) of GHG emissions is an important tool for comparing the attributes of different energy sources. Production and conversion are supply-side considerations. How efficiently can we extract oil, natural gas, or coal from a reservoir, put it in a form suitable for use, and transport it to the user? By increasing process efficiencies, energy consumed by the processes – and, correspondingly, carbon emissions – are reduced. Utilization is a demand-side consideration. By increasing the end-use efficiency, we reduce the amount of primary energy that must be devoted to the task and, accordingly, carbon emissions. The possibilities are endless: from building heating and cooling systems; ground, air, and sea transportation systems; manufacturing processes; all manner of appliances and electronic devices; and much more.

Consider a coal-fired power plant. Whatever we can do to increase the efficiency of the power plant reduces the amount of coal consumption and carbon emissions per unit of electrical energy production. In such a power plant, the gaseous products of burning the coal are used to produce steam, which, in turn, is routed through a turbine to generate electricity. The overall efficiency with which the chemical energy of the coal is converted to electrical energy depends strongly on the temperature and pressure of the steam. Most of today's power plants operate under what are termed sub-critical conditions and have efficiencies of approximately 35%. However, efficiencies can be increased to approximately 40% for newer supercritical systems operating at higher pressures and, at some point, to 45% or more for ultra-supercritical systems operating at yet higher pressures. Each such improvement reduces the amount of coal that must be burned, and hence carbon emissions, per unit of electrical generation.[1]

On the demand side, whatever can be done to improve the efficiency of using the electricity reduces its consumption and, correspondingly, coal consumption and carbon emissions. In fact, increasing the efficiency of energy utilization represents the low-hanging fruit of carbon abatement. Of all mitigation options, it is the only one that produces short-term, positive economic returns (Creyts et al., 2007). Savings associated with reduced energy consumption exceed the cost of efficiency improvements over the lifetime of an option, and carbon abatement is achieved at *negative cost*. By implementing an assortment of measures that would reduce U.S. private-sector emissions by 3% annually over seven years, the potential exists to realize $190 billion in annual savings by the end of that period (WWF, 2013).

The foregoing considerations can be applied to all sectors of energy utilization, including transportation, and were underscored in an NRC (2009) report that concluded that "the potential energy savings available from the accelerated deployment of existing energy-efficiency technologies … could more than offset the Energy Information Administration's projected increases in U.S. energy consumption through 2030," and "the life-cycle costs of most technologies are much less than the costs that would be incurred by having to purchase the saved energy." In language used by the business community, energy efficiency is a *sweet value proposition*.

The degree to which energy efficiency measures have been adopted varies between nations (Young et al., 2014). From an analysis that considered thirty-one metrics – including energy consumption in buildings, industry, transportation, and power generation – for each of the world's sixteen largest economies, some countries were found to significantly outperform others. Overall, Germany was ranked first and the United States thirteenth, behind China (fourth) and India (eleventh). The most important finding was the existence of "substantial opportunities for improvement" in all nations.

A companion to energy efficiency is *energy conservation*. Energy efficiency implies using technology to reduce the energy required to achieve a specific objective – in a sense, doing the same or even more with less. In contrast, energy conservation simply implies doing less. It has more to do with behavior than technology.

Efficiency and conservation measures are not mutually exclusive. We can improve the efficiency of heating and cooling systems while lower- ing thermostats in winter and raising them in summer. We can increase the efficiency of generating and transmitting electricity while reducing its use through both efficiency and conservation measures. We don't have to supersize our homes and we can choose to live closer to places of work, recreation, and public transportation. We can choose vehicles that provide better fuel economy, and where practical, we can take advantage of alter- native modes of transportation, public and otherwise. Such choices may be incompatible with what is often termed the *American way of life*, but by any measure, energy conservation is also a sweet value proposition.

6.2 Decarbonization of electric power: coal – the 800-pound gorilla

As indicated by the second stem of Figure 6.1, carbon emissions can also be reduced by transitioning from primary energy sources of high carbon con- tent to sources of lower or no content. Of the three fossil fuels, coal has the largest ratio of carbon-to-hydrogen atoms (C/H). It also has the largest car- bon intensity, which is a measure of the amount of carbon released to the atmosphere per unit of energy released by burning the fuel (Table B.2). At the other end of the spectrum, natural gas has the lowest C/H ratio and the lowest carbon intensity.[2] Since about 67% of the world's electricity is gener- ated from fossil fuels, these results have an important bearing on reducing emissions from power generation (EIA, 2013c). In 2011, coal provided about 40% of the world's electricity, followed by natural gas and oil at 20% and 10%, respectively. Coal also accounted for almost 45% of the world's *total* CO_2 emissions. And with coal-rich developing nations such as China and India providing most of the growth, the contributions of coal to power gen- eration and CO_2 emissions are projected to increase (EIA, 2013b).

The natural gas conundrum

For equivalent generation efficiencies and power production, use of gas instead of coal would reduce the amount of carbon dioxide released to the atmosphere by nearly 30%. Efficiencies, however, are not the same. The nominal efficiency of using natural gas in a combined-cycle power plant is 55% compared to 35% for a typical coal-fired plant. By providing both smaller carbon intensity and larger conversion efficiency, substitution of natural gas for coal reduces CO_2 emissions by approximately 55%. On this basis, since coal is the single largest contributor to the world's carbon emissions, replacement of coal by natural gas would do much to reduce emissions. And natural gas has other benefits. Because it contains little or no sulfur, nitrogen, and toxic elements, or minerals that contribute to fine particulates and ash, it is much cleaner than coal.

But, there are a number of issues that must be addressed before embracing a decarbonization strategy involving substitution of gas for coal. First, is there enough gas in the world to enable large-scale replacement of coal for power generation and to do so for much of this century? And can the gas be produced and distributed at costs competitive with those of other fossil fuels? Until a few years ago, many would have answered "no" to these questions. But there's been a game changer.

Recent years have seen explosive growth in the production of natural gas from unconventional sources such as coal beds and most notably shale rock. In the United States, shale gas went from 2% of total gas production in 2001 to 40% by 2014 (Malakof, 2014). With large shale formations throughout the United States, total consumption is projected to increase by more than 50% from 2014 to 2040, with shale gas supplying more than 50% of the total (EIA, 2014b). In combination with other sources of natural gas, total U.S resources could sustain large production rates well into the next century. And although largely untapped, abundant resources exist throughout the world, including China, Russia, and portions of Europe (Malakof, 2014; Zoback and Arent, 2014). Globally, new discoveries of conventional as well as unconventional gas have increased the resource-to-production ratio from roughly 50 years to more than 200 years.[3]

So the answer to the first question is "yes"; there is enough natural gas to enable large-scale replacement of coal for most if not all of this century. And supplies would also be sufficient to enable significant substitution of natural gas for petroleum as a transportation fuel. There are two advantages to such a substitution. Carbon emissions would be reduced by about 25%, and even in regions of the world where natural gas is expensive, it would still be cheaper than gasoline or diesel.[4]

But as the demand for natural gas increases, will prices remain competitive with other fossil fuels? The answer appears to be a qualified "yes." In 2012, abundant U.S. gas production drove U.S. prices below $2 per million British thermal units (MBtu), which is the energy equivalent of about $11 for a barrel of oil, well below a circa-2014 nominal price of $90 per barrel.

Although gas prices were three to five times higher in Europe and Asia, they were still significantly less than the price of oil and competitive with the price of coal.

Large global differences in the price of gas are due to contractual arrangements or transport requirements, both of which inhibit behavior as a true commodity. Prices have been higher in Europe because they were set by long-term contracts tied to the price of oil and highest in Asia because it depends on imports of liquefied natural gas (LNG) from a limited number of suppliers. Prices are low in North America because, lacking the infrastructure needed for liquefying the gas and shipping it as LNG, transport is limited to continental pipelines serving markets for which supply exceeds demand.

Over time the spread in global prices for gas should shrink as indexing to the price of oil fades and LNG infrastructure grows, allowing gas to be traded more freely. Gas prices will likely rise in North America as its use for power production and transportation increases and the continent becomes a net exporter, but drop for importing nations as more exporters compete for their business. With abundant supplies, prices should still remain competitive with other fossil fuels, although less so for coal than for oil.[5]

A significant transition from coal to natural gas is already occurring for power production in the United States. In the 1980s, coal-fired power plants provided approximately 60% of the nation's electricity. Largely as a result of a 1990s surge in the construction of gas-fired power plants, coal's share dropped to 50% by the 2000s, only to drop below 40% by the end of 2011 (ME, 2012). From 2005 to 2013, electricity generated from natural gas increased by 47%, as gas went from contributing 19% to 28% of the nation's total generation (Malakof, 2014). With lower construction and fuel costs for gas-fired plants, as well as reduced emissions of pollutants such as NO_x (de Gouw et al., 2014), gas is on a trajectory to one day replace coal as the principal fuel for power generation. In the United States, most post-2013 construction will consist of gas-fired power plants, providing a powerful lever for bending the curve of U.S. CO_2 emissions downward. However, the impact on global emissions will be muted if coal-rich developing nations such as China and India continue to rely heavily on coal for power generation.

Since much of the exuberance over abundant supplies of natural gas is driven by new discoveries of shale gas, it could become an important arrow in the quiver of measures for mitigating carbon emissions. But before that happens, there are two more issues that must be addressed. Are there deleterious environmental effects associated with producing shale gas? And, looking at the entire process from production to utilization, does its use in lieu of coal or oil fulfill its potential for decarbonization?

Recovering shale gas involves a combination of three technologies: (1) the ability to locate and map the boundaries of the formation; (2) use of horizontal drilling to extend the reach of a single vertical well into the formation; and (3) the use of *hydraulic fracturing*, or simply *fracking*, to extract gas from the shale. To access a formation, a vertical well is drilled to a depth of 2–3 km – well beyond water tables, which are no deeper than a few hundred meters – before turning to continue drilling in a horizontal direction through the formation for a kilometer or more. During drilling, a steel casing is inserted and cemented into the well to prevent contamination of the surroundings. Explosive charges are then used to puncture the casing at selected locations along its horizontal section, and an aqueous slurry of about 99% water and sand and 1% special chemicals is injected at pressures large enough to enlarge cracks created in the shale by the explosions – hence the term hydraulic fracturing. The process increases the permeability of the shale, enabling liberation of the gas. Typically, millions of gallons of fluid are injected into a well. The sand acts as a proppant to keep fracture sites open, and the chemicals enhance release of gas from the shale while inhibiting microbial growth and corrosion of the casing.

A wide range of environmental issues are associated with fracking (Zoback and Arent, 2014), with three in particular drawing considerable attention. One concern relates to contamination of groundwater due to migration of gas from the shale. However, because the shale and groundwater are separated by more than a kilometer of sandstone, it is unlikely the water would be contaminated by gas migrating upward from the shale. But if upper sections of the vertical well casing crack or are improperly sealed, gas can leak into groundwater. High construction standards must therefore be maintained to ensure the integrity of a well throughout its lifecycle.

Another concern is what to do with wastewater after the shale has been fractured. If it cannot be routed to deep injection wells for disposal and is instead returned to the surface, it must be properly managed to prevent contamination of surface waters. Because fracking fluids can include toxic chemicals such as benzene, hydrochloric acid, and methanol, to which radioactive elements, natural gas, and additional toxins are added as the fluid moves through the shale, wastewater returned to the surface represents an environmental hazard that must be remediated. Remediation requirements can be mitigated by reusing the water to drill new wells, but ultimately it must be treated and restored to environmentally acceptable standards. The requirements are challenging but not insurmountable (MIT, 2011; Silva et al., 2014).

For a time, environmental concerns were exacerbated by the reluctance of drillers to divulge their chemicals, citing proprietary and competitiveness issues, and by a paucity of federal and state regulations. In fact, influenced by the natural gas lobby, Congress passed a 2005 law that exempted fracking from regulation under the Safe Drinking Water Act (Barrett, 2011). However, by 2011, under the threat of lawsuits and enactment of full disclosure laws at the state level, drillers were becoming more transparent and open to regulation of their activities. There was clearly growing recognition that the stakes are too high not to get it right. The industry would have to step up to its responsibilities, and local, state, and federal governments would have to hold them accountable to the highest standards.[6]

we don't do this w/ other industries so I don't trust we would w/ them

The third environmental issue bears directly on the potential role of natural gas in decarbonizing the world's energy portfolio. As a substitute for coal in generating electricity, it would seem that increased use of natural gas – enabled by fracking – provides a significant multi-decade approach to reducing GHG emissions. Maybe not! In Section 3.2 we discussed concerns that leakage of natural gas from production and distribution systems may have been underestimated and could exceed requirements for long-term decarbonization benefits.

Two different methodologies are used to determine methane emissions due to leakage. A bottom-up approach inventories emissions associated with producing and delivering the gas, while a top-down approach uses atmospheric measurements to infer emissions associated with production wells and transmission lines. Results vary widely, with top-down estimates consistently exceeding those obtained from bottom-up procedures (Nisbet et al., 2014; Brandt et al., 2014). These differences contribute to uncertainty, and the issue has yet to be fully resolved. In one bottom-up study, an average production leakage rate of only 0.42% was reported for 190 well sites (Allen, et al., 2013). The fact that most of the wells were completed using New Source Review Standards imposed by the EPA in 2012 suggests that, at least for production, leakage can be held to acceptably low rates. The standards will be required of all new wells by the end of 2014, at which time fifteen additional studies will have been performed at every stage of the natural gas supply chain, from production and processing to transmission and distribution (Brownstein, 2013). However, another study, which considered methane emissions from agricultural and fossil fuel operations across the United States,

who knows the 2 methods.

yielded less encouraging results, namely that emissions exceeded EPA estimates by more than a factor of two (Miller et al., 2013).

Although substitution of natural gas for coal in electric power plants is a promising decarbonization option, more studies are needed before significant benefits can be claimed with confidence. Most state governments lack the expertise and resources to monitor gas production sites and pipelines, and the Environmental Protection Agency lacks the resources to launch a viable national monitoring program (EPA, 2013b). Yet, there is immediate need for rigorous and enforceable standards accompanied by an extensive monitoring system. From production well to storage reservoirs to transmission and distribution lines, a cradle-to-grave leakage rate of less than 1% is achievable if the gas industry is willing to make the appropriate investments.[7] *but we must be skeptical of this*

The foregoing concerns bring us to the following question. Rather than simply reducing emissions, wouldn't it be better to eliminate them altogether?

Nuclear and renewable energy

Because no carbon emissions are associated with nuclear power and renewables such as solar and wind energy, substitution of these energy sources for coal (and natural gas) would go a long way toward reducing emissions. But, before we get ahead of ourselves, we need to understand that coal- and gas-fired power plants can be operated without interruption. They provide what is termed *base-load* or *dispatchable* power, that is, the ability to deliver electricity on demand, anytime and almost anywhere. But the Sun doesn't shine at night and the wind doesn't always blow, making them intermittent sources of electricity. Within constraints imposed by existing technologies and infrastructure for storing, transmitting, and distributing energy, there are limits to deployment of solar and wind power. These constraints can be loosened over time, but not without advancements in technology, large capital investments, and political will. In contrast, nuclear energy provides base-load power and is also carbon-free.

Any consideration of the nuclear option must include acknowledgment of the 1986 Chernobyl and 2011 Fukushima Dai'ichi disasters and recognition that such events represent the ultimate black swan, a metaphor for a rare event of significant consequence. Fukushima was precipitated by a huge earthquake and tsunami, and failure of the Dai'ichi plant revealed weaknesses that were both technical and institutional (Bunn and Heinonen, 2011). Technically, a prolonged loss of power impaired the ability to cool the reactors and to vent accumulated gases. Institutionally, the accident revealed major deficiencies in the command-and-control structure for dealing with disasters, as well as existence of a regulatory environment that was far too lax. Fukushima provided important lessons-to-be-learned for any country that has or wishes to include nuclear power in its energy portfolio.

Although efforts were made to strengthen international standards for nuclear safety following Chernobyl, the measures were largely voluntary and accompanied by limited oversight. Bunn and Heinonen call for more stringent safety, security, and emergency response standards, with each nation agreeing to periodic review of its systems by international teams of experts. Within each nation, regulatory agencies must be independent of plant operators and empowered to take appropriate action when deficiencies are identified. And each nation must have safe and secure means of managing its nuclear wastes. Although the road to achieving these goals is strewn with geopolitical obstacles, it can be done if the will exists.

Some nations like Japan, Germany, and Switzerland are having second thoughts about the nuclear option with plans for reducing or eliminating their generating capacity. In Japan, which had obtained 30% of its electricity from nuclear power, Fukushima prompted serious discussion of a non-nuclear future, which would mean going from a once promised 25% reduction to a 3% increase in circa 2020 GHG emissions relative to 1990 (Tabuchi and Jolly, 2013). In Germany, which obtained 20% of its electricity from nuclear power, the decision was made to decommission existing plants by 2022 and to meet future needs without a nuclear option. A downside to this decision has been increased use of coal for power generation (Eddy, 2014). Although the United States had been taking measures to resume construction of nuclear power plants, progress has been slow, and it's likely that only four new plants will be completed before 2020. In contrast Britain, China, India, Russia, and South Korea are moving ahead with ambitious plans.

Today there are about 440 nuclear plants worldwide, a number that could double by 2030. Underlying it all is the fact that nuclear energy makes a significant contribution to large-scale base load power without contributing to GHG emissions. Power plants constructed over the next twenty-five years can help bridge the gap to achieving a twenty-second-century non-carbon, non-nuclear, renewable energy future. Although in the near term construction of new nuclear power plants is more costly than coal- or gas-fired plants, the ability to build plants that last for eighty years or more may well provide a long-term cost advantage.

Ultimately, the future of reducing GHG emissions to acceptable levels – for that matter, the future of energy itself – rests with renewables such as wind and solar energy. However, unlike coal, natural gas, and nuclear, which are concentrated forms of energy, wind and insolation are diffuse and require large amounts of surface area to produce comparable power. Although the world is well endowed with the open spaces needed for large-scale power generation, the best sources of wind and insolation are often found in sparsely populated (low-demand) regions. In the United States, the largest solar capacity exists in the Southwest and the lowest in the Midwest and Northeast, while the largest wind capacity exists in the Upper Midwest and Texas and the lowest in the Southeast. Such regional variations place new demands on the electric grid if the power is to be transmitted from locations of large supply but weak demand to locations of limited supply but large demand.

Despite the foregoing challenges, there has been significant growth in wind and solar power, initially in Europe, and more recently in China and the United States. For more than a decade, the installed global capacity of wind power has been growing at an average annual rate of about 25%, from 39 gigawatts of electricity (GW_e) in 2003 to 318 GW_e in 2013 (GWEC, 2014). Over the same period, solar power experienced an average annual growth of more than 40%, from less than 3 GW_e to 136 GW_e (IEA, 2014). Spurred by sharp reductions in the cost of photovoltaic panels, a record 2013 year-over-year increase of 37 GW_e in solar capacity outpaced growth of 35 GW_e in wind power for the first time. These numbers are encouraging, as is the potential for sustaining growth by advancing technologies and lowering costs. However, it is important to acknowledge that rapid growth in wind and solar power is occurring from a small base. In 2013, their combined capacity of 454 GW_e was still only 8.5% of the world's total installed capacity, while fossil fuels accounted for 70% of the total. In terms of the amount of electrical energy actually generated, their contribution was smaller yet.

The *nameplate capacity* of a power plant corresponds to the maximum rate at which it can generate electrical energy. However, over a designated period such as a year, a plant will not always be operating at full capacity, and its *capacity factor* provides a measure of actual output relative to the maximum possible output. For wind and solar systems, capacity factors range from 20% to 50%, well below the more than 80% associated with fossil fuel and nuclear power plants. Accordingly, in 2013, wind and solar systems contributed only 3.3% to the world's total electrical energy (753 TWh_e out of 23, 127 TWh_e) (BP, 2014).

Maintaining rapid growth in wind and solar power is important to reducing GHG emissions, as well as an essential requirement of transitioning to a sustainable energy future. Improving the efficiency and reducing the cost of wind and solar technologies are important enablers, along with the ability to attract steady and large capital investments. But if they are to contribute at scales commensurate with demand, eventually surpassing the contribution of fossil fuels, two limiting factors must be addressed, one dealing with intermittency and the other with transmission. The first issue can be addressed by accelerating the development of energy storage technologies, and both issues can be addressed by modernizing the electric grid.

Large-scale storage systems would enable banking some of the wind or solar energy during periods of peak generation and withdrawing it during periods of no wind or radiation. In this way, output can be better leveled to mitigate intermittency effects. Existing options include converting the electricity to mechanical energy vis-à-vis a flywheel, compressed air in large tanks or caverns, and water pumped uphill to a reservoir. The electricity could also be stored as chemical energy in a battery or by using electrolysis to produce hydrogen. When electricity is needed, it can be generated from the stored mechanical or chemical energy. Currently, there are two barriers to implementation: costs must drop, and efficiencies associated with charging and discharging the storage unit must increase. Scalable, efficient, and low-cost storage is a sine qua non if wind and solar energy are to overtake fossil fuels as the principal source of electricity, and of all the options, battery storage offers the greatest potential for getting there.

Intermittency effects can also be mitigated by expanding the electric grid to include interconnections between wind and solar farms spread across a wide region. When one area within the region is cloudy or windless, another may experience heavy winds and clear skies, enabling the grid to better provide steady power in the face of variable wind and insolation. Because regions that are well endowed with wind or solar energy are often rural and far from urban areas of large demand, the grid must also be expanded to provide transmission from one to the other. But, by itself, expansion is not enough. The grid must also have intelligence afforded by integration with modern information and communication technologies that increase its stability, efficiency, and cybersecurity. Without such capabilities, the grid will be limited in its ability to blend large-scale generation of renewable energy with base-load power and to minimize the effects of intermittency on load management. The goal is to have a grid – supplemented by large-scale energy storage systems – that maximizes the extent to which renewable power supplants traditional sources of base-load power.

A *smart grid* must also be able to manage the rapid growth of distributed power generated by rooftop solar panels on individual homes and businesses. Distributed power often involves *net metering*, which allows owners to be compensated for electricity they rout to the grid. The trend adds another layer of complexity to grid management, as does *demand response*, which uses smart meters to control a customer's appliances and machinery and to better align demand with available supply.

A cautionary note! In recent years it has become fashionable to think in terms of eventually replacing the grid with distributed power. The premise is that distributed generation will turn "tens of thousands of businesses and households into power producers," enabling them to operate off the grid (Martin et al., 2013). Distributed power should not be viewed as a means of displacing the grid but as an opportunity to integrate distributed generation with a *smart grid* in ways that maximize the use of renewable energy and the efficiency with which it is used. And another note! Distributed generators participating in net metering and enjoying the energy security provided by the grid must pay a fair share of the infrastructure costs associated with construction, maintenance, and operation of the grid.

Although the foregoing discussion has focused on wind and solar energy, other renewables, such as geothermal, wave, tidal, and hydroelectric, are contributing or will contribute to decarbonizing power generation. Hydroelectric power is currently the world's largest source of renewable energy and has the potential for significant growth, particularly in Asia, Africa, and South America. Although the generating capacity of geothermal, wave, and tidal energy is currently small and unlikely to achieve anywhere near the scales of solar, wind, or hydro, with further development they can still contribute to meeting needs in certain regions of the world.

6.3 Decarbonization of transportation

Used largely for power production, in 2011, coal provided 29% of the world's total primary energy supply (TPES) and contributed 44% of its total CO_2 emissions (EIA, 2013b). As such, it is an inviting target for curtailing emissions. But if coal wears the bull's-eye for power production, oil takes center stage for transportation. In 2011, oil contributed 32% of the world's TPES and 35% of its CO_2 emissions. Of these emissions, 63% (22% of total emissions) were due to transportation. In the United States, the transportation sector was even more influential, accounting for 34% of total CO_2 emissions from fossil fuels (Davis et al., 2014). The large dependence of the

U.S. transportation system on oil is underscored by the fact that it provided 93.2% of the total energy used by the system.

The foregoing numbers are large, and when it comes to reducing carbon emissions, the transportation sector must be viewed as another target of opportunity. The sector includes light-duty vehicles (LDVs), which in the United States include cars, SUVs, light trucks, and minivans; medium and heavy trucks; off-highway vehicles; and air, water, rail, and pipeline systems. Of these categories, the first two are the largest consumers of oil and the largest contributors to CO_2 emissions. In 2010, the global total of LDVs and medium/heavy trucks, namely highway vehicles, exceeded 1 billion for the first time. By 2012, it had risen to more than 1.1 billion, with 773 million car registrations. In the United States, personal travel was enabled by 232 million LDVs, which included 121 million cars (Davis et al., 2014).

Driven primarily by economic growth in developing nations, the world's highway vehicles have been increasing by about 3% per year. Were growth to continue at this pace, the number of vehicles would reach 2 billion by 2032, significantly increasing CO_2 emissions in a business-as-usual scenario. What can be done to curtail this increase?

Carbon emissions by the transportation sector can be reduced by reducing the amount of energy consumed by the sector. Energy consumption is reduced by increasing vehicle efficiency and by reducing vehicle miles traveled (VMT). For example, recognizing that the internal combustion engine (ICE) will remain a prime mover of vehicles for decades to come, emissions can be reduced by advancing engine technologies in ways that increase fuel economy (miles per gallon). Reductions in VMT are achieved by choosing trips more judiciously and, where feasible, opting for alternative, lower-emission modes of travel. However, more can be done by reducing the carbon content of energy consumed by the sector, in one case by transitioning from the ICE to carbon-free prime movers and in another by transitioning from oil to renewable biofuels.

Electric vehicles: back to the future

At the dawn of automotive transportation, the electric motor and ICE vied as sources of vehicle power. However, aided by the low cost and large energy density of gasoline relative to batteries, the ICE won hands down. Fast forward by a century, and the electric vehicle is back in the game.

All-electric vehicles rely exclusively on a battery or a hydrogen fuel cell,[8] while hybrids use an ICE in combination with a battery and electric motor. A decade ago, there was a good deal of optimism for prospects of using hydrogen fuel cells to power LDVs, but significant technical and economic barriers have impeded market penetration (Romm, 2004). Costs are high and, even with technical advancements and economies of scaled production, significant penetration remains problematic. The same cannot be said of battery-operated electric and hybrid vehicles.

Hybrids are of two kinds. In one version the battery is charged by the engine, and the engine and battery operate in tandem to power the vehicle. Termed a parallel hybrid, the Toyota Prius is the best-known example. In the other version (a plug-in hybrid electric vehicle, PHEV), the battery can be charged by drawing power from the grid when the vehicle is not used and from a gasoline engine when the vehicle is being used. An electric motor powers the vehicle, drawing energy from the battery until its charge drops below a prescribed level and then using the engine to maintain a charge on the battery as it continues to power the vehicle. The operating range of a PHEV is determined jointly by the storage capacity of the battery and the gasoline tank. A fully electric vehicle (EV) has no ICE. The battery is charged by connecting the vehicle to the grid, and the vehicle's operating range is determined exclusively by the battery's storage capacity.

Hybrid and fully electric vehicles have considerable potential to reduce petroleum consumption, and by 2011 virtually all of the world's major carmakers had active vehicle development programs. Battery technology is the center of attention, with efforts devoted to increasing storage capacity, and therefore vehicle range, while reducing cost and the time to charge the battery.

To the extent that the electricity is generated from low-carbon or carbon-free sources of energy, movement from today's gasoline- or diesel-powered vehicles to EVs and PHEVs would reduce greenhouse gas emissions. With advancements in battery technology, EVs and PHEVs have the potential to achieve 50% of the circa-2050 market for LDVs – about 100 million vehicles per year (IEA, 2011). Governments can encourage movement in this direction with support for research and development and with consumer incentives. In the meantime, parallel hybrids provide an excellent option for reducing fuel consumption. Currently, they provide fuel economies close to the circa-2025 U.S. target of 54.5 mpg and have potential for further improvement.

Biofuels: some better than others

Another manifestation of solar energy is its production of biomass, and during the last decade, a good deal of attention was devoted to increasing the use of renewable (bio) transportation fuels. In the United States, the Energy Policy Act of 2005 mandated annual production of 7.5 billion gallons of renewable fuel by 2012. In 2007, the Energy Independence and Security Act upped the mandate to 15 billion gallons and set a longer range target of 36 billion gallons by 2022. Basically, two fuels could be used, ethanol derived largely from corn and biodiesel derived from animal fat, agricultural oils such as soy and palm, and waste oils. Domestic, corn-based ethanol was favored and subsidized in two ways, first by providing a tax credit of $0.51 per gallon of ethanol blended with gasoline and second by imposing a $0.54 per gallon tariff on imported ethanol. By 2011, domestic production had exceeded 13 billion gallons.

Although favored by Midwest corn producers, the foregoing policy is flawed in two ways. First, corn is a vital agricultural grain, and food prices rise with use as both a food and energy source, putting a special burden on those living at the margins. Second, energy requirements of growing, harvesting, transporting, and processing feed stocks (biomass) into transportation fuels are significant and are largely provided by fossil fuels. It takes about 80 units of energy from other sources to produce 100 units of ethanol energy, providing only a modest net energy gain (Farrell et al., 2006). Moreover, life-cycle CO_2 emissions associated with using corn-based ethanol as a transportation fuel are little more than 10% less than those associated with a petroleum-based fuel. The disadvantages of corn-based ethanol outweigh its advantages, and similar arguments can be made for eliminating the use of agricultural feedstocks for biodiesel. Nevertheless, biofuels remain a promising option for reducing CO_2 emissions, particularly if production is based on alternative (non-food) sources of biomass.

One alternative involves ethanol production from cellulosic forms of biomass such as energy crops (grasses and trees), agricultural residues (corn stover and rice straws), forest residues, and waste streams such as sewage and municipal solids. Termed cellulosic ethanol, the option has the potential to reduce GHG emissions relative to gasoline by about 90% while providing a roughly tenfold energy gain relative to requirements for producing the ethanol (Farrell et al., 2006). Another option involves producing biodiesel from non-food sources such as algae. Although high costs have been a barrier to commercialization, they are coming down.

For cellulosic ethanol produced in small volumes, advances in the enzymes and catalysts needed to convert the biomass to ethanol have reduced costs from approximately $9 per gallon in 2001 to $2 per gallon in 2014 (Service, 2014). The challenge is to keep costs low while scaling production to large volumes. The first attempts to do so were made in

2014, with three plants in the U.S. Midwest projected to have a combined production of 90 million gallons per year. With continued research and development, costs of processing the biomass should eventually drop to a level for which biofuels become competitive with petroleum distillates. Although production at scales comparable to oil refineries will be limited by the logistics of collecting and transporting the biomass, biofuels can become an important contributor to reducing CO_2 emissions in transportation, as well as in other oil-based applications.

6.4 Carbon capture and sequestration: the CCS challenge

The last stem of Figure 6.1 deals with capturing and sequestering CO_2 before or after discharging it to the atmosphere. Carbon capture and sequestration (CCS) following discharge occurs naturally through photosynthesis and storage of carbon in the Earth's biomass. But carbon dioxide can also be separated from products of combustion before they are discharged to the atmosphere and sequestered. Implementation of CCS would allow coal to transition from being the largest contributor to GHG emissions to being essentially carbon free.

For coal-fired power plants, CO_2 emissions per unit of electric energy generation can be reduced by increasing the efficiency with which the chemical energy of the coal is converted to electrical energy. But even if the efficiency is raised from today's nominal value of 35% to 50%, coal would remain a significant contributor to global emissions. Apart from curtailing the use of coal, the only other option for reducing emissions is to implement carbon capture and sequestration. CCS is a three-step process for which CO_2 is (1) separated from its source (captured), (2) transported to an appropriate site, and (3) pumped underground for long-term storage as a pressurized liquid in a geological repository (sequestered). For a coal-fired power plant, capture involves removal of CO_2 from the POC before they are discharged to the atmosphere. Sequestration implies permanence, that is, little or no leakage over millennia.

Carbon dioxide can be captured by retrofitting existing coal-fired plants or by adopting alternative designs for new plants. For existing plants, CO_2 can be captured in much the same way that the products of coal combustion are scrubbed to remove sulfur dioxide. The scrubbing process would occur in an absorption tower through which a chilled amine or ammonium carbonate solution is circulated, stripping the products of CO_2. After leaving the tower, the absorbent is heated to remove the CO_2, which is then pressurized and transmitted for sequestration. Before returning to the tower, the absorbent is cooled to maximize extraction of CO_2 from the POC. The process works, but capital and operating costs, as well as parasitic energy requirements, are large.

Costs associated with different CCS options have been estimated for a large (500 MW_e) power plant with 90% CO_2 removal (MIT, 2007). Based on 2005 dollars, the scrubbing process was estimated to add approximately $30 per megawatt-hour to the *levelized cost of electricity* (LCOE) – the cost of supplying electricity to the grid – increasing it from about $50/$MWh_e$ without capture to $80/$MWh_e$ with capture. If assessed per unit mass of CO_2 removal, the cost of removal would be approximately $40 per Mt-$CO_2$ ($147 per Mt-C). Energy requirements of the separation process would reduce the plant's thermal efficiency by approximately 30%, thereby requiring a 43% increase in the generating capacity, and accordingly coal consumption, to achieve an equivalent net power of 500 MW_e. In strictly economic terms, retrofitting existing power plants with this technology would be hard to justify, unless comparable costs are imposed as a tax on emissions and/or an economic return can be obtained on the CO_2 before it is sequestered. A contributing factor to the high cost of CCS is the low relative concentration of CO_2 in the POC. At values of 15% or less by volume, low concentrations make for large absorption towers and large amounts of absorbent.

Alternative CCS technologies focus on increasing the relative concentration of CO_2 in the POC by reducing the concentration of the most prominent constituent, namely nitrogen introduced with the air used to burn the coal. In an *oxy-fuel combustion* (OFC) system, nitrogen is separated from the air before it enters the combustion chamber, and the coal is burned in a nearly pure oxygen environment (Appendix E). Nitrogen is then restricted to that contained within the coal, and the POC consist largely of carbon dioxide and water vapor. The larger CO_2 concentration in the POC and the fact that separation can now be achieved by distillation (condensing the water vapor) increase the overall effectiveness of carbon capture. Although it costs money and energy to separate nitrogen from the air, overall costs and energy requirements of CCS are projected to be lower than those associated with the scrubbing process. By using the OFC option with 90% CO_2 removal in a 500 MW_e plant, the LCOE is estimated to increase by $22/$MWh_e$, the cost of CCS would be $30/Mt-$CO_2$, and the plant's efficiency would decrease by approximately 20% (MIT, 2007).

Another CCS option involves use of an integrated gasification combined cycle (IGCC). The system combines two well-established technologies, one involving *gasification* of the coal and the other use of a combined-cycle power plant to generate electricity (Appendix E). Like OFC, the IGCC process begins with separation of nitrogen from the air, but unlike OFC, the coal is gasified, not burned.

A pulverized coal-water slurry and nearly pure oxygen enter a gasifier, where they react at very high temperatures and pressures to produce syngas (a mixture of hydrogen and carbon monoxide), carbon dioxide, and pollutants such as acid (sulfur-based) gases, fine particulates, and mercury. The carbon monoxide can be converted to additional hydrogen and carbon dioxide through a water-gas shift reaction, and the pollutants can be removed. With its relative concentration significantly increased, the carbon dioxide in the resulting gas can be captured at lower cost. The remaining hydrogen-rich gas is burned, and the POC are used to power a highly efficient combined cycle involving integrated gas and steam turbines. Without CO_2 removal, the LCOE from an IGCC plant is approximately \$50/MWh$_e$. With removal, it's estimated that approximately \$14/MWh$_e$ would be added to the LCOE, the cost of carbon removal would be approximately \$20/Mt-CO_2, and the efficiency would be reduced by about 35% (Socolow, 2005; MIT, 2007).

Additional information on CCS options and related costs is provided in Appendix E. Although results suggest that the IGCC system is the lowest-cost option, the MIT study was clear in stating that it's far too early to pick winners and losers and that selection of the most technically and economically viable option would have to be preceded by experiences gained from the development and operation of large-scale (100 MW$_e$ or more) power plants. Because the OFC and IGCC systems with CCS are relatively new, there is likely to be upside potential for cost and efficiency improvements. However, there must first be large-scale demonstration projects that establish baseline conditions. At this point estimates of CCS options are just that, *estimates*. Confidence in the estimates must await construction and operation of the first large (commercial) scale power plants.

As of September 15, 2014, worldwide there were twenty-six CCS projects for power plants larger than 100 MW$_e$ and twenty-four pilot projects for plants of 1 to 48 MW$_e$ (MIT, 2014). Although these numbers suggest that CCS is well on its way to becoming a viable option for reducing carbon emissions, implementation at meaningful scales remains problematic. Of the twenty-six large-scale projects, only three were under construction, with the remainder in various stages of planning for which construction may never occur. Of the pilot projects, two were in planning and six had been terminated after one to two years of operation. To get a sense of the challenges, consider the following projects.

In 2009, the American Electric Power Company (AEP) initiated a pilot project at its 1,300 MW_e coal-fired Mountaineer power plant in West Virginia. The objective was to demonstrate use of a chilled ammonia system to capture 2% of the more than 4 Mt-CO_2 discharged per year. Once captured, the CO_2 was pressurized and pumped 2,500 meters below ground for sequestration in a porous rock formation shrouded by cap rock. Following successful completion of the pilot project, plans were to scale it for sequestration of more than 20% of the plant's emissions. However, in 2011, AEP scrapped the entire effort, citing the "uncertain status of U.S. climate policy" and a weak economy (Wells and Elgin, 2011).

Another project, termed FutureGen, involved a broadly based government-industry partnership that sought to create the world's first emissions-free coal-fired power plant, in this case a 275 MW_e unit based on IGCC technology. But, after taking almost a decade to design the plant and to select a suitable site, the project was deemed too costly. In 2010, its scope was changed to one of using OFC technology to retrofit an existing 229 MW_e power plant. But, with the partnership fraying and tight time constraints placed on access to government subsidies, the project remains in planning and its future in doubt (Wald, 2011; MIT, 2014).

What would it take to complete a large-scale CCS project? In 2012, one such project was seemingly moving forward. Termed the Texas Clean Energy Project (TCEP, 2011), plans called for construction of a 377 MW_e IGCC plant, with 214 MW_e routed to the Texas Grid, 121 MW_e used for plant operations, including CO_2 capture and compression, and 42 MW_e used to produce urea for fertilizer. The goal is to capture 90% of CO_2 emissions – about 2.5 million tonnes per year – and to sequester the CO_2 in nearby oil fields for enhanced oil recovery (EOR).[9] The plant is also designed to capture 95% or more of the sulfur, nitrogen oxides, mercury, and particulates in the POC. If all goals were met, it would merit designation as a clean-coal facility. But, even if successful, requirements for activating the project point to the barriers that would await attempts to replicate it.

First, there's the price tag. At an estimated cost of $2.8 billion to supply about 250 MW_e of electricity to the grid, capital costs come to $11,000 per kW_e, a staggering amount when compared with costs of about $2,000/$kW_e$ estimated in the MIT (2007) study (Table E.1) or the cost (~$1,000/kWe) of a natural gas combined-cycle plant without CCS. To secure private capital for such a project, it must make economic sense to potential investors. To make TCEP a more attractive investment, the federal government awarded $450 million under its Clean Coal Power Initiative and the American Recovery and Reinvestment Act, augmented by $63 million of investment tax credits and additional tax incentives from the State of Texas. The project also benefits from having four commercial products: electricity for the grid, CO_2 for EOR, urea for fertilizer production, and sulfur for producing sulfuric acid. Nevertheless, despite generous financial subsidies and several co-products, the project – scheduled for completion in 2014 – remains in a planning stage (MIT, 2014). Difficulties associated with launching a project under such favorable conditions suggest that prospects may be poor for wide adoption of CCS.

In Kemper County, Mississippi, another large IGCC project with CCS is faring better than TCEP. At 582 MW_e, the plant will capture 65% of CO_2 emissions (about 3.5 Mt/yr) for EOR and reduce SO_2, particulate, and mercury emissions below normal levels (MIT, 2014). Construction began in 2010 at a projected cost of more than $2.4 billion (more than $4,000/$kW_e$). Like TCEP, government incentives were provided through a $270 million grant and $133 million in tax credits. And in October 2014, a 110 MW_e plant in Saskatchewan, Canada became the first operational coal-fired plant with CCS (90%) and EOS at a cost of more than $1.2 billion ($11,000/kW_e) with a $355 million government subsidy.

The rationale behind providing generous subsidies for first-generation projects such as the TCEP, Kemper County, and Saskatchewan plants is that they would not otherwise attract private investment and that such projects are needed to realize cost reductions anticipated for subsequent (nth generation) plants. However, the extent to which costs can be reduced is yet to be determined, and it's questionable whether generous government subsidies can continue for future projects. Also, a key economic requirement for matching an emission source with a storage site is the need to accommodate more than fifty years of emissions, and it's questionable how many such matches exist for sites that also provide for EOR.

Coal-fired power plants provide an important target for CCS, but they are not the only target. Other large point (concentrated) sources of CO_2 include gas-fired power plants, petrochemical refineries, and facilities for producing natural gas, ammonia, cement, and steel. Globally, Dooley et al. (2006) identified approximately 8,000 point sources, each generating more than 0.1 Mt-CO_2 per year and providing a suitable candidate for CCS. In 2005, fossil fuel-fired power plants comprised 60% of the point sources and accounted for more than 70% of their total emissions. Facilities for processing natural gas and cement plants were the next largest contributors at 10% and 6%, respectively, while refineries and steel mills each contributed 5%.

There is certainly no dearth of concentrated sources to which CCS technologies can be applied, and globally there is sufficient subterranean (geologic) capacity to store hundreds of years of anthropogenic CO_2 emissions. Moreover, evidence accumulated to date suggests that long-term retention can be achieved.[10] Yet, from assessments of the status of CCS (Haszeldine, 2009; WorleyParsons, 2009; MIT, 2014), it is hard to be sanguine about future prospects. The fact remains that CCS is a high-cost approach to reducing CO_2 emissions. In an economic analysis of mitigation options, Creyts et al. (2007) estimated new plant costs in 2005 dollars to be approximately $110 and $165 per tonne of carbon for a coal-fired plant with and without EOR, respectively. Costs were higher yet for retrofitting existing coal-fired plants with CCS, and the LCOE exceeded costs for wind, photovoltaic, and nuclear systems.

By 2012, concerns about the viability of CCS were leading to a wave of project cancellations in Europe and the United States (Handwerk, 2012; Cala, 2012; Plumer, 2012). In Europe, concerns are for the geological risks of storing pressurized CO_2 underground, as well as for the high costs of capture. Two projects were cancelled in 2011, and of twelve demonstration plants scheduled for start-up by 2015, no more than six were likely to begin by 2020. In the United States, there was little to show for the nearly $7 billion in federal funds authorized from 2005 to 2012 for advancement of the technology, and daunted by escalating cost concerns, utilities were showing little appetite for investment in large-scale projects.

Current strategies for stabilizing atmospheric CO_2 below 450 ppm include extensive implementation of CCS, particularly for coal-fired power plants, and under existing conditions, there may be some niches where combining CCS with EOR makes economic sense.[11] But if CCS is to become a viable mitigation option and to be widely used, there would have to be a major breakthrough in CCS technology, one that significantly reduces costs. At this time, the likelihood of using CCS to meet expectations for significantly reducing carbon emissions is not high.

6.5 A sobering perspective: the notion of wedges

Curbing carbon emissions is a nontrivial task. Yes, there are many options, but in the main they come with large costs and significant departures from business-as-usual (BAU). Pacala and Socolow (2004) highlight the magnitude of the problem and the scale of measures needed to address it. Adopting circa-2004 global fossil fuel emissions of approximately 7 Gt-C/yr as a baseline, they asked the following question: *What mitigation measures would be needed to keep emissions at this level over a fifty-year period, while allowing for population growth and economic development?* In a BAU

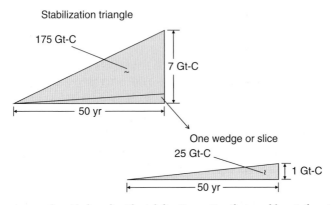

FIGURE 6.2. Carbon dioxide stabilization option that would curtail emissions by 1 Gt-C at the end of fifty years (a wedge) and seven options that would collectively curtail emissions by 7 Gt-C (a triangle). Adapted from Pacala and Socolow (2004).

scenario corresponding to a 1.4% annual increase in emissions, a doubling to 14 Gt-C/yr would occur over fifty years. Allowing for growth, what could be done during that time to hold emissions at 7 Gt-C/yr?

Pacala and Socolow couched the problem in terms of a stabilization triangle (Figure 6.2). The hypotenuse of the right triangle represents a BAU scenario for which emissions are assumed to increase linearly from 7 to 14 Gt-C/yr over fifty years. The base of the triangle is the scenario for which emissions remain at 7 Gt-C/yr, and the area (175 Gt-C) is the total amount of carbon emissions that would be averted over fifty years. They then examined mitigation options that would have a significant effect, which they defined as achieving 1 Gt-C/yr in averted emissions per option by the end of fifty years. Each option, termed a wedge, was also represented as a right triangle, with a base of fifty years and a height of 1 Gt-C/yr. Each wedge would avert 25 Gt of carbon emissions over fifty years, and seven wedges would combine to form the stabilization triangle. Fifteen different mitigation options (wedges) were identified.

One option addressed the LDV segment of the transportation sector. Globally, the segment was projected to grow from 540 million vehicles in 2003 to about 2 billion over fifty years. As a reference case, Pacala and Socolow assumed a fifty-year BAU condition for which the average LDV would operate on conventional fuel (gasoline/diesel), have an average fuel economy of 30 mpg, and be driven 10,000 miles per year. For these conditions each LDV would emit 1 tonne of carbon per year, and two billion of them would emit 2 Gt-C/yr. What could be done to take 1 Gt-C/yr from these emissions?

Options included achieving an average global fuel economy of 60 mpg within fifty years or reducing annual driving to 5,000 miles. The first is certainly doable with advanced engine and hybrid vehicle technologies, while the second is more problematic, at least for some nations such as the United States. Land development and community-planning strategies would have to be altered to reduce dependence on the automobile and provide greater access to efficient and user-friendly public transportation, as well as more opportunities for movement by foot and bicycle. Although electric vehicles and biofuels also provide options for reducing carbon emissions, there are some caveats. Unless the electricity is produced from a carbon-free source of primary energy, it is a misnomer to talk about zero-emission electric vehicles. And to achieve significant reductions in GHG emissions while eliminating competition with needs for food, a biofuel such as ethanol would have to be produced from cellulosic matter such as crop residues and natural vegetation, currently an expensive proposition. That said, in general, the transportation sector provides one of the more promising opportunities for achieving giga-scale reductions in carbon emissions.

Biological options for reducing atmospheric CO_2 concentrations involve improved land-use and forest management practices such as preserving existing forests, planting new forests (afforestation), and restoring previously forested land (reforestation). Pacala and Socolow estimated that 1 Gt-C could be removed from net emissions by eliminating deforestation entirely and reforesting 250 million hectares in the tropics or 400 million hectares in temperate zones. But the most promising regions for maintaining large forest growth exist largely in developing nations that obtain much needed revenue by clearing forested land and developing it for other purposes. To discourage such practices, financial incentives would have to be provided, largely by wealthier developed nations. Although a small group of nations, led by Norway, has pledged $4.5 billion for this purpose, as much as $10 billion per year in recurring funds would be needed to have the desired impact (Economist, 2010). With the current state of the global economy, maintaining such a commitment is problematic. Another option involves capturing and sequestering CO_2 from large-scale fermentation or gasification processes that convert biomass to a biofuel or from co-firing biomass with a fossil fuel for generating electricity. The option would actually provide *negative* emissions by essentially removing CO_2 from the atmosphere (Benson, 2014). But with nominal costs of $100 per t-$CO_2$ for CCS, prospects for implementation are poor.

Bigger challenges are associated with other options, among them the decarbonization of electric power. When it comes to power production, the biggest opportunity (and challenge) is to curtail CO_2 emissions associated with coal-fired plants. There is much to like about coal. It is an abundant, low-cost resource in many regions of the world, including Asia, Europe, and North America. And historically it has not been subject to the kind of price volatility and/or disruptions in supply that have characterized oil and gas markets. In 2013, it accounted for approximately 37% of power generation in the United States, 50% in Germany, 70% in India, and 80% in China. Within certain regions of the United States, it fueled more than 85% of power generation, as in states such as Ohio (86%), Indiana (94%), Wyoming (95%), and West Virginia (98%). Replacement of coal-fired power plants is therefore a tempting target for reducing CO_2 emissions.

As a fifty-year reference case, Pacala and Socolow considered coal-fired plants with a cumulative generating capacity of 700 gigawatts (700 GW_e), an efficiency of 50%, and an annual capacity factor (percentage of time in operation) of 90%. A 50% efficiency corresponds to the projected capability of a USCPC plant. What would it take to eliminate one wedge corresponding to the reference case? Assuming that gas-fired, combined-cycle power plants would reduce emissions per unit of power production by 50%, a slice could be achieved by replacing 1,400 GW_e of coal-fired capacity with an equivalent amount of gas-fired capacity. Although a large amount of natural gas would be needed to fuel the power plants – about 5 billion cubic meters per day or fifty times the output of a large natural gas pipeline – it could be doable in view of explosive growth in the production of natural gas. And the transition from coal to gas would have other benefits.

Contrary to television commercials appearing in the United States, coal is not a *clean fuel*. In addition to carbon dioxide, emissions include oxides of sulfur and nitrogen, particulates, and heavy metals, all of which are deleterious to human health and the environment. Were the external costs of these emissions internalized, the price of coal-based power could more than double in the United States and increase by much more in nations such as China and India, where restrictions on emissions are more lax (Economist, 2011a). Coal is also disadvantaged by the risks it poses for those who mine it. In contrast, lacking sulfur, nitrogen, heavy metals, and precursors to particulate matter, natural gas is a much cleaner fuel. Of course, like so many aspects of climate change, there are caveats. As we've noted, natural gas is not without its own environmental issues. Production can contaminate surface and ground waters, and global warming can be exacerbated by CH_4 leakage across the supply chain.

To reduce CO_2 emissions, better options would involve replacement of coal-fired power plants by power plants with zero emissions. A reduction of

1 Gt-C per year could be achieved by adding 700 GW_e of nuclear power plants in lieu of new coal-fired capacity, almost doubling the world's circa-2010 nuclear capacity. Nuclear power has several appealing attributes. It releases no CO_2 or pollutants such as mercury and oxides of sulfur and nitrogen, and it enhances a nation's energy security by diversifying its energy portfolio. However, satisfactory means must be established for processing and storing radioactive waste materials, and absolutely failsafe processes must be implemented for preventing proliferation and use of the materials for terrorism. Over fifty years, the global accumulation of waste materials from the nuclear option would be about three times the amount accumulated through the year 2010. And at construction costs of more than $5,000 per kW_e of generating capacity – or more than $3.5 trillion for 700 GW_e – access to capital markets would require government assistance.

In lieu of or in addition to nuclear power, coal-fired power generation could also be replaced by renewable, zero-emission technologies such as hydro, wind, and solar power. Assuming an equivalent capacity factor, a wedge could be realized by replacing 700 GW_e of coal-fired power with an equivalent amount of hydropower. However, to account for their intermittent nature and smaller capacity factors, Socolow and Pacala estimated that 2,300 GW_e of peak wind or solar power would be needed to displace 700 GW_e of coal generation, which would require approximately 38 million hectares or 2.3 million hectares of land, respectively.

At first glance, the foregoing requirements appear daunting. Circa-2013 global wind and solar PV capacities of 318 and 150 GW_e would have to increase more than seven- and fifteen-fold, respectively, to produce 2,300 GW_e. However, renewable sources of electricity are growing rapidly, and in 2013 they accounted for 56% of all the new (year-over-year) increase in global generating capacity (REN, 2014). Driven by steep drops in the cost of solar panels and land-based wind turbines, growth in wind and solar power has been strong, and for solar it occurred at an average annual rate of more than 50% from 2008 to 2013. There is no doubt that renewable forms of power offer significant potential for reducing CO_2 emissions. But there are some caveats.

Apart from hydropower, generation from renewables is growing from a small base, which in 2013 supplied only 2% of the world's total energy consumption (REN, 2014). Even with sustained growth, it will be some time before they can meet a significant fraction of the world's needs, particularly if accompanied by a 2% annual growth in energy demand. And meeting a significant fraction depends on advancements in energy storage and grid technologies. Storage is needed to mitigate the intermittent nature of wind and solar energy, as is expansion of the grid and endowing it with greater intelligence. And it will all require trillions in capital investments. The reality is that it will be years before renewable energy will provide a major share of the world's electricity, and absent implementation of an ambitious plan for nuclear power, the world will continue to rely heavily on coal- and gas-fired power plants.

The 2004 study by Pacala and Socolow provides a sobering perspective of the scale of changes that would have to be made to the world's energy portfolio if emissions are to be held in check while the global economy continues to grow. Since then, has the world struck the appropriate balance between curtailing emissions and growing the economy? The answer is an unqualified "no!" Implementation of mitigation measures has been incremental, and the challenges have become more daunting. In 2011, the notion of wedges and a stabilization triangle was updated (Socolow, 2011), with nine instead of seven wedges needed to fill a stabilization triangle extending from 2011 to 2061. And even if achieved, stabilization would have to be followed by a precipitous reduction in emissions. If the rise in atmospheric GHG concentrations is to be stemmed, mitigation efforts must be more audacious.

Davis et al. (2013) argue that aggressive action must be taken over a much shorter timeline, with emissions sharply curtailed within a few decades rather than fifty years from now. They also argue that, with emissions having risen from 7 Gt-C in 2001 to 10 Gt-C in 2013 and the likely continuation of an ascendant trajectory, twenty-one rather than nine wedges would be needed to fill the stabilization triangle and that thirty-one wedges would be needed to achieve zero emissions within fifty years. In the face of demands for economic growth in much of the world, implementing nine, much less twenty-one or thirty-one, wedges may be an insurmountable challenge.

6.6 Economic factors

To recap, the foregoing mitigation strategy calls for: (1) pursuing efficiency and conservation measures across all areas of energy production and utilization; (2) reducing the use of coal for power production by increasing the use of natural gas and nuclear energy as twenty-first-century bridge fuels and by accelerating deployment of renewable sources of energy; and (3) by reducing the use of oil for transportation by moving to electrification and/ or the use of biofuels for the world's fleet of ground and air transportation vehicles. What's to prevent aggressive implementation of these measures?

After all, many of the requisite technologies already exist with considerable upside for improvement. But technology is only one part of the solution. Apart from energy efficiency and conservation, the foregoing measures require significant capital investments.

The cost of capital is critical, particularly for utility-scale power plants, and those choosing to invest want to be assured of adequate returns on their investments. Key economic parameters include the rate of return expected by investors, profitability vis-à-vis other investment options, the weighted average cost of capital obtained from investors and debt providers, and the effective rate that must be paid on the debt. These factors determine whether to proceed with a project, and for power plants they affect the levelized cost of electricity (LCOE).

Using nominal values for relevant economic parameters, Lazard (2013) determined the LCOE for new U.S. power plants based on conventional and alternative sources of energy. Not surprisingly, coal- and gas-fired power plants without CCS were at the low end of the cost spectrum. For an advanced SCPC power plant with coal priced at \$1.99 per MBtu, the LCOE was \$65/MWh_e; for a combined-cycle plant with gas priced at \$4.50 per MBtu, it was \$61/MWh_e. However, with 90% capture and compression of CO_2 emissions, the LCOE more than doubled to \$145 and \$127 for coal and gas, respectively. The additional cost of \$80 for coal more than doubled earlier estimates of \$30 per MWh_e (MIT, 2007).

The foregoing estimates are, of course, sensitive to fuel costs. Since coal is abundantly available and traded as a global commodity, its price is not likely to vary much between nations, nor is it likely to experience significant upward pressure. That was certainly true of global coal markets in 2013, as coal dropped to less than 40% of U.S. power generation from 50% a decade earlier and China banned new coal plant construction in certain regions while encouraging greater use of natural gas, nuclear, and renewable energy. That is not to say that global coal consumption is experiencing an inexorable decline, only that prices will remain low enough to maintain its status as a low-cost means of power production. The extent to which global consumption increases (or decreases) will depend strongly on choices made by developing nations, especially China and India (Krauss, 2013).

Conditions differ for natural gas. In the United States, where fracking has significantly increased production, prices in the early 2010s ranged from \$2 to \$4 per MBtu, making it an extremely attractive option for domestic power production. Although gas priced at \$10 or more in Europe and Asia makes the value proposition less attractive, in time the situation will change. The world has plenty of natural gas, and as new LNG facilities and exports grow, price differentials will narrow, with natural gas likely to remain a relatively low-cost power production option for several decades in North America and becoming a more attractive option in other regions of the world. As demand for energy increases, it is easy to envision a multi-decade scenario for which gas and to a lesser extent coal continue to supply a significant fraction of the world's electricity.

In the United States, the low cost of natural gas has significant implications for other sources of energy. While natural gas can reduce emissions by displacing coal-fired power plants, its low cost can indirectly increase emissions by discouraging investment in carbon-free sources of electricity, as well as energy efficiency measures (Wald, 2014). Consider nuclear power. Without subsidies and accounting for decommissioning costs, the LCOE for a new nuclear plant is estimated to be $115 per MWh$_e$ (Lazard, 2013), almost twice that of a gas-fired plant. So, while other nations, particularly in the Far East, move forward with nuclear plants, only four new plants are planned for the United States. And even when the plants come on line, the nation's nuclear capacity will likely decrease due to decommissioning of existing plants. Utilities are hard-pressed to make the case for nuclear power, and elected officials are generally averse to approving rate requests that raise costs for their constituents. But is that perspective short-sighted? What is the level of confidence that the price of gas will remain low for more than a few decades, particularly in the face of rapidly rising demand? A new nuclear power plant with a capacity factor of 90% provides a bulwark of base-load power and would have an operating life of more than sixty years. Thinking long term, could it one day – well before its decommissioning – become a low-cost source of electricity? Because nuclear energy can do much to reduce twenty-first-century carbon emissions while providing a large and reliable source of base-load power, it should play a prominent role through much of this century.

For advocates of renewables, the news concerning the cost of electricity is good and getting better. Wind is already cost-competitive in many regions, and solar is approaching competitiveness in regions of high insolation. Three trends are converging to improve the outlook of solar photovoltaics (M. O'Sullivan, personal communication, 2013): (1) increasing energy conversion efficiencies, (2) decreasing panel and installation costs, and (3) the willingness of investors to accept a lower rate of return. Assuming a cost of $1.50 per watt and a single-axis tracking mechanism for solar panels, the LCOE for utility-scale generation in the U.S. Southwest ranges from about $65 to $90 per MWh$_e$ without subsidies (Lazard, 2013). And the LCOE will continue to drop as panel costs go to $1.00 per watt by the end of the decade. Although the LCOE for distributed generation by rooftop PV systems is much higher – from about \$150 to \$200 per MWh_e without subsidies – a range of incentives has stimulated rapid growth in the United States and elsewhere. The same cannot be said of utility-scale solar thermal systems that rely on large arrays of mirrors to operate a steam turbine by concentrating solar energy. Costs exceed \$125 per MWh_e, and apart from circa-2014/15 systems under construction, new builds are unlikely.

The economics also work well for wind energy. With technology improvements yielding higher capacity factors, circa-2013 costs of electricity generated in wind-rich regions extending south and west from the Dakotas to West Texas were as low as \$45 per MWh_e and no higher than \$80 in states like Michigan, Indiana, and Ohio (Lazard, 2013; M. O'Sullivan, personal communication, 2013). If the usual production tax credit of \$20 per MWh_e is included, the LCOE ranges from \$25 to \$60 and is less than the cost for coal or gas. With an LCOE of about \$155, the same cannot be said for offshore wind. However, offshore winds provide an abundant resource close to regions of high demand and are less affected by intermittency. Offshore wind farms have been deployed in Europe, and plans are to install the first U.S. farm off the Rhode Island coast. They will become more cost competitive over time, and unlike natural gas or nuclear power, they are more than a twenty-first-century bridge to a sustainable energy future. Offshore wind will become an integral part of the future.

If GHG emissions are to be significantly dampened, contributions of renewable and nuclear energy to meeting the world's needs for electricity must increase in relative as well as absolute amounts. With growing demand, it is not sufficient for them to increase without capturing an ever larger share of the demand and in the process reducing dependence on fossil fuels. Large investments in research and infrastructure are needed to get there, investments that increase the efficiency and reduce the cost of converting primary energy to electricity, provide cost-effective and efficient systems for low (residential and commercial) to large (utility) scale storage of electrical energy, and create smart grids with net metering to optimize integration of electricity generated by utilities with distributed generation by homes and businesses.

6.7 Adaptation

Mitigation is the first line of defense against global warming and should be pursued vigorously and persistently. But what if efforts are unable to reduce atmospheric GHG concentrations to acceptable levels, a scenario that is becoming ever more likely? The second line of defense involves preparing for such an eventuality by implementing adaptation measures to reduce the impact of climate change. Today it's clear that mitigation and adaptation must occur concurrently, the first to reduce the extent to which the second is needed and the second to protect against deficiencies in the first.

Adaptation deals with increasing the resilience of a region or country to climate change. In essence it is an exercise in risk management, in this case complicated by the uncertainties of climate change (Schwartz, 2010). There is uncertainty in the trajectory of future GHG emissions and atmospheric concentrations, which will depend on a range of uncertain technical, economic, political, and social factors (Section 3.4). There is also uncertainty in related climate response mechanisms and their impact on different regions of the world. Hence, adaptation must be based on a planning process that applies the principles of risk management to a range of scenarios that vary from one region to another.

Risk assessment begins by identifying possible outcomes of climate change for a particular region and system. The region could be the U.S. Northeast or South Asia, and the system could pertain to agricultural production, human health, or infrastructure. Specific climatic events could involve intense and prolonged heat waves or drought, severe storms with extreme precipitation and flooding, and/or rising seas and storm surges. In each case, metrics and probabilities would be assigned to the event. How much is the sea level expected to rise over a prescribed time? How many days per year are temperatures expected to exceed a prescribed level? What probabilities are assigned to the expectations? For example, what is the probability that, by 2050, the City of Houston could experience temperatures exceeding 110°F (43.3°C) for ten or more days? Would it be low, medium, or high?

The next step would be one of assessing the vulnerability of the system of interest to the event. Would adverse consequences be low, medium, or high? A worst case would correspond to highly adverse effects caused by a highly probable event, while the best case would correspond to a low-impact, low-probability event. It is the probability of the event multiplied by the severity of its consequences that determines its risk and prioritization for appropriate action. Action would, of course, be influenced by economic considerations. Could today's costs to enhance a system's adaptive capacity be justified in terms of higher costs that would be incurred in subsequent decades by ignoring the problem?

For adverse consequences to which a high probability is assigned, the next step is to identify and implement adaptation options. To protect against rising seas and storm surges in coastal regions, how should new infrastructure be built and existing infrastructure reinforced? Where and to what extent should large seawalls, floodgates, and diversions be built? Should natural barriers such as salt marshes and beaches be restored or reinforced? And if these systems fail, what provisions should be made for providing a quick response to needs of the survivors, particularly when disaster strikes developing countries ill-equipped to cope with the consequences? Barbier (2014) proposes establishment of permanent regional or global disaster task forces that would be first responders capable of providing immediate and effective assistance.

A three-step planning process has been developed for dealing with the impact of climate change on infrastructure (NRC, 2008). The process considers (1) stresses imposed by climate change and the probability of effects on the infrastructure, (2) the potential for damage and disruption of services, and (3) measures that can be taken to increase resilience. But the process can be applied to more than infrastructure. Where and to what extent should adaptation include implementation of drought-, temperature-, and/or salt-resistant crop varieties? What measures must be taken to ensure adequate food supplies in response to disruptions occurring anywhere in the world due to persistent drought or flooding? Where and to what extent should plans be developed for curtailing the spread of disease vectors? And what procedures should be established for dealing with large dislocations caused by migration of people from regions severely and irreversibly impacted by climate change? The final element of any risk assessment/management process involves monitoring, assessing, and adjusting adaptation measures as needed. A comprehensive treatment of adaptation methodologies is provided by the National Research Council (NRC, 2010).

To be effective, assessment and management strategies must be data-driven, and although much is yet to be done, decision-making tools are being developed to assist governments in gauging their *vulnerability* to climate change and their *readiness* for deploying resources to improve *resilience*. In one such system (ND-GAIN, 2014), vulnerability metrics include food and water security, ecosystems, habitat, and infrastructure, while readiness accounts for economic, political, and social conditions. Of the 176 countries included in the assessment, well over half are poorer developing nations that combine poor readiness with high vulnerability and have the greatest need for assistance – technical and financial – in identifying and implementing adaptation options. But whether developing or developed, poor or rich, nations must provide specificity in their planning processes.

Talbot (2007) calls for the development of reliable regional climate models, or what Schwartz (2010) would call a "finer-scale understanding of climate change." Where will flooding be most acute? Where will drought and water shortages be most pronounced? Where will climate change spawn the propagation of vector diseases? How should coastal regions be managed, communities developed, food and water security maintained, and human health preserved? As the frequency and intensity of such events increase, where and how does it make sense to invest in adaptation measures? Measures adopted on the U.S. East Coast would differ from those in the Midwest and Southwest and certainly from those in sub-Saharan Africa and South Asia.

For much of the world's population living in urban areas that would bear the brunt of climate change – particularly those in low-lying coastal regions – adaptation is an existential consideration. In the United States, New York City has undertaken a comprehensive, risk-based analysis of the vulnerability of its low-lying infrastructure to climate change, and results are being integrated into the city's processes for managing existing infrastructure and future development (NPCC, 2010; Yohe, 2010). In Boston, a cost-benefit analysis has been performed for investments designed to protect a highly developed coastline from the effects of a rising sea level and more intense storms (Yohe, 2010). Decisions to allocate billions, if not trillions, of dollars to adaptation are not easily made, particularly with uncertainties in predictive methods and the time spans associated with dire consequences. But the decisions may well reduce to a *pay me now or pay me later* conundrum. For 2012, the estimated cost of eleven extreme weather events in the United States was $114.6 billion, led by Hurricane Sandy at $65 billion (Hampton and Cross, 2014).

The potential impact of climate change on displacement and migration of people deserves special attention. Warner et al. (2009) identified regions of the world where displacement due to climate change has already begun, largely attributable to the increased frequency and intensity of storms, floods, and drought. The longer-term effects of changes such as rising sea levels were also considered. A one-meter rise in sea level could displace almost 24 million people along the Ganges, Mekong, and Nile deltas and remove more than 1.5 million hectares from agricultural production. By 2050, total global migration could exceed 200 million people. Mass migrations are a catalyst for political instability, as well as human suffering. If manifested, the impact on social and political systems and on human suffering would make events like Hurricane Katrina appear mild by comparison. How would governments deal with such conditions?

6.8 Geoengineering

A few years ago, a cartoon making the rounds showed the Earth encased in a red box with a glass front and the instructions, "In Emergency Break Glass." The implication was that, if all else fails and global warming exceeds a dangerous threshold – a level beyond which mitigation and adaptation are no longer credible options – humankind can look to geoengineering for relief. Also termed "climate engineering" or "climate remediation," it refers to large-scale engineering endeavors designed to counteract the warming effects of GHG emissions. Following mitigation and adaptation, it is the third and last arrow in the quiver of measures for dealing with the effects of global warming. But unlike the first two arrows, for which implementation is long-term and costly, geoengineering would provide a relatively quick fix at a comparatively low cost.

Geoengineering measures fall into one of two categories: those that reduce the amount of solar radiation absorbed by the Earth-atmosphere system, and those that extract CO_2 from the atmosphere. Each alters the global radiation balance described in Section 2.3, but in different ways. The first approach involves solar radiation management (SRM) through human control of the Earth's albedo. Simply put, human agents are used to reflect more solar radiation back to space, thereby providing a negative radiative forcing to counter the positive forcing of larger atmospheric GHG concentrations. The other approach – termed carbon dioxide removal (CDR) – alters the radiation balance by removing carbon dioxide from the atmosphere, thereby reducing its positive forcing and increasing the transfer of terrestrial radiation to space.

Solar radiation management

One way to increase the reflection of solar radiation is to inject large amounts of aerosols (small particles or droplets) or aerosol precursors such as SO_2 and H_2S into the upper atmosphere. Atmospheric aerosols occur naturally from volcanic eruptions and anthropogenically from the use of fossil fuels. In 1991, the eruption of Mount Pinatubo (a natural phenomenon) discharged approximately 20 million tonnes of SO_2 into the atmosphere, creating a negative forcing that reduced the global average temperature by 0.6°C over fifteen months (NASA, 2006). In Section 4.1 we noted that sulfur discharged from proliferating coal-fired power plants (an anthropogenic source) could have been a significant contributor to the global cooling experienced in the 1950s and may in part be responsible for the post-1998 temperature plateau. Hence, if we must, could we not impose a large negative forcing by injecting aerosols into the upper atmosphere? The question was raised by Crutzen (2006) and answered in the affirmative by Wigley (2006).

Using a global climate model, Wigley showed that a sustained forcing of −3 W/m² would counter much of the GHG warming expected for the twenty-first century and that such a forcing could be maintained by annually transporting 5 million tonnes of sulfur into the stratosphere, well below that produced by Mount Pinatubo and only 7% of circa-2006 SO_2 emissions from fossil fuel combustion. The negative forcing associated with Pinatubo was short-lived, and within two years higher temperatures were restored as the forcing decayed while GHG emissions continued to rise. Likewise, periodic injection of SO_2 would curtail warming, but were injection to cease, the aerosol layer would dissipate and the warming effect of atmospheric GHGs would be fully restored. Moreover, other effects of increasing atmospheric CO_2 concentrations, such as ocean acidification, would not only continue but be enhanced by the fallout of acidic aerosols (BPC, 2011).

The goal of reflecting more solar radiation could be achieved in other ways. Large mirrors could be placed in orbit outside the earth's atmosphere, but high costs render the option impractical. A less costly approach involves seeding clouds with aerosols to *brighten* them by increasing their reflectivity (Russell, 2012). As discussed in Section 3.1, aerosols increase the number and reduce the size of cloud droplets, making them less likely to coalesce and descend as precipitation. A corresponding increase in cloud size would be accompanied by an increase in solar reflectivity. But, as also discussed in Section 3.1, there is still a lot we don't know about the effect of aerosols on clouds.

Carbon dioxide removal

A different strategy calls for removing CO_2 from the atmosphere and sequestering it. Think of it as an after-the-fact form of CCS. Of course, such sequestration occurs naturally through photosynthesis and could be enhanced by planting more trees. It could also be enhanced by accelerating the growth of oceanic phytoplankton. Oceans could be fertilized with iron to enhance algal growth near the surface, drawing more CO_2 from the atmosphere, and when the algae die, they would presumably descend to the sea bed where the carbon would be sequestered. But algal blooms are an attractive food source for other aquatic life, and it's likely that much of it would be consumed and its carbon content returned to the environment before it sinks very far. Moreover, afforestation and accelerated plankton growth are long-term strategies that would not provide the quick response desired of geoengineering solutions.

The most intriguing CDR option involves the use of separation technologies akin to those proposed for removing CO_2 from the flue gas of a power plant. In this case the technologies, which involve special sorbents or membranes, would capture CO_2 directly from the atmosphere (Keith, 2009; ScienceDaily, 2012a, 2012b). It is certainly doable, but at what cost? A simple calculation suggests that economic viability is more than problematic.

Today's atmospheric CO_2 concentration of approximately 400 ppm corresponds to a volume fraction of 0.0004 or 0.04%. Contrast this concentration with a nominal value of 15% for the flue gas of a coal-fired power plant, and remember that for a power plant the cost of extraction *increases* with *decreasing* concentration. It's this inverse relationship between cost and CO_2 concentration that motivates the use of IGCC and OFC power plants. But extracting CO_2 from air, where its concentration is 375 times smaller than the nominal value for a flue gas, takes us in the opposite direction. Even if the atmospheric CO_2 concentration were to reach 1,000 ppm, it would still be low (0.10% by volume) relative to other atmospheric constituents. The driving potential for separating the CO_2 from the air would remain small, and system size and cost requirements needed for large-scale capture would continue to render the option impractical. Short of a major breakthrough in separation science and technology, costs are likely to remain unacceptably high.

The foregoing conclusion was also reached in an assessment performed by the American Physical Society (APS, 2011), which found that chemical capture of CO_2 from the atmosphere would cost approximately $600 per tonne of CO_2 ($2,200 per tonne of carbon), about ten times the cost of removal from the POC of a coal-fired power plant. It was also affirmed by a thermodynamic and economic analysis (House et al., 2011) that predicted a cost of more than $1,000 per tonne of CO_2 compared to less than $100 per tonne for existing power plant scrubbers. At this point, it's difficult to view capture from air as anything more than a diversion from the challenges of adopting more realistic and effective mitigation measures.

Ramifications and risks

The degree to which geoengineering options are being taken seriously and the surrounding controversy are reflected by a spate of books, articles, and reports on the subject (Kerr, 2006d; Buesseler et al., 2008; Tilmes et al., 2008; Hegerl and Solomon, 2009; Robock et al., 2009; Fleming, 2010; Goodell, 2010; Kintisch, 2010; Parkinson, 2010; Ricke et al., 2010; BPC, 2011; Hamilton, 2013; Keith, 2013). The underlying message is: *beware of unintended consequences*. Implemented on the massive scale needed to counter the effect of greenhouse gas emissions, other effects on life forms and regional climates are uncertain and fraught with risk.

A stratosphere laden with sulfate particles could deplete it of ozone and would certainly increase acid rain and ocean acidification. It could also reduce global precipitation – as did the eruption of Mount Pinatubo – alter monsoon rains, and increase regional drought. By rendering sunlight incident on the Earth's surface more diffuse, stratospheric aerosols and cloud brightening could also affect plant growth and the general well-being of terrestrial ecosystems. Serious concerns also exist for the effect of massive fertilization on ocean ecosystems. And if GHG emissions were to continue in a business-as-usual fashion, what happens if unintended consequences were to force suspension of geoengineering measures? Throughout implementation of the measures, the radiative forcing associated with GHG emissions would continue to increase, and with the measures no longer in play, warming would rapidly resume as if they had never been implemented.

And if geoengineering measures were to be implemented, who would be making the decisions? What's to keep one nation – even a multinational company or a wealthy individual – from altering the environment in ways that yield benefits for one person, corporation, or region while triggering adverse effects for others? How would effects be monitored? Who would be liable for damages inflicted to specific regions and populations? Currently, there are no laws prohibiting projects such as fertilizing the oceans or injecting aerosols into the upper atmosphere. Yet, large-scale geoengineering affects everyone, and the stakes are large.

Parson and Keith (2013) make the case for international cooperation and governance of geoengineering and provide several recommendations for developing comprehensive international standards. Recognizing that a time may come when geoengineering becomes necessary, they advocate research that addresses the efficacy and global ramifications of different approaches. If the world must act, better to do so from a knowledge base that enables more informed decisions. The research should be governed by more than normal scientific review processes and existing national laws. It should be conducted with a high level of transparency that encourages public discourse and oversight of appropriate research goals and experiments, while discouraging "reckless interventions," including rogue experiments, subject to "low-probability, high-consequence risks." And the research should be coordinated internationally. Similar recommendations have been made in a report published by the U.S. Government Accountability Office (GAO, 2011), which emphasized the need for risk assessment and international cooperation, and by the Oxford Geoengineering Programme (Oxford, 2014), which emphasized regulation as a public good, public participation, transparency, and robust governance.

The use of geoengineering measures to counter the effects of GHG emissions on global warming will no doubt receive greater advocacy if efforts to mitigate emissions and adapt to climate change flounder. Should it get to the point that implementation is unavoidable, humankind will have reached an unfortunate and dangerous state. Mitigation and adaptation should be the pillars of climate policy.

6.9 Summary

By addressing the root cause of global warming – namely greenhouse gas emissions – mitigation provides the first line of defense against warming. In this chapter, the scope of the problem was framed in terms of wedges, where one wedge corresponds to emission of 1 billion tons of carbon per year (1 Gt-C/yr). In 2001, global emissions were about 7 Gt-C/yr, and Pacala and Socolow (2004) posed the following question: Over a fifty-year period, what can be done to maintain emissions at this level while allowing for anticipated population and economic growth? In a business-as-usual scenario, sustaining growth would double emissions to approximately 14 Gt-C/yr at the end of fifty years. The problem was then one of identifying mitigation options that could eliminate 7 Gt-C/yr (7 wedges) of emissions by 2051.

For each mitigation option, the question was asked: What would it take to cut emissions by a single wedge? To remind ourselves of the magnitude of the problem, let's revisit one of the options – reducing emissions from coal-fired power plants. As a baseline, Pacala and Socolow ascribed a circa-2051 wedge to power plants having a total generating capacity of 700 GW_e and operating at an efficiency of 50% for 90% of the time. What could be done to eliminate that wedge by replacing coal-fired plants with alternative sources of electricity? To fully appreciate the scale of such a replacement, consider the fact the total circa-2011 U.S. generating capacity was about 1,050 GW_e.

Pacala and Socolow reasoned that the target of 1 Gt-C/yr could be achieved by replacing 1,400 GW_e of coal-fired power plants with an equivalent amount of gas-fired plants. But the replacement would increase the global consumption of natural gas by about 5 billion cubic meters per day, or fifty times the output of a large gas pipeline. Replacement of 700 GW_e of coal-fired capacity with an equivalent nuclear capacity would almost double the world's circa-2012 capacity of 375 GW_e. Replacement with wind or solar energy would require about 2,300 GW_e of peak power in each case, five times the world's combined circa-2013 capacity of wind and solar power. And the wind and solar farms would cover about 94 million and 5.7 million acres of surface area, respectively. Each of the above is doable but not without a commitment of significant resources and a major transformation of the world's power generation portfolio.

Things have changed since the 2004 Pacala and Socolow study and not for the better. By 2013, global emissions had risen to 9.6 Gt-C, 37% higher than the benchmark used in the study. The task of getting to 7 Gt-C/yr by 2051 is now even more challenging. And were it to be achieved, emissions would still have to be rapidly reduced to about 2 Gt-C/yr to keep the atmospheric CO_2 concentration and Earth's average surface temperature from rising to unacceptable levels. Current emission reduction measures are simply not working, and there is growing urgency to do more and to do it quickly.

A cornerstone of mitigation measures should be a relentless drive to increase the efficiency of energy utilization and to promote a global mind-set that assigns a high priority to the principles of energy conservation. We must improve the efficiency with which primary energy is used to generate electricity while transitioning from the use of fossil to carbon-free forms of energy. The same can be said for transportation. Opportunities remain for increasing the fuel economy of combustion and hybrid vehicles, while concurrently increasing the electrification of transportation and the use of carbon-free energy to generate electricity. Efficiency improvements are also possible in all manner of devices and processes used in residential, commercial, and industrial sectors of the economy. The costs of reducing emissions through energy efficiency measures are largely negative – simply put, *efficiency saves money.*

Energy conservation also saves money, and by leveraging effective educational programs, conservation can once again become a *core value*, rooted in people and nations throughout the world. We've passed the point where admonitions for conservation are only relevant for developed nations. It's clear that, in emerging economies, people want what Westerners have long enjoyed. A world in which ever increasing numbers of people are consuming ever more is not sustainable. That is not to say humanity must move to a state of deprivation. It is possible to lead comfortable and fulfilling lives while moderating our tendencies for self-indulgence and overconsumption. Living big is not a prerequisite to living well.

Development and deployment of carbon-free sources of energy such as wind and solar must be accelerated and accompanied by investments in energy storage and a smart grid. But even under the best conditions, it will be decades before these sources can meet a significant fraction of the world's energy needs. During this time, the abundance and low cost of natural gas make it an excellent bridge fuel for decarbonizing the economy, so long as increased usage is accompanied by curtailment of leakage and water pollution. And if we are serious about reducing carbon emissions, we can't ignore the nuclear option. As a carbon-free source of base-load power, nuclear energy also provides a bridge to the future, but it must be accompanied by stringent procedures for managing and securing radioactive wastes.

Of all the obstacles to mitigation, none looms larger than the world's large coal reserves and new discoveries of oil and natural gas. On strictly economic grounds, fossil fuels will remain tough competitors, and vested interests in this multitrillion-dollar industry, abetted by those convinced of the inherent good of unfettered markets, will do what they can to continue to supply a large share of the world's primary energy needs. Enormous capital investments have been made in power plants fired by fossil fuels, as well as in facilities for recovering, transporting, and refining the fuels. These facilities will not be decommissioned until appropriate returns on investment have been realized. And without a major breakthrough in carbon capture technology, CCS is not likely to meet expectations associated with reducing emissions from coal-fired (or gas-fired) power plants.

The need for adaptation is a fait accompli, and efforts to increase the resilience of built and natural systems are under way. However, adaptation must proceed in concert with – not in lieu of – mitigation. Even if mitigation measures fail to achieve desired outcomes, to the extent that they *bend the emissions curve* downward, they reduce the extent to which adaptation measures must be implemented. The need for a balanced approach to mitigation and adaptation was articulated in a report of the U.S. President's Council of Advisors on Science and Technology (PCAST, 2013). Quoting from the report: "Mitigation is needed to avoid a degree of climate change that would be unmanageable despite efforts to adapt. Adaptation is needed because the climate is already changing and some further change is inevitable."

Geoengineering is the last line of defense against the effects of warming, one that would be accompanied by significant risks and should be avoided.

CHAPTER 7

Public policy options

Public policy measures can do much to accelerate deployment of carbon mitigation measures, and there are basically three options. One approach is to put a price on emissions, which can be done through a cap-and-trade system or an outright tax. The approach is motivated by the following premise: if the goal is to put a large dent in GHG emissions, a price tag must be put on the emissions. A second approach is to mandate reductions through the regulatory process. Forms of government regulation could include corporate average fuel economy (CAFE) standards for automotive transportation, renewable portfolio standards (RPS) for power production, and efficiency standards for buildings and home appliances. The third approach is to provide financial incentives to ease the cost of mitigation. Incentives can be provided as outright grants, tax credits for producing carbon-free energy, or preferential treatment for sale of the energy.

In each of the three approaches, a critical issue, both politically and economically, is cost management. If costs are too high, economic growth is stifled; if they are too low, emission reductions and innovation are stifled.

7.1 Cap-and-trade

Lawmakers are generally averse to increasing taxes. Hence, if they believe that a price should be placed on carbon emissions, they're inclined to favor a cap-and-trade system, even though, like a tax, the cost of implementation is ultimately borne by the consumer. In principle, the system works as follows. Governments impose caps (limits) on GHG emissions from large central sources such as power plants, oil refineries, natural gas producers, and manufacturers of energy-intensive products such as concrete, steel, and glass. Initially, the caps are high to allow time for adjustment but are gradually reduced until atmospheric GHG concentrations drop to desired levels. Commensurate with a prescribed cap, permits (allowances) are granted and/or auctioned to the emitting entities, with one permit corresponding to a unit of annual emissions such as 1 t-CO_{2eq}.

Whether auctioned or granted, permits are traded on a market, which can be regional, national, or international. Permits are sold by those able to economically reduce emissions below their cap and bought by those finding such purchases to be a more cost-effective approach to compliance. If a company finds it too costly to meet its prescribed limit by reducing its emissions, it can purchase permits traded by a company that is within its cap. The system may include the use of offsets by companies unable to operate within their limits. A common offset involves reforestation or averted deforestation. Carbon credits would be provided according to the amount of atmospheric CO_2 assimilated by reforestation or the amount by which emissions are reduced by averting deforestation. If the cost of purchasing credits associated with reducing net GHG emissions by reforestation or averted deforestation is less than the cost of reducing the company's emissions, it can reduce the price of compliance by purchasing the credits. If applied in tropical regions of the world, costly emission reduction measures in developed nations could be supplanted by lower-cost programs in developing nations.

In principle, the cap-and-trade system provides a market-based approach to reducing emissions, with permit prices driven by the dynamics of supply and demand. Prices could increase over time, as the caps and the number of permits are progressively reduced, but would also depend on market conditions such as the relative cost of different energy sources and whether the global economy was experiencing growth or contraction. Market prices would also depend on the nature of allowable offsets and the extent to which technology and innovation made it easier for emitters to operate within their caps.

By invoking market principles, the cap-and-trade system ostensibly provides a mechanism for achieving emission reduction targets at the lowest possible cost. Ideally, entities would reduce their emissions only so far as the marginal costs of doing so did not exceed the market price of the permits (or offset credits). Above that threshold, additional permits would be purchased. Although emitting entities needing permits could purchase them directly from entities with excess permits, most transactions would be conducted on electronic exchanges involving third parties including banks and hedge funds. The system can take on layers of complexity that are susceptible to speculation and volatility.

To reach its full potential, a cap-and-trade system would have to be implemented globally. However, due to large disparities in per-capita energy consumption and economic growth between developed and developing nations, achieving global agreement on a system is difficult at best. To what degree should caps differ between developed and developing nations? Should allocations be weighted more heavily in accordance with emissions per capita or per GDP? Even if developing nations agreed to participate in a cap-and-trade system, they would likely insist on an allocation formula that provided allowances on a per-capita basis, which, in the face of a substantive cap on global emissions, would result in a huge unidirectional sale of allowances by developing nations to developed nations. Quoting Mankiw (2007), in the case of China, such a system "would amount to a massive foreign aid program to one of the world's most rapidly growing economies." Convincing developed economies to support such a program would be a nonstarter, even in good economic times.

Other issues relate to how caps should be set and whether emission allowances should be awarded or auctioned. Should limits be restricted to utilities, refineries, and other *concentrated sources* of emission, or should they be applied to a wider swath of economic sectors? If allowances are auctioned, how should the revenues be used? Since the costs are ultimately borne by the consumer, should the process be made revenue neutral, at least by providing tax relief for those of low and middle incomes? Should some of the revenues be used to accelerate development and implementation of GHG mitigation and adaptation measures?

The system has other issues. Prescription of specific caps would not occur without intense lobbying by special interests, while enforcement would require yet another government bureaucracy. Consider the intense lobbying and the political and corporate pressures that would precede establishment of a cap, an allocation formula, and permissible offsets. Then think about implementation and whether the rules could be manipulated by participants.

Despite its complexities, adherents of cap-and-trade believe it is the best way to capture free-market efficiencies. It is favored by politicians who are open to GHG curtailment measures, by a significant number of emitters (assuming generous caps and flexible trading arrangements), and by segments of the financial community that would benefit from a global permit-trading process. But, how does it compare with other options?

7.2 A carbon tax

Although carbon taxes do not impose specific emission caps, Economics 101 tells us that raising the price of a consumable is likely to do two things: reduce demand and encourage the pursuit of alternatives. By taxing GHG emissions from carbon-based fuels, society would be motivated to use them more efficiently, to capture and sequester emissions where feasible, and to advance the development and implementation of noncarbon energy sources.

Carbon taxes need not be viewed exclusively in terms of raising revenue for governments. As with revenues obtained by auctioning allowances in a cap-and-trade system, carbon taxes could be made revenue neutral by reducing other taxes. Since those most impacted by a carbon tax would be of low-to-moderate income, the impact could be mitigated by a graduated reduction in taxes paid on income below a certain level and/or a reduction in payroll taxes. As advocated in the early twentieth century by the British economist Arthur Pigou, taxes can also be used to remedy societal problems. A Pigovian tax would compensate for social costs not included in prices determined exclusively by supply and demand.

Pigou (2013) believed that when the interests of individuals (or corporations) harm the larger society, governments should intervene by imposing a tax commensurate with the harm. In the context of climate change, harm is associated with environmental degradation, resource depletion, and risks to human health and welfare. Although other issues have long contributed to divergence of individual and societal interests, climate change provides a more recent and intractable dimension.

Pigou was an advocate of market economies but believed they would only be sustainable if the full cost of goods and services was immediately recognized. Drawing from Pigou, Hawken (2005, p. 75) underscores the point with the statement that, "Today we have free markets that cause harm and suffering to both natural and human communities because the market does not reflect the true costs of products and services." Pigou and Hawken, along with many, largely heterodox economists, believe that external costs of the harm – those not otherwise included in the transaction between buyer and seller – must be internalized. But how are the costs to be determined, for future as well as current generations?

Models for estimating the economic impact of climate change must be superimposed on models used to predict global warming and its effects on climate. For extreme weather events or rising sea levels, how does one account for the loss of man-made and natural capital and the effect of the losses on current and future productivity? How does one deal with the effect of chronic drought on food production or the spread of vector diseases on human capital? Are common economic parameters such as discount and growth rates still useful when dealing with the vagaries of climate change?

In a report commissioned by the British government (Stern, 2006), an economic case was made for adopting strong and early action to reduce global warming by imposing a large enough tax on GHG emissions to stabilize atmospheric concentrations at no more than 550 ppm CO_{2e}. Absent such action, the Stern Review projected that climate change would provide a persistent drag on the global economy, reducing GWP by 5–20% per year. In contrast, the nominal cost of stabilization was predicted to be 1% of circa-2050 GWP within a –2% to +5% range.

The Stern Review did not go unchallenged and was criticized on three fronts: (1) for understating the real costs of reducing GHG emissions, (2) for using an unduly low discount rate to weigh the relative well-being of future and current generations, and (3) for discounting the role that adaptation measures, such as flood prevention, could play in reducing the economic costs of climate change (Lomborg, 2006; Nordhaus, 2007). Criticism of the manner in which the discount rate was treated illustrates how a seemingly arcane but important economic parameter can complicate the climate change debate.

The discount rate is commonly used to assess the effect of today's actions on future benefits according to the time value of money and the belief that today's dollar is more valuable than a future dollar because of benefits derived from its investment. In the context of climate change, measures to reduce GHG emissions have present costs and future benefits. What is the present value of those benefits, and how does it compare with present costs? Does siphoning some of today's dollars away from other investments to fund mitigation measures reduce the benefits that would have otherwise accrued to future generations? Answers depend on many assumptions, including selection of the discount rate.

With adverse effects of climate change increasing over time, a low discount rate increases the present value of mitigation measures and benefits for future generations. Conversely, a high discount rate would reduce the present value of mitigation measures, tilting the argument in favor of a business-as-usual approach that foregoes mitigation measures and encourages economic growth. The implication is that returns on the growth would better enable future generations to deal with the effects of warming, as for example, by developing drought- and heat-resistant cereal grains, barriers to protect against rising seas, and medicines to protect against new diseases. Rather than devote today's resources to curb global warming, the argument calls for investment in measures that would enhance the resource base for future generations to deal with the effects of warming.

Jamieson (2014) compares approaches taken by Stern and Nordhaus in developing economic models of climate change. While Stern chose a discount rate of 1.4% – well below historic values and one that would not discriminate against future generations – Nordhaus chose a more representative value of 5.5% over fifty years and an average of 4% for the century. Stern and Nordhaus both support a global carbon tax but in different amounts. Stern advocates a steep tax of $85/t-CO$_2$ to immediately and significantly curb emissions. In contrast, motivated by concern for the economic impairment that could accompany a large tax, Nordhaus recommends a tax that would begin at $7/t-CO$_2$ and ramp to $25/t-CO$_2$ by 2050. While Stern advocates an *all-in* approach to reducing emissions, Nordhaus recommends a conservative approach that reduces economic risks associated with the tax but only yields a 25% reduction in emissions by 2050 and 45% by 2100 – not enough to stabilize GHG concentrations below 550 ppm CO$_{2e}$ – while projecting costs approaching 3% of GWP by 2100. It is not surprising to see such divergent results.

Climate change is the ultimate challenge for economic modeling. The models represent the climate system as a capital asset that is diminished by GHG emissions. A major objective is to determine an optimum cost (tax) for emissions, one that achieves a suitable trade-off between reducing degradation of the climate system and risks to the economic system. But the models involve many parameters and assumptions and allow for considerable latitude in specific choices. Add the large uncertainties associated with the choices to those of the climate models, and there is ample room for widely divergent results, as well as good reason for skepticism in the value of the models (Economist, 2013c; Nordhaus, 2013; Pindyck, 2013; Stern, 2013). Pindyck believes that results are so dependent on arbitrary assumptions as to be useless, while Stern feels that the models are biased toward underestimating economic impairment, which could be enormous in the case of abrupt climate change (Section 5.8). That said, many mainstream economists feel that a carbon tax is the best way to deal with climate change. So, how to proceed?

A carbon tax should be large enough to stimulate meaningful emission reductions, yet not so large as to inflict significant damage to the economy. Consider a tax of $30/t-CO$_2$ ($110/t-C), which corresponds to the low end of estimated costs for including CCS in a new, coal-fired power plant. The tax would add approximately $60/tonne to the cost of coal and, if the plant were operated under supercritical conditions, approximately $0.025/kWh$_e$ to the cost of electricity. Although many would complain, the tax would not bring a developed economy to its knees. In contrast, if applied to gasoline, the same tax on emissions would only add about $0.27 to the cost per gallon. A tax on coal comparable to the cost of CCS would have the intended effect of reducing emissions by encouraging CCS for coal-fired plants and by accelerating development of carbon-free sources of power. A 27 cent gasoline tax would have little effect on fuel consumption. The implication is that, if applied to emissions, taxes should be differentiated according to fuel source and application. Similar conclusions were reached in a study conducted by the MIT Joint Program on the Science and Technology of Climate Change (TR, 2009). If the goal is to reduce carbon emissions, a common tax would not fit all applications.

Tax revenues could be used in several ways: (1) to reduce other taxes, making the carbon tax revenue neutral; (2) to accelerate progress toward a sustainable, decarbonized energy future by investments in public transportation, the electric grid, and technologies that increase energy efficiency and the use of renewable energy; and/or (3) to reduce federal budget deficits. In a survey of the U.S. public (NSEE, 2014), only 34% of the respondents supported a tax with unspecified use of the revenue and only 38% supported using it to reduce the budget deficit. However, 56% supported the tax if it is revenue neutral and 60% supported it if revenues are used to support renewable energy. An argument against any tax is that money is withdrawn from the economy, slowing economic growth. But if a carbon tax is revenue neutral, money would not be withdrawn from the economy. If the tax is used to support mitigation measures, it would contribute to economic development through manufacturing and service industries built around energy efficiency and carbon-free energy sources.

7.3 Cap-and-trade or a carbon tax?

If a price is placed on carbon, would it be better to implement a cap-and-trade system or to simply impose the tax? In a study conducted by the U.S. Congressional Budget Office (CBO, 2008), a carbon tax was determined to be the most efficient of several incentive-based options for reducing CO_2 emissions. Defining the most efficient option as one that "can best balance the costs and benefits of the reductions," a steadily rising carbon tax would eliminate fluctuations in the cost of emissions and allow both producers and consumers of energy to more confidently determine when and to what extent emissions should be reduced. Of variants on the cap-and-trade option, imposition of an inflexible annual cap on emissions was the least efficient approach. Preferable options were those that provided flexibility in the form of a *safety valve* (a ceiling on the price of emission allowances), a *circuit breaker* (an adjustment to the cap), and either *banking provisions* or a *price floor* to prevent the cost of emission allowances from dropping too low. Banking would allow firms to defer the use of allowances when costs of meeting emission requirements are low and draw on them when costs are high. However, even with such flexibility, an outright tax on emissions was deemed to be the most efficient approach. From a broad range of 103 options, a tax of $25 per ton of CO_{2eq} was determined to be the most effective means of reducing the U.S. budget deficit (CBO, 2013). With a 2% annual increase, the tax could raise $1 trillion over a ten-year period while encouraging efficiency and conservation measures that reduce emissions by 10%.

One advantage of a carbon tax is the certainty of related costs. Assuming implementation over an extended period, energy providers would be able to more effectively make long-term investments in energy systems. The tax would send the clearest possible signal to energy markets and would encourage market-based solutions by stimulating the development of improved and innovative energy technologies. The tax could start low and be incremented gradually to provide time for adaptation by energy producers and consumers. Although a carbon tax does not explicitly cap emissions, annual increments could be fine-tuned to achieve the desired emissions trajectory.

A tax would also be simpler to administer, particularly if it were applied at a limited number of sources, such as coal mine heads, gas pipelines, and oil refineries. Once implemented, it would be less vulnerable to political manipulation and lobbying by special interests. Although conventional wisdom suggests that it would be more difficult to implement globally than a cap-and-trade system, suitable mechanisms have yet to be vetted. Taxes could initially be varied across developed and developing nations, larger for the former and smaller for the latter. Differences could be reduced over time, as developing economies mature.

The single most important barrier to prescribing ambitious targets for reducing GHG emissions is the cost of implementation and its impact on business and the consumer. For cap-and-trade, it is cost containment that drives caveats such as offsets and ceilings on the market price of an emission credit. If offset allowances are too generous and/or the ceiling on the price of an emission credit is set too low, prospects for maintaining the cap are diminished. In principle, a cap provides emissions certainty, but if trade is regulated by limiting the price of emission credits and/or providing generous offsets, achieving the cap becomes problematic. Ideally, one would like to find a sweet spot that provides a good mix of emissions certainty and price certainty. A carbon tax would achieve price certainty and, if high enough, would bring down emissions. An outright carbon tax provides the simplest and most efficient means of reducing emissions, with recent studies pointing to negligible adverse effects on the economy (GCEC, 2014; IMF, 2014).

7.4 Regulatory options

In the 1970s and early 1980s, the world's supply of petroleum was sharply curtailed by conflicts involving embargos imposed by producing nations. The United States needed to respond by discouraging domestic consumption of transportation fuels, but it was reluctant to do so by increasing fuel taxes. Instead, it chose to implement CAFE standards for LDVs. Enacted by the Nixon administration and to be achieved by 1985, the standards prescribed separate fleet averages of 27.5 mpg and 19.5 mpg for cars and light trucks (minivans, pickups, and SUVs), respectively. The target for light trucks was subsequently raised to 22.2 mpg, with a recommendation by the G.W. Bush administration that it be raised again to 23.5 mpg by 2010. However, over a thirty-five-year period since their inception, CAFE standards had failed to adequately curb fuel consumption, and by extension carbon emissions. In 2005, average U.S. fuel economy was actually 22.2 mpg for cars and 16.9 mpg for light trucks, well below the mandated values.

Motivated by a desire to reduce emissions and viewing existing CAFE standards as inadequate, thirteen states and cities challenged the status quo in court, advocating for stricter standards. The Ninth Circuit Court of Appeals in San Francisco agreed and ruled that the federal government had undervalued the benefits associated with reducing GHG emissions and had overestimated the costs associated with imposing higher standards (Ball, 2007b). The court rejected existing standards, and the government was expected to revise them. Although the Bush administration did not act on the ruling, things changed in 2009 when the Obama administration secured an agreement for a standard of 34.5 mpg across all LDVs by 2016. In 2011, the bar was set higher, with a standard of 54.5 mpg prescribed for 2025. For the first time, standards were also prescribed for larger trucks, with improvements in fuel economy ranging from 10% to 23% based on the size of the truck.

It remains to be seen whether the foregoing objectives will be achieved, and at this point it is highly unlikely.[1] Even if they are met, it is sobering to note that the United States will continue to lag behind many nations in

which stricter standards and high fuel taxes have shaped consumer preferences for fuel-efficient vehicles and manufacturers have responded accordingly. In the European Union, standards are based on vehicle (tailpipe)
CO_2 emissions. From a 1995 average of 186 grams of carbon dioxide per
kilometer (g-CO_2/km), standards for 2008 and 2012 were reduced to 140 and
120 grams, respectively. The 2012 standard is approximately equivalent to 50
mpg. In Japan and China, fuel economy standards are couched in terms of
vehicle weight, with circa-2007 standards equating to 46 mpg and 36 mpg,
respectively (Gallagher et al., 2007). Although CAFE standards provide an
instrument for reducing carbon emissions in the transportation sector, a
meaningful carbon tax of $1 or more per gallon would do far more to drive
consumer preferences for fuel economy.

In the electric power sector, renewable portfolio standards (RPS) dic-
tate production of a certain amount or fraction of energy from renewable
sources (solar, wind, biomass, hydro, and/or geothermal). Germany, which
has ambitious 2020 targets of obtaining 35% of its electricity and 18% of its
total energy from renewable sources, is well on its way to meeting its goals.
Although similar targets have been established by the European Union,
it is problematic whether they will be met by all members, particularly in
the face of weak economic growth and the debt-driven financial difficulties
experienced by some nations. Although generally less aggressive, renewable portfolio standards are being established in other regions of the world.

Out of concern for its impact on consumer energy prices, the U.S.
Congress has not enacted a national RPS and is not likely to do so. However,
many states have adopted their own standards, which typically take the
form of achieving a certain percentage (10% to 40%) of power production
from renewables by a certain year (2015 to 2030). As of 2014, thirty-one
states and the District of Columbia had established standards, with seven
more having established aspirational targets (C2ES, 2014). Aggressive standards have been set by Maine (40% by 2017) and California (33% by 2020),
with other states seeking to achieve 20% or more. In many states the ability
to meet prescribed standards is enhanced by financial incentives, trading
of *renewable-energy credits* (RECs), and/or receiving credit for improved
energy efficiency.

State renewable portfolio standards have been a major contributor to
rapid growth in large-scale wind and solar power systems. However, sit-
ing issues, including NIMBY (*not in my back yard*) considerations, and
transmission bottlenecks are looming impediments to sustained growth.
From large solar farms in Sun-rich regions of the Southwest to wind farms
in the Northeast, NIMBY has become a formidable obstacle to securing

site permits. Transmission bottlenecks exist between regions of the United States with the greatest potential for wind and solar energy, such as the Upper Midwest, Southwest, and population centers with large demand. To reduce the bottlenecks, there is a growing need for cooperation between states in transmitting and sharing renewable energy and in issuing and tracking RECs. The federal government could play an important role by coordinating state initiatives, supporting expansion of the grid, and adopting consistent, long-term approaches to maintaining tax credits for renewable energy.

Although policy makers have given far more attention to renewable power, targets and support policies also exist for the heating and cooling sector. By 2014, at least twenty-four countries had adopted heating (and cooling) targets through building codes, financial incentives, and other measures (REN, 2014).

7.5 Financial incentives

Financial incentives come in different forms, two of which are tied to the tax code. One involves the amount of electricity produced from renewable energy. In the United States, the subsidy is termed a production tax credit (PTC). For each unit of energy, the federal tax liability of the producer is reduced by about $0.02 per kWh_e. The other incentive involves an investment tax credit (ITC). Depending on the type of renewable energy, credit for up to 30% of the cost of development is provided when the facility becomes operational. Although the economics of wind and solar power are becoming more competitive with fossil and nuclear energy, retention of the credits would do much to sustain a strong trajectory of growth for the renewables.

In Europe, production subsidies are more commonly provided in the form of feed-in tariffs (FIT), which obligate utilities to purchase renewable energy at a set price over a prescribed duration. To encourage development of renewable sources, prices are typically set at a premium, no more so than in Germany, which mandated prices of approximately $0.55/$kWh_e$ and $0.084/$kWh_e$ over twenty years for solar and wind energy, respectively. The utilities, in turn, are allowed to pass the increased cost of electricity to the consumer. While the costs are high per unit of renewable energy, adverse economic effects are mitigated if the energy makes a comparatively small contribution (e.g., less than 20%) to the total energy portfolio. Led by Germany and Spain, European countries that have adopted FITs have experienced the largest growth rate in using renewables for power generation, with concomitant development of renewable energy industries that have stimulated job growth and technology export opportunities.

Financial incentives in the form of loan guarantees or outright cash awards can also be used to encourage the development of related infrastructure. Such incentives are being used worldwide, driven in part by the goal of reducing GHG emissions but also by the desire to gain competitive advantage in emerging clean energy industries. In the United States, the Energy Policy Act of 2005 authorized the Secretary of Energy to provide loan guarantees for projects whose accelerated development would provide a more secure energy supply. Under the American Recovery and Reinvestment Act of 2009, guarantees and grants were used to enable large solar and clean coal initiatives.

7.6 Summary

If energy efficiency and renewable energy are to have a growing and significant impact on reducing GHG emissions, public policy must play a prominent role and policy instruments must continue to evolve. In the heating/cooling and transportation sectors, policy instruments must do more to reduce the use of fossil fuels. In the power sector, new measures are needed for adapting the grid to deal with increasing levels of renewable electricity, including support for development and installation of large-scale energy storage systems as well as net metering and demand control technologies.

Moving forward, it will be important to retain a mix of regulatory and financial measures, but one point cannot be emphasized too strongly. Putting a price on carbon emissions is the best way to stimulate adjustments in energy consumption patterns and to trigger appropriate market responses. This view is shared by many mainstream economists and members of the business community. Quoting Jeffrey Immelt, CEO of General Electric (Friedman, 2007), "the multi-billion dollar scale of investment that a company like GE is being asked to make in new clean-power technologies or that a utility is being asked to make in order to build coal sequestration facilities or nuclear power plants is not going to happen at scale – unless they know that coal and oil are going to be priced high enough for long enough that new investments will not be undercut in a few years by falling fossil fuel prices." And, quoting from an assessment of climate change and fiscal policy made by the International Monetary Fund (IMF, 2014),

Stabilizing atmospheric concentrations of greenhouse gases will require a radical transformation of the global energy system over coming decades. Fiscal instruments (carbon taxes or similar) are the most effective policies for reflecting environmental costs in energy prices and promoting development of cleaner technologies, while also providing a valuable source of revenue (including, not least, for lowering other tax burdens). The message is that carbon emissions must be monetized.

At the end of the day, the extent to which mitigation measures are implemented at meaningful scales depends a good deal on political considerations – global, national, and regional. An argument made against strong mitigation measures is that they are too costly and impede economic growth. There is no better testimony to the fact that the argument is overstated than the economic prowess of Germany. Energy efficiency and conservation are ingrained in the German culture and mindset, and no nation has invested more in developing its renewable energy portfolio. Despite the increased costs associated with very generous FITs, the nation remains an industrial juggernaut, the world's leading exporter of manufactured goods, and a leader in developing future energy technologies.

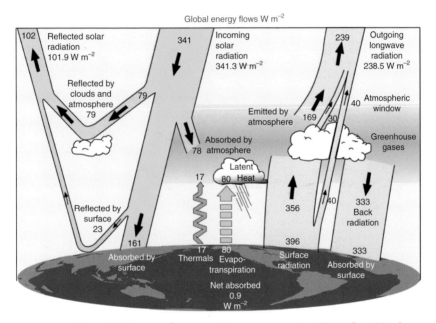

Global energy flows W m⁻²

FIGURE 2.3. Mean global energy fluxes in watts per square meter (W/m²) from March 2000 to May 2004 for the Earth-atmosphere system. From Trenberth et al. (2009).

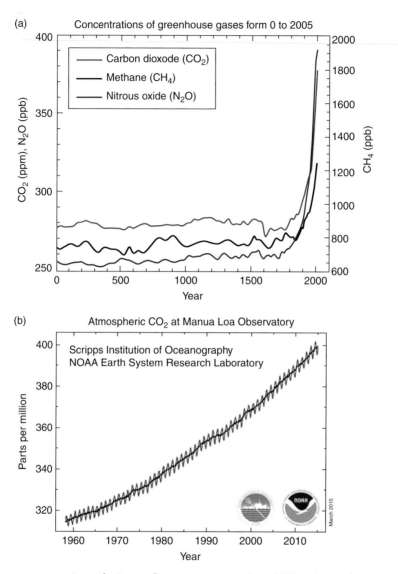

FIGURE 3.1. Atmospheric greenhouse gas concentrations: (a) Variations in the concentrations of CO_2, CH_4 and N_2O over the last two millennia (IPCC, 2007b). (b) Variation in the CO_2 concentration measured at the Mauna Loa Observatory from 1958 to December 2013 (NOAA, 2014a).

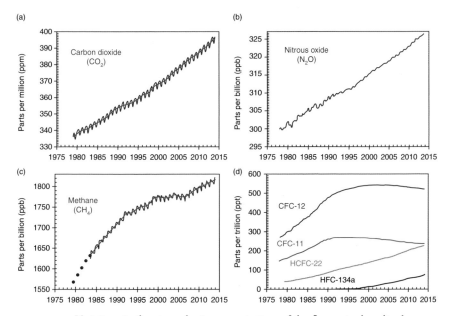

FIGURE 3.2. Variations in the atmospheric concentrations of the five major long-lived greenhouse gases from 1979 to August 2014 (NOAA, 2014b).

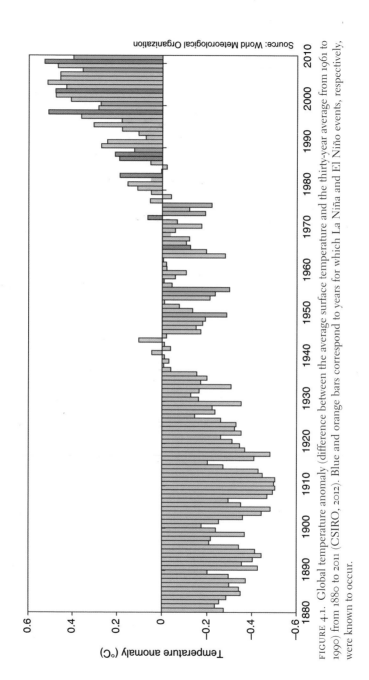

FIGURE 4.1. Global temperature anomaly (difference between the average surface temperature and the thirty-year average from 1961 to 1990) from 1880 to 2011 (CSIRO, 2012). Blue and orange bars correspond to years for which La Niña and El Niño events, respectively, were known to occur.

Source: World Meteorological Organization

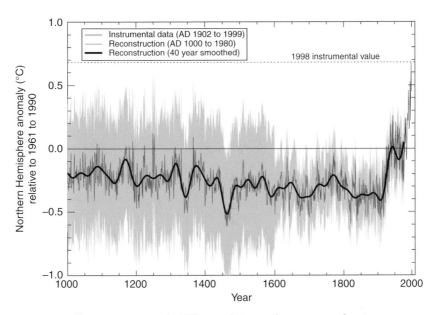

FIGURE 4.2. Temperature anomaly (difference between the average surface temperature and the thirty-year average from 1961 to 1990) for the Northern Hemisphere from 1000 to 2000 (IPCC, 2001b).

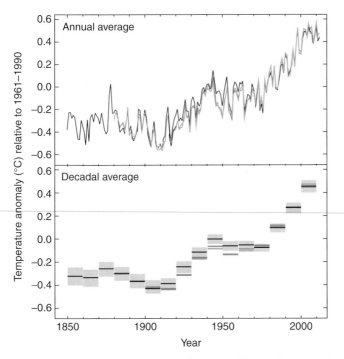

FIGURE 4.3. Annual and decadal globally averaged land and ocean surface temperature anomaly from 1850 to 2012 (IPCC, 2014a).

Thermohaline circulation

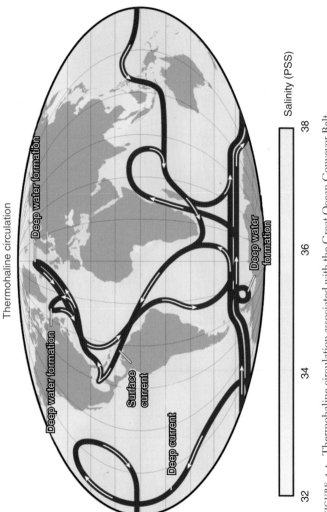

Salinity (PSS)

32   34   36   38

FIGURE 4.4. Thermohaline circulation associated with the Great Ocean Conveyor Belt. Courtesy of the National Aeronautics and Space Administration.

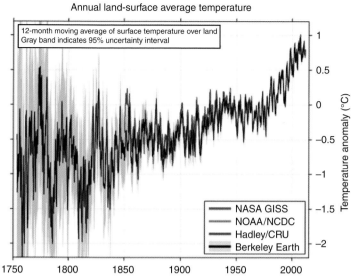

FIGURE 4.5. Global *annual* average land temperatures relative to the 1950–79 mean. Results reported by the Hadley Centre of the UK Meteorological Office and the Climate Research Unit of the University of East Anglia (HadCRU), the Goddard Institute of Space Studies (GISS) of the National Aeronautics and Space Administration, the National Oceanic and Atmospheric Administration (NOAA), and the Berkeley Earth Surface Temperature Project (http://berkeleyearth.org/resources.php). Accessed January 15, 2015.

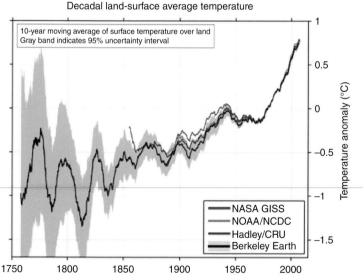

FIGURE 4.6. Global *decadal* moving average land temperatures relative to the 1950–79 mean. Results reported by the Hadley Centre of the UK Meteorological Office and the Climate Research Unit of the University of East Anglia (HadCRU), the Goddard Institute of Space Studies (GISS) of the National Aeronautics and Space Administration, the National Oceanic and Atmospheric Administration (NOAA), and the Berkeley Earth Surface Temperature Project. (http://berkeleyearth.org/resources.php). Accessed January, 15, 2015.

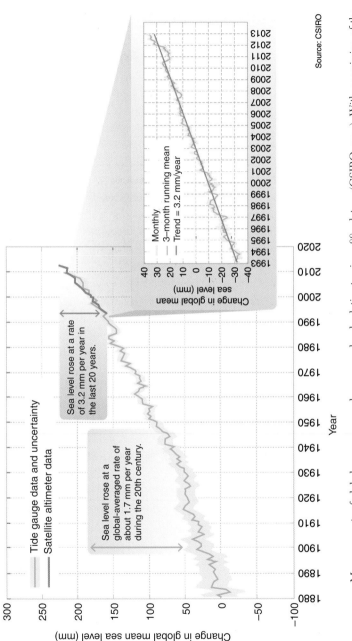

FIGURE 5.1. Measurements of global-average annual mean sea level relative to circa-1880 datum (CSIRO, 2014). With permission of the Commonwealth Scientific and Industrial Research Organization of Australia.

CHAPTER 8

The politics of global warming: a history lesson and future prospects

Consider the atmosphere from any point on the Earth's surface. Move east, west, north, or south; the atmosphere has no boundaries. If a CO_2 molecule is discharged from a point source in the United States, it can move in any direction, and one year hence it could be over India, Brazil, or anywhere else in the world. Likewise, GHG emissions in India and other nations make their way to the United States and to every other nation. All nations contribute to GHG emissions, some more than others, and all nations are recipients of the emissions. Geographically, the atmosphere is nondiscriminatory.

To the extent that GHG emissions contribute to global warming and, in turn, warming contributes to climate change, inherent problems are not local, regional, or national. They are transnational. The problems cannot be solved by any one nation – although some can contribute more than others – but must be addressed by collaboration among all nations. Even the smallest of nations that contribute little to GHG emissions must be involved, for they may well be among those most adversely impacted by climate change and least able to adapt.

Whether the problem is perceived as potential or omnipresent, the fact that it is international is indisputable. And the international community has been paying attention. In this chapter, we will examine efforts to address the problem and the mine fields that have impeded substantive progress. As applied to climate change, the chapter is a history lesson in geopolitics, highlighting how difficult it is to achieve consensus among nations of widely disparate cultures and economic conditions.

8.1 The Intergovernmental Panel on Climate Change

International agreements have long been used to address environmental issues for segments of the planet as diverse as the atmosphere, the open seas, and Antarctica. A relatively recent agreement is the *Montreal Protocol* of 1987, which calls for protection of the stratosphere's ozone layer through elimination of industrial chemicals termed chlorofluorocarbons (CFCs) and hydrochlorofluorocarbons (HCFCs).[1]

Although interest in global warming can be traced to the late nineteenth century (Weart, 2003), it was not until 1988 that the issue began to receive serious attention. Under auspices of the United Nations Environmental Programme (UNEP) and the World Meteorological Organization (WMO), the Intergovernmental Panel on Climate Change (IPCC) was established and charged with addressing a broad range of issues. In addition to assessing the scientific basis for anthropogenic climate change and scenarios for future change, the IPCC was asked to consider the potential effects on ecological, economic, and social systems, as well as mechanisms for mitigating and adapting to change.

With active participation of approximately 400 natural/social scientists and economists representing approximately 120 nations, the IPCC was well resourced and was charged with periodically publishing reports based on its findings. Five major reports have so far been published – in 1990, 1995, 2001, 2007, and 2013–14. Despite efforts to acknowledge uncertainties and a cautious approach to stating conclusions, the IPCC has not been immune from criticism. For example, the methodology used to forecast future economic growth and attendant GHG emissions was believed to have an upward bias (Castles and Henderson, 2003). The simplicity of the climate models and an assertion in the 1995 report that "the balance of evidence suggests a discernible human influence on the global climate" were subjected to scorn (Singer, 2004). And conclusions that average temperatures experienced in the Northern Hemisphere during the twentieth century likely exceeded those of any comparable period of the last millennium were contested (Soon and Baliunas, 2003), igniting a hostile review in the U.S. Congress (Monastersky, 2003).

Whatever criticisms are levied at its assumptions and methodologies, which are subject to continuous internal and external assessment and refinement, the IPCC represents humankind's best effort to understand the consequences of GHG emissions. It also plays a role in setting the international agenda and policies for action. Its input was instrumental in adopting a global climate treaty termed the United Nations Framework Convention on Climate Change (FCCC), which was ratified at the 1992 Earth Summit Convention in Rio de Janeiro. The treaty called for *stabilization of GHG concentrations at a level that would prevent dangerous anthropogenic interference with the climate system* but stopped short of setting quantitative targets for emissions and stabilized concentrations. However, it did call for regular meetings of its adherents, termed the Conference of Parties (COP), to advance its agenda, and it was at the 1997 meeting in Kyoto, Japan, that specific goals were established.

8.2 The Kyoto Protocol

For more than a decade, the Kyoto Protocol (KP) was the centerpiece of international efforts to deal with global warming. Its key provision called for industrialized nations to reduce their GHG emissions to 95% of 1990 levels by the year 2012, while exempting emerging economies from compliance. Led by China and India, nations with rapidly expanding economies weren't about to jeopardize economic growth by curtailing their use of fossil fuels. The exemption was not received well by some industrialized nations, nor was selection of 1990 as the baseline for targeting emission reductions.

A case could be made for exempting developing nations on the basis of climate equity. Since developed nations contributed disproportionately to GHG emissions from the time of the Industrial Revolution to the end of the twentieth century, it could be argued that they should bear most of the burden of curtailing emissions, as well as provide technical and financial resources to help developing nations curb their own emissions and adapt to the effects of climate change. But moving into the twenty-first century, the argument ignores the large economic strides made by many developing nations. With globally competitive industrial sectors and ever increasing GHG emissions, nations such as Brazil, China, and India have taken large steps toward movement into the developed category. Although it was well intended and a cornerstone of the KP, a two-tiered system of charging developed nations with the responsibility for curbing GHG emissions and exempting developing nations was problematic.

Before it could take effect, the KP would have to be ratified by industrialized nations that collectively contributed more than 55% of global, year-1990 CO_2 emissions. In 1990, the United States, Russia, and the European Union (EU) contributed 21%, 17%, and 15%, respectively, to global emissions, and within this group selection of 1990 meant that compliance would be easier for some than for others. For example, following reunification in 1989, the closure or modernization of many inefficient production facilities in the former East Germany significantly reduced GHG emissions of the new Germany. In Russia, emissions plummeted due to a severe economic downturn in the 1990s. The United Kingdom also benefited from timely conversion of many of its coal-fired power plants to natural gas. Accordingly, these nations experienced significant post-1990 reductions or, at worst, modest growth in their emissions, making it easier for them to meet the Kyoto requirement. In contrast, there were no comparable effects on U.S. emissions, which increased steadily from 1990 in line with economic growth.

Although the United States agreed to the 1990 baseline, the Clinton administration, anticipating congressional disapproval, proposed to meet its target, not through real reductions in emissions, but by receiving credit for carbon sequestered in its vast forests and by purchasing emission allowances from nations such as Russia, whose emissions were below those imposed by the KP. Nevertheless, concerned with the Protocol's exemption of developing countries and with potential damage to the economy, the Senate was quick to voice its opposition to any agreement that risked economic impairment.

Although a meeting of the COP at The Hague in 2000 was intended to achieve ratification of the KP, discussions were contentious, and the meeting concluded without reaching the 55% requirement. As a condition for ratification, the United States continued to argue that the Protocol should be amended to allow for credits obtained by emissions trading between nations, as well as for benefits accrued from sequestration of carbon by net growth of a nation's forests. Emissions trading would allow a nation failing to meet its target to do so by purchasing credits from another nation that was within its limit. A sequestration provision would yield credits commensurate with carbon sinks provided by net conversion of CO_2 to biomass in a nation's forests. Both amendments were roundly rejected, particularly by the Europeans, and the meeting concluded without the United States, Russia, and Japan as signatories. A U.S. view that it should not reduce its emissions without limits on developing nations was also rejected. These nations refused to be held to any limits, arguing that their per capita emissions were nominally less than 5% of U.S. emissions. This argument was supported by many of the industrialized nations, which viewed the United States as a large consumer of fossil fuels and capable of reducing consumption without adverse effects on its economy.

With concessions to allow an emissions-trading system made at a 2001 meeting of the COP, Japan joined the EU in ratifying the KP. Russia, which would benefit from such a system, appeared inclined to follow suit. However, discounting the significance of global warming, the new administration in Washington, under President George W. Bush, declared its opposition to the Protocol. Although Washington had legitimate reasons for concern, it was the abrupt nature of its disengagement that elevated European views from disappointment to outright anger (Prestowitz, 2003).[2]

Pursuing its own course, in 2002, the Bush administration articulated a goal of restraining emissions without impairing economic growth (White House, 2002). Rather than committing to a reduction in annual emissions, the nonbinding goal was stated in terms of reducing the GHG intensity of the U.S. economy (the ratio of GHG emissions per GDP). By realizing an 18% reduction in GHG intensity by 2012, year-2012 emissions would be reduced by 100 Mt-C_{eq} (million tonnes of carbon equivalent) and by a cumulative amount of 500 Mt-C_{eq} over the ten-year period. The goal would be achieved by increasing fuel economy standards by 1.5 mpg for sport utility vehicles and light trucks and by creating a $3 billion Climate Change Technology Program to accelerate development of carbon sequestration, energy efficiency, and non-fossil energy technologies. However, U.S. GHG intensity had been decreasing throughout the second half of the twentieth century and decreased by 17.5% in the decade of the 1990s (EIA, 2001). With GHG intensity already projected to decline by 13.8% from 2002 to 2012 (EIA, 2001), the administration's target of 18% struck many as little more than business-as-usual. In any case, absolute GHG emissions would not decrease and relative to 1990 were projected to increase 30% by 2010 (Clausen, 2004).[3]

Despite approval by approximately 120 countries, representing 44% of 1990 CO_2 emissions, without adoption by the United States, approval by Russia became critical to achieving the 55% threshold. By the end of 2003, Russia was having second thoughts (Ball, 2003). Without the financial gains it had expected to accrue from emissions trading, it was less inclined to participate. In the words of its chief delegate to a 2003 meeting of the COP, "Our position is pragmatic. We want to participate in Kyoto with benefits." Concurrently, the chief science advisor to Russian President Vladimir Putin and the head of the Russian Academy of Sciences were beginning to openly question "the science behind the Kyoto initiative." Nevertheless, in May 2004, the pragmatic view prevailed. After years of negotiation, the EU agreed to support Russia's entry to the World Trade Organization (Chazan, 2004). In return for this support, President Putin indicated that Russia would "speed up" its ratification of the KP, despite the opposition of advisors who continued to challenge its scientific basis and express concern for adverse effects on the growing Russian economy. In November 2004, President Putin signed the Kyoto Protocol, making Russia the 126th nation to do so and clearing the 55% threshold requirement.[4]

So, what is the import of the Kyoto Protocol? This question was posed in an article by Revkin (2003). Quoting David Sandalow, a guest scholar at the Brookings Institute, "the standard of success isn't whether the first treaty out of the box sails through … (it) is whether this puts the world on a path to solving a long-term problem." Revkin suggested that "the process of moving the world toward limiting releases of the (GHG) gases after more than a century of relentless increases has clearly begun," but regrettably without a plan for the future role of developing nations and with the United States looking much like an obstructionist. As we'll see, the KP had some success, but by the time it expired in 2012, it was problematic whether the world had been put "on a path to solving a long-term problem."

8.3 Implementation of the Kyoto Protocol

The Kyoto Protocol was an agreement with two phases running from 2005 to 2012. A cornerstone of Phase I, which ran from 2005 to 2008, was a cap-and-trade system (Section 7.1) modeled after that developed and successfully implemented by the United States to curtail SO_2 emissions from coal-fired power plants. As applied to GHGs, large industrial users of fossil fuels would be allotted a certain number of emission credits. If an entity's emissions exceeded its allocation, additional credits could be purchased from surplus credits created by other entities or acquired as offsets by funding emission reduction projects in other nations. The largest portion of the offset market was the Clean Development Mechanism (CDM), which sponsored emission reduction projects in developing nations and was operated by the United Nations. Its goals were twofold: to reduce the cost of GHG mitigation projects and, by providing financial incentives, to encourage emission curtailment in poorer nations. But, as we'll find, although well intentioned, the CDM would be manipulated in ways that undermined emissions reduction and the cap-and-trade system itself.

The EU was quick to take a leadership role by formulating an Emissions Trading Scheme (ETS), which did three things: establish emission limits (a cap), develop a protocol for distributing allowances to member nations, and provide guidelines for acquiring offsets. An initial cap of 1.8 Gt-CO_2 per year was prescribed for six industrial sectors (electric utilities, oil refineries, concrete, steel, glass, and paper/pulp) responsible for about half of the EU's total emissions. With one allowance per tonne (t) of CO_2, 1.8 billion permits were awarded – not sold or auctioned – to member nations, and each nation was left to develop a plan for distributing its allowances. Targets could be achieved by improving the efficiency of existing systems, by introducing alternative energy conversion systems, and/or by offsets.

In the ETS, offsets could be acquired by using the CDM to fund projects in developing nations exempted from the Kyoto accords and where emissions could be reduced at lower cost. A large target was China, where investments in clean energy technologies could do much to mitigate that nation's contribution to GHG emissions. The offsets were perceived to be a win-win proposition. For the EU, they would reduce the cost of curtailing emissions; for China and other developing nations, they would enhance economic development while reducing the consumption of fossil fuels.

With ratification of the KP and acceptance of cap-and-trade procedures, private investment funds and exchanges quickly emerged to manage trading of emission credits.[5] In 2004, it was estimated that annual trades would soon reach two billion tonnes of GHGs (Fialka, 2004) and that monetization of trading could reach $10 billion by the time the Protocol's mandates were fully implemented in 2008 (Ball, 2004a). The ETS began with allowances priced at €7 per tonne of CO_2. Prices climbed to €25 by the middle of 2005 and to €31 by April of 2006. But, within a matter of days, prices plummeted to approximately €10, and by early 2007 the allowances had become nearly worthless (Brahic, 2006). The ETS was in danger of failing to meet the sine qua non of a cap-and-trade system, namely to cap emissions. With carbon priced so low, there was no incentive to deploy more expensive, low-emission technologies or to reduce emissions by deviating from business-as-usual in other ways. What had gone awry?

The market had simply become glutted with allowances, reflecting the fact that, at least for some nations, the caps had been set too high. Under intense lobbying by industry, the EU had been too generous. With emission credits accumulating, a surplus of supply guaranteed a reduction in price. Ironically, receiving allowances at no cost, power companies still raised rates – much to the benefit of their bottom lines – arguing that emission reduction targets had increased the cost of generation (Kanter and Mouawad, 2008).

The imbalance between supply and demand for emission credits in the ETS was exacerbated by the CDM, which allowed companies to sponsor relatively low-cost emission reduction programs in developing nations and to thereby acquire additional credits that could be sold. The CDM spawned many projects representing billions of euros, in some cases with unintended consequences including corruption (Economist, 2007; Tyndall, 2011). Consequently, EU emissions actually increased from 2005 to 2007.

One particularly egregious manifestation of the CDM involved reducing emissions of HFC-23, a by-product of producing refrigerant HCFC-22 with a 100-year global warming potential nearly 15,000 times that of CO_2. The Montreal Protocol called for replacement of HCFC-22, which has a 100-year GWP of 1,800, with more benign refrigerants, and production was declining in most developed nations. But in China and other developing nations, it was expanding, and with windfall CDM revenues for capturing and incinerating the HFC by-product, strong incentives remained for continued production. Developing nations had no incentive to join the developed world in eliminating HCFCs when they could derive dual benefits from continued production, one from sale of the product itself and the other from emission credits generated by the disposal of HFC-23.

By the end of 2006, HFC-23 disposal accounted for almost 50% of the credits sold by developing nations (Ball, 2007c). In 2006, China alone received approximately €3.5 billion from CDM projects funded by the EU. Although the problem posed by HCFCs was acknowledged by developing nations, which agreed to freeze production at 2013 levels and to move the time for complete elimination from 2040 to 2030 (Kintisch, 2007), nineteen plants continued to profit from HFC-23 credits, eleven of them in China and five in India (Rosenthal and Lehren, 2012).

Although fewer emission allowances were granted in Phase II of the ETS, which ran from 2008 to 2012, continued implementation of the CDM sustained the purchase of credits from less developed nations, depressing their market value. In the words of a Green member of the Swedish Parliament (Winfield, 2007): "These credits have already been exposed as highly flawed, and often fraudulent.... They don't always reflect absolute reductions in emissions, whilst many of the credits are generated from projects ... that would have happened anyway." Although the cap was lowered by 6.5% and 10% of the allowances were auctioned, by then a global recession was in full bloom, and with emissions falling due to reduced economic activity, there was again no motivation to deviate from business-as-usual. And things weren't about to get better.

From 2008 to 2010, spot prices fluctuated between €15 and €20 per tonne, but by 2012 they fell to about €7 and to less than €3 in 2013. The EU was still awarding too many allowances, creating a glut of more than 1.5 billion tonnes or approximately a year's worth of EU emissions (Economist, 2013a). A plan to increase the market value of carbon permits by temporarily removing 900 million tonnes of permits from the market – and restoring them when demand increased – was rejected by the European Parliament (EP). The plan was opposed by EU countries that rely heavily on coal, by energy-intensive industries fearing a loss of competitiveness due to the higher cost of carbon emissions, and by those objecting to any interference with a market-driven system. Although the EP subsequently reversed its decision, removal of 900 million credits still left the market with a large surplus, and the slight rebound in price to about €4.70 was well shy of the €30 or more needed to achieve objectives of the ETS (Reed, 2013). Whether the EU can implement the structural changes needed to significantly reduce surplus credits and increase the price per credit remains to be seen.

Although the EU has been in the vanguard of efforts to reduce emissions, its resolve may be wavering. The low price of carbon allowances and the high price of natural gas, as well as the security of gas supply, are making a return to coal-fired power plants more appealing. And with estimates of the cost of CCS ranging from €60 to €130, why invest in CCS when the cost of allowances is much lower? Because the EU has led efforts to implement cap-and-trade through its ETS, lack of success impairs prospects for eventually achieving a global market and jeopardizes use of cap-and-trade as a viable market mechanism.

Are there lessons learned from the EU's experiences? For cap-and-trade to be a viable option, emission caps must be lowered and offsets more carefully scrutinized. Caps should also be prescribed in accordance with long-term policies. The three-year (2005–8) window for Phase I of the ETS was simply too short for companies to make significant capital investments in carbon reduction, preferring instead to wait for longer-term policies that would reduce uncertainties associated with the investments. And moving forward, if cap-and-trade is to be an effective component of an international agreement to reduce emissions, the caps must be administered globally. For that to happen, the world's two largest emitters, China and the United States, would have to find common ground that could be shared by other developing and developed nations. Finding that ground would not be easy!

8.4 Beyond Kyoto

Following activation of the KP, the Eleventh meeting of the COP in December of 2005 began the process of assessing progress and planning for the future. Developing nations indicated a willingness to consider how they might contribute to reducing future emissions but remained opposed to mandated limits, citing their already low emissions per capita. The United States agreed to participate in assessing "strategic approaches for long-term cooperative action to address climate change," while continuing to stress voluntary approaches and the need for more research. But, in the main, views remained divergent. China, India, and other developing nations continued to argue that they should have the same opportunities historically enjoyed by developed nations to grow their economies, while the Bush administration would not commit the United States to reducing emissions without commitments from the developing world.

In its early stages, signatories to Kyoto were experiencing mixed success in achieving their targeted reductions. The top achievers, Britain and Germany, were already well below their 1990 targets, aided in one case by the large-scale 1990s replacement of coal by natural gas for power production and in the other by modernization of the former East Germany's industrial base. However, other nations, such as Canada and Spain, were a long way from meeting their goals. Canada, which had agreed to a 6% reduction from 1990 emissions by 2012, was exceeding the 1990 level by 19% and the 2012 target by 25%. The 2012 target was simply beyond reach, particularly with ever increasing oil production from Alberta's tar sands. Basic elements of the KP would have to be revisited.

Establishment of absolute, legally binding, short-term emission targets with uncertain costs and implications for economic growth was simply not tenable. It is one thing to apply a cap-and-trade system to reduce SO_2 emissions from coal-fired power plants in the United States, but it is far more difficult to monitor and enforce GHG emissions globally, even if restricted to a subset of concentrated sources. Is a cap-and-trade system the best vehicle for doing so? At what level would the cost of emission credits for cap-and-trade, or an outright carbon tax, provide sufficient incentives for reducing emissions without economic impairment? And how should the system be administered globally?

As an alternative to the KP, a group of non-signatories (Australia, China, India, and the United States) and signatories (Japan and South Korea) coalesced to form the Asia-Pacific Partnership on Clean Development and Climate. The partnership was founded on the primacy of economic development and human well-being, rejection of mandated emission targets and the notion that energy consumption must be curtailed, and on the development of clean energy technologies. However, little of substance emerged from the Partnership, and Australia subsequently elected to ratify the KP. This decision left the United States as the only industrialized nation not to sign the treaty.

At the previous meeting of the COP in December 2004, efforts had been made to engage the United States in a discussion of future measures to curb GHG emissions. However, along with developing nations and members of OPEC, it declined to participate, and the meeting concluded with frustration over what was perceived to be U.S. obstructionism (Rohter, 2004). There was little change at the 2007 COP meeting in Bali, Indonesia, where the United States resisted efforts to establish a road map for negotiating a sequel to the KP, prompting anger and derision from both developed and developing nations. In the final hours of the conference, the United States relented and agreed to participate in negotiations to establish a new treaty by the end of 2009. But motivated by concern for impairment of its economy, it remained opposed to mandatory reductions (Fuller and Bowley, 2007). China's position was that developed nations should lead by cutting their emissions and by providing financial and technical assistance to developing nations. However, companies having relevant state-of-the-art technologies were understandably leery of China's tendency to violate intellectual property rights by copying Western technologies (Spencer and Wright, 2007). Also, with consistent double-digit economic growth and accumulation of large foreign reserves, China's designation as a developing nation was becoming arguable. The acrimony of the two-week Bali conference provided clear signals that the road to a global consensus would be contentious. Nevertheless, some nations were determined to move forward.

Influenced by the Stern Review, Britain ratified the Climate Change Act of 2008, calling for reductions of 26% and 80% in emissions by 2020 and 2050, respectively, and becoming the first nation to enact a GHG emission law. In 2008, the European Commission (EC) committed to a 20-20-20 plan. Relative to 1990, member nations would reduce emissions 20% by 2020, with renewable sources meeting 20% of total energy consumption. It went a step further by committing to a 30% reduction if other developed nations made comparable reductions and developing nations contributed commensurate with their abilities. Targets were extended to sectors not covered by the ETS, such as buildings and transportation, and the cost of achieving the plan was estimated to be 0.5% of GDP by 2020, or about €150 per person per year (EC, 2008). The plan also called for revision of the ETS, with a new Phase III to extend from 2013 to 2020. For the power sector, 100% of the permits would be auctioned over the entire period; for other sectors the amount of auctioned permits would begin at 20% and increase to 100% by 2020. Additional controls and restrictions would also be placed on the use of credits and offsets.

Of course, ratification of measures such as the British Climate Change Act and the EU 20-20-20 plan does not guarantee successful implementation. A case in point is Japan and its failure to meet its obligations under the KP. Instead of achieving a 6% reduction from 1990 emissions, its 2009 emissions exceeded the 1990 level by 14%. Another is the willingness of the EP to allow the ETS to limp along in 2013, falling far short of its intended role. How far will nations go to achieve reduced emissions at the expense of economic interests? The evidence to date suggests *not far*. Policies can be modified in response to ever changing conditions.

With the KP expiring in 2012, hopes were high that a December 2009 meeting of the FCCC in Copenhagen, Denmark would lead to a binding treaty that replaced the KP. The EU was poised to provide leadership, but would the United States and developing nations be able to find common ground? By early 2009, it was becoming evident that harmonization of a global policy would be very difficult. While most nations acknowledged the existence of a climate change problem, economic interests were making it difficult to find common ground for a solution.

8.5 Copenhagen and subsequent meetings

The Copenhagen meeting involved participation of all 192 members of the FCCC, and even before it began, it was recognized that such a large and unwieldy group with a wide range of agendas would make it difficult to achieve quantifiable and binding commitments. Again, issues centered on divisions between developed nations, whose per capita emissions ranged from approximately 10 to 25 tonnes of CO_2 equivalents per year ($t\text{-}CO_{2eq}$), and developing nations, whose emissions were no higher and, in most cases, much less than 5 $t\text{-}CO_{2eq}$.

Outcomes of the meeting were incremental at best. It was agreed that the rise in the Earth's average temperature from preindustrial conditions should not exceed 2°C. However, without binding commitments to significantly reduce emissions, the goal was simply aspirational. To get there, at least a 50% reduction in global emissions from 1990 levels would have to be achieved by 2050, with developing nations expecting richer nations to bear most of the burden. Developing nations minced no words in claiming that developed nations should shoulder the costs of reducing emissions and requested assistance of approximately $500 billion per year. One could argue that the developed nations have an obligation to provide assistance since they are responsible for most of the atmosphere's existing GHGs and that developing nations need assistance to reduce their emissions. But to which nations, in what forms, and at what levels? Should nations such as China and India, with burgeoning economies, be recipients? With slowly growing economies and, in some cases, large levels of debt, how much support could developed nations provide without impairing their own economies?

At Copenhagen, such issues were contentious and largely unresolved. Developed nations did agree to provide $10 billion per year over three years to help the least-developed nations reduce their emissions and adapt to climate change. A target of increasing allocations to $100 billion per year from public and private sources by 2020 was also established. For their part, China and other developing nations agreed to provide greater transparency in reporting emissions. But all pledges were nonbinding and, at best, no more than expressions of good intentions.

If progress was made on any issue, it was on deforestation, which was widespread in the tropical forests of developing nations like Indonesia, the Congo, and Brazil. In 2010, it contributed about 18% of global GHG emissions. If these emissions are to be curtailed, there must be economic incentives. Curbing tropical deforestation comes at the expense of lost revenues for developing nations and will not occur without appropriate compensation. Part of the $10 billion per year committed by developed nations would be used to slow net deforestation, with the intent of establishing measures for complete cessation over an appropriate timeline. However, success depends on the manner in which economic assistance is provided, and there are pitfalls, not the least of which involves the use of carbon credits (Economist, 2009a). Although inclusion of credits for deforestation in cap-and-trade schemes could be an important source of funding, we've seen that it can come at the risk of reducing the market price of a credit and correspondingly incentives for adopting other carbon reduction measures.[6]

Although developing nations, led by China, were initially opposed to establishing a globally uniform system of monitoring and reporting emissions, they did agree to a compromise in which each nation would allow for international review of its measurement, reporting, and verification processes. This matter is consequential since evidence was mounting that atmospheric concentrations of GHGs such as methane and sulfur hexafluoride were exceeding levels associated with reported emissions (Carey, 2010). The ability to accurately report results on a global scale is jeopardized by the absence of uniform standards for quantifying, documenting, and verifying emissions. However, achieving standardization and high levels of confidence in the results will be elusive.

While not agreeing to binding commitments, many nations did agree to submit expressions of intent by January 31, 2010. For developed nations, the focus was on reducing emissions and for developing nations on reducing the rate of growth in emissions. Fifty-five nations met the deadline, including China and India, the twenty-seven members of the EU, and the United States. China and India pledged to reduce the *rate of growth* in their emissions by at least 40% and 20%, respectively, while the EU pledged an *absolute reduction* of 20% – relative to 1990 levels – by 2020 and 30% if other nations deepened their cuts. Using 2005 emissions as a reference, the U.S. administration pledged a 17% reduction by 2020, but without enactment of enabling legislation by Congress. In contrast, South Korea targeted a 30% reduction from 2005 levels and enacted legislation committing 2% of its GDP to energy efficiency and low-carbon technologies. However, due to changing conditions – political, economic, or otherwise – some nations were unlikely to meet their Copenhagen commitments. With newly elected governments more supportive of producing and exporting fossil fuels, Australia and Canada were unlikely to do so. And, with its nuclear power industry decimated by repercussions of the Fukushima disaster, the same could be said for Japan.

Fast forward one year to early December 2010, when the COP reconvened in Cancún, Mexico for another round of climate talks. Unlike Copenhagen, expectations were subdued going into the meeting, and discussions were much less contentious. The Cancún Agreements were largely formalizations of the Copenhagen Accord (Stavins, 2010). Mechanisms for monitoring, verifying, and reporting GHG emissions and mitigation measures were clarified, as were procedures for transferring clean-energy technologies to developing nations. The Agreements also affirmed the commitment of public and private funds to assist developing nations with mitigation and adaptation measures. Emphasis was placed on reducing emissions due to deforestation by developing means of measuring forest resources and market-based mechanisms to compensate developing nations for the reductions. Although developing nations such as Brazil, China, and India made nonbinding commitments to reduce carbon intensities, they resolutely supported the continuation of the two-tiered system that exempts them from binding commitments. What would happen next, at least under the auspices of the United Nations, wasn't clear. Even less clear was what would happen to the KP when it expired in 2012. Replacement by a comprehensive agreement with binding commitments would be dead on arrival in the United States, China, and many other nations.

The 2011 and 2012 COP meetings in Durban, South Africa and Doha, Qatar affirmed the goal of establishing a $100 billion annual fund to assist developing nations but failed to specify how the money would be raised or how it would be dispersed. There was also agreement that the KP should be extended through 2019, with a replacement treaty formulated for approval at the 2015 meeting in Paris and scheduled for implementation in 2020. However, at the 2013 meeting in Warsaw, Poland, there was little to suggest that agreement on a new accord to replace the KP would be reached by 2015. Although China and the United States could find common ground on reducing the use of HFCs, divisive issues remained, and apart from the EU there was little indication of the extent to which mitigation measures would be implemented by other nations. The state of affairs brings to mind the global climate treaty ratified at the 1992 Earth Summit Convention in Rio de Janeiro. Although the treaty called for *stabilization of* (atmospheric GHG) *concentrations at a level that would prevent dangerous anthropogenic interference with the climate system*, it stopped short of setting quantitative targets for emissions and stabilized concentrations. More than two decades later, a global consensus on measurable and enforceable targets had yet to be achieved, with developing nations still favoring KP provisions that exempt them from emissions cuts and many developed nations reluctant to make binding commitments. And although the KP was renewed for seven years, it was a shadow of its former self. With nations such as Canada, Japan, and Russia opting out, the renewal included nations accounting for only 15% of global emissions, far less than the 55% requirement of the original measure.

An underlying message of the Copenhagen through Warsaw meetings is that the role of the United Nations in coordinating a global effort is limited. Through the IPCC, the UN has done its most important work by vetting the underlying science, assessing potential mitigation and adaptation measures, and increasing awareness of the problem. But the FCCC has been unsuccessful in forging international agreements that substantively curb emissions. Consider the fact that emissions of 36 Gt-CO_2 for 2013 were 61% higher than the 1990 baseline of the KP (GCP, 2013). The geopolitical and economic dimensions of climate change are complex, and the divide between developed and developing nations remains large. Bullis (2014) considers actions since Kyoto as "largely symbolic," although a better descriptor may be incremental. He also criticizes the UN for focusing on specific emission limits and timetables. Without confidence in predicting their economic impact and in many cases lacking the ability to monitor and regulate emissions, governments are becoming more reluctant to make such commitments.

Moving forward, it's easy to be skeptical about the ability to achieve a balanced and effective international agreement on measures for reducing GHG emissions. That is not to say efforts to do so shouldn't continue, and the 2015 COP in Paris provides a good opportunity. The Synthesis Report of the IPCC (2014c) conveyed a sense of urgency for seizing the opportunity, and once again the EU provided leadership by announcing what it would bring to the table. In October 2014, its twenty-eight member nations agreed to reduce emissions by at least 40% relative to 1990 and to do so by 2030 (Kanter, 2014).[7]

But global cooperation remains the most important piece of the climate change puzzle, and the divide between developed and developing countries has been the largest obstacle to cooperation. Nothing could do more to bridge that divide than cooperation and leadership by the world's two largest emitters, China and the United States – one developing with hundreds of millions of people yet to be elevated from poverty and the other developed with low levels of poverty. Henry Paulson, a former U.S. Secretary of the Treasury, expressed it well (Paulson, 2014): "This problem (climate change) can't be solved without strong leadership from the developing world. The key is cooperation between the United States and China – the two biggest economies, the two biggest emitters of carbon dioxide and the two biggest consumers of energy.... The two nations must come together on climate."

And it appears to be happening. On November 12, 2014, a historic agreement was reached between the two nations. For its part, the United States committed to a 26–28% reduction in GHG emissions from 2005 levels by 2025, essentially doubling the rate of reduction from an average of 1.2% per year projected from 2005 to 2020 to between 2.3% and 2.8% percent per year from 2020 and 2025 (Revkin, 2014). The United States also announced that it would contribute $3 billion to help poor nations adapt to climate change (Davenport and Landler, 2014). In turn, China committed to restraining the increase in its emissions such that a peak would be reached no later than 2030, at which time 20% of its electricity would be generated from carbon-free sources of energy – about 1,000 GW_e and comparable to the total generating capacity of the United States. Embedded in its commitment is the goal of capping annual coal consumption at 4.2 Gt by 2020 – reducing it to 62% of primary energy consumption – compared with 3.6 Gt and 66% of consumption in 2013 (Wong, 2014).

Although the commitments made by the United States and China are steps in the right direction, they are neither drastic nor sufficient to keep global warming below 2°C. They do, however, provide a template for how developed and developing nations can work together and an incentive for other nations to follow. Collectively, the commitments made by the EU, United States, and China, which accounted for 55% of total 2013 emissions (BP, 2014), set a positive tone and increased prospects for a productive Paris meeting. That said, the commitments are not binding, and it's appropriate to question whether they can be met, particularly by the United States. The state of climate policy in the two nations is examined in the following sections.

8.6 The Washington debate

In meetings of the COP throughout the 2000s, the United States took positions that were contrary to those held by many of the world's developed nations. But that doesn't imply existence of a unified view within the U.S. political establishment. It's instructional to examine the debates that occurred in Washington during that period, if for no other reason than to underscore difficulties associated with achieving consensus within a nation, much less between nations.

Tilting at congressional windmills

In Washington, winds of change appeared to be stirring in 2006, when the Senate considered three different global warming initiatives. Even senators once opposed to the KP were beginning to have second thoughts about imposing carbon constraints in a cap-and-trade system and accelerating implementation of carbon-free nuclear, clean-coal, and renewable energy technologies. Fialka (2006) characterized transition of the Senate debate from "whether there is, in fact, a threat of global warming" to "how" the nation should deal with the threat. With the November 2006 elections returning control of the House and Senate to the Democratic Party, legislative proposals for curtailing GHG emissions were being viewed more favorably.

In the Senate, the Lieberman-Warner Climate Security Act of 2008 became the first GHG cap-and-trade bill to secure approval by a Congressional committee. The bill would subject 87% of U.S. emissions to cap-and-trade provisions, with goals of achieving 19% and 71% reductions from 2005 levels by 2020 and 2050, respectively. If successfully implemented, the bill would achieve a 65% reduction in total GHG emissions by 2050. Initially, 75% of the permits would be granted, and the percentage of auctioned permits would gradually increase from 25% to 60% by 2032.

Cost containment measures included the use of offsets as well as the ability to bank or borrow permits. Funds from the auctioned permits would be used to assist low-income households in meeting the increased costs of energy and to accelerate the development of clean energy technologies and energy efficiency measures.

By 2007, segments of the U.S. energy sector were becoming more inclined to accept a cap-and-trade system, but with one important caveat, namely that a ceiling be placed on the market price of a carbon credit (Ball, 2007a). As a concession to the energy sector, Senators Bingaman and Specter crafted a competing bill (the Low Carbon Economy Act) that would initially place a $12/t-CO$_{2eq}$ limit on the price of a traded permit and thereafter annually increase the limit by 5% above the rate of inflation. Termed a safety valve, it would allow producers to purchase additional permits at the prescribed limit. Targets were less ambitious than those of the Lieberman-Warner bill, with emissions maintained at the 2006 level through 2020, reduced to the 1990 level by 2030 and to 60% below the 1990 level by 2050. Initially, 75% of the permits would be granted, with the balance sold at $8/t-CO$_{2eq}$ and revenues used to assist low-income families and to develop clean energy technologies.

There were two problems with the Bingaman-Specter measure. First, it undermined fundamental market principles. Cap-and-trade is promoted as a system that can best capture the efficiencies of free markets. But if market efficiencies are to be fully realized, there should be no constraints on price. Second, if such constraints are to be imposed, a safety valve of $12/t-CO$_{2eq}$ is too low to encourage the use of fossil fuels in ways that significantly reduce carbon emissions. For coal-fired power plants with costs of $50/t-CO$_2$ to implement CCS, why would any utility beholden to its shareholders incur such a cost when it could buy additional permits for much less?

Despite their limitations, passage of the Lieberman-Warner or Bingaman-Specter bill would have sent two important signals to the nation and the world, first that the United States was taking the problem of global warming seriously and second that it was willing to contribute to a solution. But it was not to happen. Lobbying in opposition to the bills was intense, as were pressures to maximize the amount of free allowances and the ability to purchase additional credits that would work against meeting provisions of the cap. Even if a version of the bills had been approved by Congress, it would have likely been vetoed by then President George W. Bush.

With inauguration of President Barack Obama in January 2009, the quest for a climate change bill was renewed, this time spearheaded by Congressmen Henry Waxman and Edward Markey and again rooted in a cap-and-trade system. In May, the Energy and Commerce Committee of the House of Representatives approved a resolution (H.R. 2454) that called for, respectively, 17% and 83% reductions relative to 2005 emissions by 2020 and 2050. The percentage of auctioned permits would increase from 15% to 100% by 2030. The resolution also imposed a ceiling of $28/t-$CO_{2eq}$ on the cost of emissions and required states to meet 20% of their electricity demand from renewable energy, with allowance for a 5% offset from increased efficiency. The final bill, termed the American Clean Energy and Security Act (ACESA) of 2009 (Waxman, 2009), was narrowly approved by the full House in June 2009.

However, simplicity was not a hallmark of the ACESA, which exceeded 1,400 pages. With numerous handouts and offsets, it was laden with concessions to special interests that would likely have been amplified by any Senate version. It also fell short of dealing substantively with the problem. Although it would have been a departure from business-as-usual, a 17% reduction in emissions from the 2005 baseline would only bend the emissions growth curve incrementally, and the United States would continue to contribute significantly to global emissions.

Critics of ACESA who wished to see no constraints on carbon emissions pointed to the financial burden it would impose on the American public, despite evidence to the contrary. Additional annual energy costs for the average American household were estimated to be $134 and $339 by 2020 and 2030, respectively (EIA, 2009). Comparable estimates were made by the Environmental Protection Agency (EPA) and the Congressional Budget Office. If the intent was to curtail emissions, the problem with ACESA was not that it imposed a large financial burden on American consumers, but that it attached too low a price on carbon emissions.

On September 30, 2009, Senators Boxer and Kerry released a bill termed the Clean Energy Jobs and American Power Act. It applied a cap-and-trade system to utilities and large industrial sources contributing about 70% of total GHG emissions and called for a 20% reduction in emissions from the 2005 benchmark by 2020. On the same day President Obama authorized the EPA to release a proposed rule under the Clean Air Act that would begin the process of regulating GHG emissions from sources that produced more than 25,000 tons per year and collectively accounted for 70% of total emissions. The unstated implication of the president's message was clear: if Congress was unwilling to deal with the issue, his administration would. Although a provision of ACESA would restrain the EPA's authority to regulate emissions from stationary sources, the Boxer-Kerry bill included no such restrictions.

With no progress made on the Boxer-Kerry bill, a similar measure was proposed by Senators Kerry and Lieberman in 2010. Termed the American Power Act, it would apply to 85% of all carbon emissions, with reductions from 2005 levels of 17% and 83% by 2020 and 2050, respectively. Starting in 2013, the Act would price carbon at $16.5 per tonne, with gradual escalation to $55.4/t-C by 2030. But in the face of solid opposition from Republicans, many of whom continued to view climate change as a nuisance, if not a hoax, and by Democrats from states tied to coal and oil production, the measure never made it to the Senate floor.

In July 2010, lacking sufficient votes to overcome a filibuster, the Senate abandoned efforts to pass a climate bill. Despite support of cap-and-trade legislation by two of the 2008 presidential candidates (Barack Obama and John McCain) and intentions to reduce U.S. emissions expressed in the run-up to the 2009 Copenhagen meeting, 2010 ended without any U.S. climate legislation. One consequence of Congress's failure to put a price on carbon was the drag it placed on private-sector investments in carbon-free sources of energy, investments that bear on the nation's energy future as well as economic development and job creation (Pooley, 2010b). Quoting Pooley, by failing "to pass a comprehensive climate-and-energy bill, the country risked losing the clean energy race to China – sacrificing the jobs of the future in a timid, ill-fated effort to preserve the jobs of the past." China had no problem seeing the future and was acting boldly to secure its position as the world's leading manufacturer of clean energy technologies.[8]

Although some proponents argue that passage of U.S. climate legislation in almost any form would at least begin to address the problem, the collapse of efforts to craft legislation could be viewed as almost a blessing. Laden with concessions to the fossil fuel industry, including generous free allowances and marginalization of the Clean Air Act, the legislation could be criticized for giving too much to get too little. In any case, Congressional inaction was yet another indicator that the United States was not ready to assume a leadership role in addressing climate change. However, the White House still had a card it could play, one enabled by provisions of the Clean Air Act and a 2007 Supreme Court ruling empowering the EPA to regulate emissions if they posed a danger to human health.

Another lever to pull: enter the EPA

By late 2006, approximately twenty climate change lawsuits were pending in state and federal courts, with defendants ranging from the EPA to major oil and automotive companies, electric utilities, and coal producers. Cases against utilities were intended to limit emissions by forcing use of clean-coal technologies for new plant construction. The case against the EPA, which made its way to the Supreme Court, sought to compel the agency to establish federal standards for tailpipe emissions from cars, trucks, and buses under the Clean Air Act of 1970. The plaintiffs, which included twelve states led by Massachusetts, argued that: (1) the CAA empowers the federal government to regulate any pollutant that "threatens public health or welfare," (2) climate change due to GHG emissions represents such a threat, and (3) the EPA is empowered to regulate such emissions (NYT, 2006a). The Bush White House countered that the Act was only intended to deal with the immediate threat of pollutants such as oxides of nitrogen and sulfur and therefore did not encompass GHGs.

On April 1, 2007, in a narrow 5-to-4 decision, the Supreme Court ruled that the EPA could not ignore its authority under the CAA to regulate GHG emissions. The Court rejected the government's arguments that GHGs are not pollutants and that, because they are not specifically cited in the CAA, the EPA lacks the authority to regulate them. It also rejected the government's argument that the states have no basis for filing a suit and asserted their right to protect themselves against the adverse effects of climate change. Landmark features of the ruling are: (1) because they contribute to global warming and climate change, carbon dioxide and other GHGs are pollutants, and (2) the EPA is empowered by existing authority under the CAA to regulate GHG emissions. But, under the Bush administration, there was no interest in regulating emissions, and the EPA circumvented the Court's decision by indicating it would have to study the matter before rendering a judgment.

Views changed in 2009 when the EPA – under a new Obama administration – concluded that GHG emissions did represent a danger to human health (the so-called endangerment finding) and identified six offending gases and categories (CO_2, CH_4, N_2O, HFCs, perfluorocarbons, and sulfur hexafluoride). Although the finding was challenged by fourteen states and the coal, oil, and gas industries, it was upheld in a 2012 ruling of the United States Court of Appeals for the District of Columbia (Wald, 2012). The court also affirmed EPA's authority under the CAA to regulate GHG emissions from cars, trucks, and stationary sources. Together, the 2007 Supreme Court decision and the 2012 ruling of the D.C. Circuit Court provided a legal framework for regulatory action by the EPA, giving the Obama administration considerable license to act on its goal of achieving a 17% reduction from 2005 emissions by 2020.

Empowered by the 2007 and 2012 court rulings, the EPA and the Department of Transportation (DOT) – with the cooperation of automotive manufacturers – imposed LDV CAFE standards of 34.5 mpg and 54.5 mpg by 2016 and 2025, respectively. However, with the House of Representatives reverting to Republican control in the midterm elections of 2010, comparable regulations for stationary sources became problematic. Most Republicans remained dismissive of global warming, and the Republican majority in the House made clear its intentions to assert greater control over the EPA by restricting its ability to regulate emissions.

Mandated by law to regulate emissions on the one hand and confronted by a troubled economy and a Congress hostile to regulation on the other, the EPA found itself between the proverbial rock and a hard space. So, in a display of pragmatism, on December 23, 2010, it announced plans to establish GHG emission standards at a measured and moderate pace (EPA, 2010). Fifteen months later, it submitted for public comment its first-ever New Source Performance Standards (NSPS) for GHG emissions from electric generating units (Eilperin, 2012). The standards would require *new* power plants to emit no more than 1,000 pounds (~450 kg) of CO_2 per megawatt-hour of power production. The standards would have little effect on gas-fired, combined-cycle plants, which release less than 400 kg-CO_2/MWh$_e$, but they would place a significant burden on coal-fired plants, which nominally produce more than 800 kg-CO_2/MWh$_e$ without CCS. The rule was criticized by advocacy groups for failing to limit GHG emissions from *existing* power plants, while strongly opposed by House and Senate Republicans who viewed it as a death knell for new coal-fired plants.

On September 20, 2013, following the public comment period, the EPA released its NSPS, calling for limits of 450 and 500 kg-CO_2/MWh$_e$ for gas- and coal-fired plants, respectively (Wald and Shear, 2013; EPA, 2014d). Without CCS, there is no way that coal plants can meet the requirement, even with improvements in efficiency projected over the next two decades. The underlying premise of the NSPS that CCS technologies can be deployed to meet the requirements is problematic at best. Remember that CCS has yet to be demonstrated at requisite scales, and implementation would add 30% or more to plant construction and operating costs. Even at locations where captured CO_2 can be sold for EOR, the cost of natural gas would have to more than double for coal plants to become competitive. With gas-fired plants already dominating the market for new power generation, absent a significant cost-cutting breakthrough in carbon capture technology, it is difficult to imagine conditions under which construction of a new coal-fired unit with CCS would be viable.

The status of coal as an energy source for power generation was further jeopardized by pending EPA guidelines for existing plants. In anticipation of the guidelines, the Tennessee Valley Authority (TVA) announced plans to close six old coal-fired plants and convert two others to natural gas (Wines, 2013b). During their peak, the plants had annual emissions of 22 Mt-CO_2, as well as emissions of mercury and sulfur dioxide that exceeded limits to be imposed in 2016. The closures are part of a plan to reshape the TVA's energy portfolio by reducing coal to 20% of the total while increasing renewables to 20%. Similar decisions were being made across the country, motivated in no small way by the low cost of natural gas and the related benefits associated with replacing coal-fired generation at older plants by more efficient and far cleaner gas-fired facilities. From 2002 to 2012, the number of coal-fired plants in the United States had dropped 12%, with the share of coal's contribution to the nation's electricity decreasing from about 50% to 38%. The lost generating capacity was more than matched by increases in the contributions of natural gas – from approximately 20% to 30% – and renewables – from 9% to 12%.

By early 2014, the issue of power plant emissions remained contentious, with many in Congress viewing pending EPA regulations as a "war on coal" and a "job killer." The House was proposing legislation that would restrict implementation of the September 2013 standards, arguing that they cannot be achieved with existing CCS technology. For its part, the EPA was proceeding cautiously in its efforts to finalize the standards, indicating openness to compromise by providing a menu of options for compliance (Davenport, 2014a).

Released on June 2, 2014, the EPA standards – termed the Clean Power Plan (CPP) – called for a 30% reduction in power plant carbon emissions from 2005 levels by 2030, more than a third of which had already been achieved, largely through replacement of coal-fired plants by natural gas. In effect, the EPA plan amounted to an 18% reduction from 2012 emissions by 2030. EPA targets were set in terms of CO_2 emissions per MWh_e and varied among states according to their dependence on coal and the economic impact of reduced use. States such as Kentucky and West Virginia, which rely heavily on coal, would only have to reduce carbon emissions by 20% while Washington, which has only one coal-fired plant and depends primarily on hydropower, would have to reduce emissions by 84% (Davenport and Baker, 2014). The states were also given a good deal of flexibility in meeting their targets, allowing for selection of low-cost options. In addition to replacing aging coal-fired plants with natural gas and carbon-free sources of energy, they could also receive credit through interstate emission trading programs and for reducing demand through energy efficiency measures. Choosing from a menu of options, states would have to develop implementation plans for meeting their targets by 2016.

Although the EPA standards are modest and flexible, they were met with a firestorm of criticism – couched in terms like "job killer," "death of coal," and "ballooning costs of electricity" – particularly from states heavily reliant on coal. In fact, the rhetoric was grossly exaggerated. Coal use would continue to decline, but only at a slightly accelerated pace. It was still projected to supply 30% of the nation's electricity by 2030, with newer and more efficient plants continuing to operate (Barrett, 2014). A case in point is Xcel Energy, a utility that had reduced emissions by 19% since 2005 and was well on its way to a 31% reduction by 2020. It was doing so by retiring aging coal-fired plants and replacing the lost capacity with natural gas, wind, and energy efficiency measures, while still constructing a new 750 MW$_e$ coal plant (Davenport and Baker, 2014). Collins (2014) summarizes actions taken by the EPA since the Supreme Court decision of April 2007 through June 2014, along with a timeline for future actions through 2020.

The CPP was central to the Obama administration's Climate Action Plan, and if properly implemented, overall benefits should exceed costs (Fowlie et al., 2014). Its biggest benefit could well be the example it sets for other nations. The world could no longer view the United States as a laggard in addressing climate change, and nations that have been resisting commitments to reduce emissions would no longer have a convenient excuse for doing so. But it's yet to be determined whether the CPP will survive the inevitable legislative and court challenges and, if enacted, whether state-by-state agreements can be successfully negotiated. Because the dynamics of climate policy in the United States involve a triad of policy makers – Congress, the White House, and the judicial system – policies can change, particularly after an election cycle. The outcome of the 2016 presidential election may be pivotal in determining the staying power of EPA decisions and whether President Obama's commitment of at least a 26% reduction in emissions by 2025 will be achieved.

8.7 China: serious about emissions

As the United States wrestled with its approach to climate change, where did China stand? In 2013, the two nations jointly contributed 44% of global

CO_2 emissions from fossil fuels, with China accounting for 27% of the total (BP, 2014). But while U.S. emissions generally trended downward since 2008, China's had been steadily increasing for more than forty years. From 2008 to 2013, U.S. emissions declined by 6.3%, while China's increased by 41.4%.

In 2009, China released a document describing its position for the forthcoming meeting in Copenhagen. It called on developed nations to reduce their emissions by 40% from 1990 levels and to contribute at least 0.5% of their GDP to aid developing nations in coping with the effects of climate change. It also articulated the need for China and other developing nations to continue on a trajectory of economic growth without constraints imposed by reduced emissions. In essence, China placed the burden of reducing GHG emissions on developed nations and was adamant that developing nations should not be expected to contribute.

At the Copenhagen meeting, China underscored its position by announcing it would not consent to binding reductions in emissions. Instead, its primary goal would remain one of rapid economic development to reduce poverty and increase living standards for the Chinese people. China's position was shared by other developing nations, who rightly proclaimed that their emissions per capita were a small fraction of those for developed nations. This position certainly dampened prospects for reaching global consensus on a new treaty. Yet, despite its seemingly intractable stance, China was beginning to take climate change seriously.

China is by no means oblivious to the consequences of global warming (Zeng et al., 2008). A one-meter rise in sea level would inundate nearly 100,000 square kilometers of lowlands, including three major industrial centers, and rising temperatures would exacerbate drought and flooding in different regions with adverse effects on agriculture. Deglaciation of the Himalayan Mountains and the Tibetan Plateau would reduce flows to major rivers such as the Mekong and Yangzi, while weakening monsoons would increase flooding in Southeastern coastal regions and drought in North Central regions. China also recognizes the economic opportunities afforded by addressing global warming. By investing in renewable energy technologies, it can reduce its emissions while becoming a leading exporter of the technologies. For its own large market, it targeted 150 GW_e and 20 GW_e of new wind and solar generating capacity, respectively, by 2020. Consistent with increased use of carbon-free energy sources, its goal was to achieve a 17% reduction in its carbon intensity by 2015 (Li and Stone, 2011). Today, China views emission reduction as an important component of economic development.

Like many nations – developing and developed – China is unwilling to commit to binding reductions in GHG emissions. But it is not unmindful of the need for reductions, and it is proceeding along two paths, one involving energy efficiency measures and the other carbon-free sources of energy. Despite average annual economic growth of approximately 10% for more than two decades, which increased China's emissions from 2,400 Mt-CO_2 in 1990 to 9,525 Mt in 2013 (BP, 2014a), efficiency measures enabled it to reduce its carbon intensity (emissions per GDP) by more than 50%. China has mandated some of the world's most stringent energy efficiency standards in its building and transportation sectors, ranking first and fifth, respectively, in the world (Young et al., 2014). As part of a 2020 goal to reduce its carbon intensity by as much as 45% relative to 2005, China has also established cap-and-trade markets in two provinces and five large cities as a prelude to announcing a national program by 2016 (WRI, 2014; Yuan, 2014).

In addition to improving its energy efficiency, China is well on its way to becoming a global leader in carbon-free energy technologies, from solar, wind, and nuclear energy to hydropower. It has the world's largest installed base of wind energy and is installing solar photovoltaic power at a pace exceeding that of any other nation (Pew, 2013a). China is in fact becoming ever greener, driven in no small way by recognition of the economic benefits of becoming a world leader in exporting green technologies. Although extensive use of fossil fuels will continue to underpin economic growth, China's intention to reach peak emissions by 2030 is achievable, as is its goal of obtaining 20% of its generating capacity from carbon-free sources of energy.

8.8 From the bottom up

If national governments are unable to effect substantive reductions in GHG emissions, all is not lost. There are other agents of change, and the United States provides an excellent case study. The inability of the legislative branch to craft a plan for reducing GHG emissions and its ability to encumber efforts of the president have not meant that other elements of government and society had given up on the matter.

State and city governments

Despite inaction by the federal government during years preceding and following ratification of the KP, the issue received growing attention at the state level. States were becoming more assertive in legislating energy efficiency and renewable portfolio standards (RPS) and in establishing fixed targets for reducing GHG emissions. True to form, the most aggressive measures were being taken by California.

Impatient with the reluctance of the last Bush administration to impose more stringent CAFE standards, California was determined to press forward. Relative to 2005 levels, the state mandated a 30% reduction in emissions by 2016. The mandate was prescribed under provisions of the Clean Air Act, which authorized California to set its own vehicle emission standards and allowed other states to adopt the California standards in lieu of existing federal regulations. However, there was a caveat. Before California could implement its standards, it had to be granted a waiver by the EPA.

If granted a waiver, the California mandate would have been emulated by sixteen other states, largely in the Northeast and Far West. The mandate was vigorously opposed by the auto industry, claiming it would result in the loss of 65,000 jobs nationwide, undermine auto safety, and cause irreparable hardship for the industry. The industry also argued that the mandate intrudes on the purview of the federal government, which has sole authority to set fuel economy standards. Such resistance to regulation of automotive fuel economy and emissions had been a gut reaction of the auto industry since the 1970s.

In 2007, automakers filed suit in Vermont, one of the states supporting the mandate, to block implementation, arguing that it was a backdoor attempt to regulate automotive fuel economy and would therefore usurp a prerogative of the federal government. However, the Federal District Court of Vermont ruled in favor of the states, saying that the California measure went beyond fuel economy by including other GHG emissions, such as those associated with refrigerants used in automotive cooling systems. The decision augmented that rendered by the U.S. Supreme Court a few months earlier, which ruled that the EPA had the authority to grant the waiver being requested by California, and was affirmed a few months later by a second Federal District Court, this time in California. The court rejected the auto industry's assertions that only the federal government had the authority to regulate fuel economy standards and affirmed the right of California and other states to impose tougher standards to reduce GHG emissions.

As 2007 drew to a close, the EPA, acting on behalf of the White House, rejected the authority of California and other states to set their own automotive emission standards, which were more restrictive than those signed into law by the president a day earlier (Broder and Barringer, 2007). Although the states were prepared to appeal the decision, the issue was resolved in 2009 when, under the Obama administration, the EPA announced stricter standards consistent with those advocated by the states. By 2016, light-duty vehicles would have to achieve a fleet average of 35.5 mpg, at least 39 mpg for cars and 30 mpg for light trucks.

But California had other emission targets. In November 2006, Governor Arnold Schwarzenegger signed a Global Warming Solutions Act (AB 32) that established economy-wide limits on GHG emissions and a return to 1990 levels by 2020. It also required that renewable energy provide one-third of California's electricity by 2020. The law targeted large industrial sources, such as oil refineries and utilities, and included a cap-and-trade system for which permits can be purchased by emitters or investors and traded in a futures market.[9] Both the cap and the number of permits would decrease over time. The law received a good deal of political support from governors of other states and corporations such as DuPont and BP. However, it was not without its detractors, and in June 2010, Proposition 23, termed the California Jobs Initiative, was certified for the November ballot.

Proposition 23 was largely funded by two Texas oil companies, each with two refineries in California and each reluctant to invest in the emission control equipment needed to comply with AB 32. If approved, Prop 23 would require suspension of California's emissions reduction program until its unemployment rate dropped to 5.5% – from its 2010 level of 12% – for four consecutive quarters. In counterpoint, proponents of AB 32 argued that approval of the proposition would decrease, if not reverse, the growth of green jobs. On November 2, 2010, Prop 23 was soundly defeated by California voters, and in October 2011, the California Air Resources Board gave its unanimous approval to implementation of AB 32. In view of the state's long-standing commitment to environmental quality and the large investments being made by its high-technology sector in clean energy, the outcome was not surprising. But it did stand in sharp contrast to the lack of progress on climate issues in Washington and the negative impact that the 2010 midterm national elections would have on future efforts to curb emissions.

In its cap-and-trade system, California was careful to avoid problems plaguing the EU's ETS by carefully setting and pricing its caps and regulating offsets. As previously noted, the ETS granted emission permits to utilities and allowed them to earn credits from a wide range of poorly regulated offset projects. The initial effect was market volatility and windfall profits for the emitters, followed by a depressed market for emission credits.

At its first auction in November 2012, California *sold* all of its 23.1 million permits, each at $10.09 per tonne of carbon and slightly above a preset floor of $10. Three months later the second allotment of 12.9 million permits sold at $13.62/t-C (Doan, 2013). Both allotments were for 2013 and were oversubscribed by factors of 2.5 to 3, with proceeds used to support emission reduction projects throughout the United States. The fact that they were oversubscribed and were trading at about $15 in futures markets boded well for the sustainability of the California system. With expansion of the program to include mobile emission sources, the system is expected to generate $5 billion a year by 2017, with revenues earmarked for increasing low-carbon transportation options and water and energy efficiency measures (Sustainable Business, 2014). To date, California has been successful in avoiding volatility in the price of a carbon credit and has expanded its market by linking it with the Province of Quebec.

Other states were also moving to reduce non-vehicular GHG emissions. By June 2006, twenty-two states, concentrated in the Northeast, upper Midwest, Southwest, Far West, and the District of Columbia, had adopted renewable portfolio standards (Rabe, 2006). As discussed in Section 7.4, the standards prescribe thresholds for power generation from renewable energy sources such as wind, solar, and biomass, as well as timelines for reaching the threshold. The threshold can be specified as a percentage of electricity generation or consumption within a state or as an absolute amount of generation/consumption. California's AB 32 RPS dictates that 33% of the state's power must be generated from renewable energy by 2020, while Texas prescribes a generating capacity of 5,580 MW_e by 2015.[10]

In 2007, a group of ten Northeast and Mid-Atlantic states agreed to launch a program termed the Regional Greenhouse Gas Initiative (RGGI) – a cap-and-trade system for controlling emissions from the region's 230 power plants. The system implemented lessons learned from the ETS by *auctioning* most of its permits and by closely regulating offsets (Fairfield, 2007). Its first cap of 165 million credits – each corresponding to one ton of carbon dioxide – was implemented in 2009, and proceeds were used to support a portfolio of energy initiatives. Utilities would be penalized for releasing more than the amount of carbon for which they had allowances, with penalties determined by market dynamics of the auction and subsequent trading of allowances. Offsets were allowed but limited to 3.3% of a utility's total emissions and to the following options: methane recovery from landfill gases and/or livestock manure, new forest plantings, reducing leaks of sulfur hexafluoride from power plant equipment, and/or reducing building energy consumption associated with heating, cooling, and lighting.

Following start-up, two issues became problematic. Faced by growing budget deficits, some states channeled revenues to deficit reduction instead of energy efficiency and clean energy measures. And in 2010, elections shifted the political orientation of some state governments, which were less disposed to mitigating GHG emissions. However, while New Jersey's governor pulled his state out of RGGI, attempts made by the Delaware and New Hampshire legislatures failed, and RGGI was left with its current contingent of the six New England states, Delaware, Maryland, and New York. Although the market value of allowances was low – ranging from approximately $2 to $3 per ton of CO_2 – revenues from quarterly auctions through 2012 enabled investments projected to return more than $2 billion in lifetime energy savings and to reduce CO_2 emissions by approximately 8 million tons (RGGI, 2014a). And at the first auction in 2014 following a 45% reduction in the supply of allowances from 165 to 91 million tons per annum, allowances were priced at $4/t-$CO_2$, (Doom, 2014). From 2009 to 2014, the nine states had reduced their emissions by 18% while growing their economies by 9.2% – in contrast with averages of 4% and 8.8%, respectively, in the other forty-one states (Fairfield, 2014). The data suggest that RGGI had little, if any, effect on slowing economic growth. Subsequently, RGGI (2014b) announced the sale of 18 million allowances at $5.02, returning more than $90 million for reinvestment.

Although contributions made by California and RGGI to reducing carbon emissions are modest, they do affirm the ability of state governments to implement cap-and-trade systems effectively. And states were becoming proactive in other ways. In May 2009, governors of thirty states formed a Governor's Energy and Climate Coalition Group to seek common ground on a national strategy to curb climate change. In New York, a 1921 securities law was used to investigate energy companies planning to construct coal-fired power plants (Barringer and Hakin, 2007). Normally used to investigate corruption, the law was used to determine if the companies had provided investors with adequate information concerning potential financial liabilities linked to CO_2 emissions and climate change. The liabilities would be incurred if future regulations curbing emissions increased the cost of coal-fired power generation.

Involvement of government in reducing GHG emissions and/or preparing for adaptation to climate change became yet more local when 132 cities established a coalition in 2005 to address the problem (Sanders, 2005). By 2010, almost 600 local governments had enrolled in a program to better understand the implications of climate change and adaptation options (ICLEI, 2011). The communities are large (e.g., New York City, Los

Angeles, Chicago, San Francisco, Seattle) and small (e.g., Hurst, Texas, Chula Vista, California, and Bellevue, Nebraska), with political leadership from both major parties. Motivations ranged from concern for the effect of drought on agricultural production in the heartland and forest fires in the West to increased flooding in coastal regions. And U.S. cities weren't alone. By 2013, thousands of cities and towns worldwide had policies for advancing renewable energy that in many cases were more ambitious than national policies (REN, 2014).

In 2014, many of the foregoing initiatives were ongoing in the United States, but not without pushback from those opposed to climate legislation and regulation. With the midterm elections of 2010 increasing the number of governors and state legislators opposed to regulating GHGs, there was growing resistance to climate change legislation. Efforts to thwart further implementation of renewable energy were strongly supported by the American Legislative Exchange Council (ALEC), a group heavily funded by the fossil fuel industry and the Heartland Institute. Although it would be better to have a more uniform national system, it remained unlikely that Congress and the White House would be able to reach agreement on specific measures. Climate change had become a highly partisan and divisive issue, with rejection of global warming a sine qua non of the Republican Party (Rosenthal, 2011).

The corporate sector

Generally, it cannot be said that U.S. industry supports regulatory constraints on GHG emissions. However, despite strong opposition from groups such as the Competitive Enterprise Institute, the Chamber of Commerce, and the American Petroleum Institute, chinks in the armor had begun to develop in the early 2000s and have been growing steadily. Consider the following quote from an advertisement for the Royal Dutch/ Shell Group that appeared in the August 4, 2003 issue of *Business Week*. The ad posed the question: "Is the burning of fossil fuels and increased concentration of carbon dioxide in the air a serious threat or just a lot of hot air?" It responded by saying:

> Shell believes that action needs to be taken now, both by companies and their customers. We are actively managing greenhouse gas emissions in our world-wide operations such that by 2010 they are still 5% or more below 1990 levels, even as we grow our business. We are also working to increase the supply of natural gas, a cleaner option for electricity

generation and home heating, and offering alternative energy sources such as solar, hydrogen and wind power. It's all part of our commitment to contribute to sustainable development, balancing economic progress with environmental care and social responsibility.

Balancing **economic progress** *with* **environmental care** *and* **social responsibility**! Consider these words carefully. The three goals comprise the essence of a balanced strategy. Strive for economic advancement while remaining mindful of the need for environmental protection and greater social goods. But was the ad simply a public relations ploy, an example of *green washing*? If not, Shell was not alone. From the corner offices of major multinational corporations such as GE, DuPont, and Alcoa, as well as large producers of electricity such as Exelon and Duke Energy, support for mandatory restraints on GHG emissions was emerging. In anticipation of GHG emissions eventually being capped, many U.S. corporations were beginning to plan for a carbon-constrained world, some acting voluntarily and others responding to pressure from shareholders (Seelye, 2003; Ball and Fialka, 2003; Ball, 2004b).

In 2004, Cinergy – now part of Duke Energy, one of the United States's largest operators of coal-fired power plants – announced its support of a national cap on GHG emissions, and although their motives may not have been entirely altruistic (WSJ, 2004), both Cinergy and American Electric Power (AEP) indicated intentions to construct advanced IGCC power plants to more cleanly convert coal to electricity (Ball and Regaldo, 2004). Because Cinergy and AEP would have to decommission aging coal-fired plants in the near future, they would derive additional benefits from new IGCC plants with CCS, if a GHG cap-and-trade system were in place. Fast forward to 2014, however, and a cost had yet to be placed on emissions. Although Cinergy and AEP shelved their plans for building IGCC power plants, their willingness to consider means of reducing emissions was a harbinger of changing views. By 2005, signs were emerging that key industry segments had begun to take the problem seriously. Corporations whose technologies could contribute to reducing emissions saw opportunities to market their products and services, while others saw measures to reduce emissions as a way of improving profits.

Utilities such as Duke Energy and Exelon began to break ranks with their industrial sector by advocating nationally mandated emission limits and market-based incentives for controlling emissions. Such measures were deemed inevitable, and existing uncertainties in emission standards were cited as a deterrent to planning new plants with projected lifetimes of forty or more years. Corporations place a high priority on managing risk, and particularly when planning large, long-term infrastructure investments, they want clear guidelines. Despite the absence of such guidelines, in 2013, Duke still completed conversion of an old coal-fired plant in Indiana to a much larger IGCC plant. Although the plant did not include CCS, it significantly reduced sulfur, nitrogen, and particulate emissions and by virtue of its higher efficiency reduced carbon emissions.

Increasingly, there were also those who argued that commitments to reducing GHG emissions need not require trade-offs between environmental protection and economic development. In some cases, costs associated with climate protection could be negative (the cost of doing business decreases), and new technologies, while welcome, may be unnecessary. Companies that implemented existing energy efficiency options were finding that they can increase profitability (Lovins, 2005; Aston and Helm, 2005; Mouawad, 2006). Relative to 1990 levels, DuPont reduced its 2005 energy use and GHG emissions by 7% and 72%, respectively, while saving more than $2 billion. Johnson and Johnson reduced emissions by more than 11% since 1990, while growing its business by 350%. Significant GHG reductions and economic savings were also being achieved by corporations such as Alcoa, BP, IBM, and International Paper, while companies as diverse as 3M, AMD, GE, and Wal-Mart were pledging to do so through increased energy efficiency. Among U.S. utilities, NextEra Energy emerged as a leader in renewable energy – with 900 and 11,500 MW_e of solar and wind capacity in place by the end of 2014 and expansion to 2,000 and 13,000 MW_e projected by 2016 (M. O'Sullivan, private communication, 2014) – while NRG committed to a 50% reduction in its emissions by 2030 and 90% by 2050 (Cardwell, 2014).

In 2006, as it opened the door to considering mandatory controls on emissions, the Senate Energy Committee invited input from the corporate sector, which responded with a combination of downstream and upstream approaches. Downstream controls, which apply to users of carbon fuels, could include implementation of a CO_2 cap-and-trade system for utilities and large manufacturers, while upstream controls, which apply to producers of carbon fuels, could include higher taxes for transportation fuels. Companies such as Southern (an Atlanta-based utility), Wal-Mart, and DuPont advocated use of both approaches, depending on the emission source, while others, such as Duke Energy, advocated use of upstream measures for all sectors. Such an approach would tax all fuels on their carbon content and was recommended by Yale University's William Nordhaus on the basis of its relative ease of implementation and difficulty to game (Glenn, 2006). Whether applied upstream or downstream, the price assigned to carbon emissions must vary according to the nature of the fuel if it is to have the intended effect of curbing emissions.[11]

Corporate concerns for global warming rose to a new level in 2007 when ten major companies, representing a cross-section of the economy, formed the United States Climate Action Partnership and called for a national commitment to reduce GHG emissions (Barringer, 2007; Ball, 2007a). Charter members of the coalition included technology and manufacturing leaders (Alcoa, Caterpillar, DuPont, and GE), electric utilities (Duke Energy, NextEra Energy, Pacific Gas and Electric, PNM Resources), a multinational oil company (BP), and a now-defunct financial services company (Lehman Brothers). Among its recommendations, the coalition called for a 10% reduction in U.S. emissions within fifteen years, a 60–80% reduction by 2050, limited construction of coal-fired power plants lacking provision for carbon capture, and a national cap-and-trade system. Their proposals were strongly motivated by interest in a single set of national emission standards – rather than an assortment of state standards – and implementation of market-based approaches. The partnership represented a sea change in corporate views, and by 2009, divisions had developed between corporations seeking to curb GHG emissions and those resisting curbs. Resistance was led by groups such as the U.S. Chamber of Commerce, the American Petroleum and Heartland Institutes, and the National Association of Manufacturers (Carey, 2009).

By 2014, lines of demarcation between corporate sectors had blurred considerably, as more companies were viewing climate change through a lens of economic self-interest (Davenport, 2014b).

Motivated by disruptions to their supply chain due to extreme weather events and concerns for rising costs of energy due to constraints on GHG emissions, there was growing awareness of the many ways in which climate change could affect the cost of doing business. Increasingly, conservative economists – including former advisors to President Reagan and two Republican presidential candidates – were endorsing a carbon tax, and more companies were including the tax as a line item in projecting future costs (Davenport, 2013). Prices assigned to carbon for corporate planning purposes ranged from $40 to $60/t-CO$_2$ for oil companies such as BP, Shell, and ExxonMobil to about $10 for companies like Google and Microsoft (Economist, 2013c). A World Bank declaration calling on all nations to enact laws placing a price on carbon emissions was signed by more than 1,000 businesses, as well as 74 countries – including China – and 7 U.S. states (Davenport, 2014c). Citing concerns for delays in climate change policies that would increase investment risks, 350 of the world's leading institutional investors with $24 trillion under management also called for carbon pricing that would enable them to channel their investments better in the face of climate change (Ceres, 2014).

Some words of caution, however! Today it is common for corporations to represent themselves as *environmentally virtuous*, whether or not their commitment to environmental protection is genuine. *Greenwashing*, a pejorative term, refers to measures taken by corporations to create an illusion of environmental citizenship, while acting in ways that are detrimental to the environment. It has been defined as *the dissemination of misleading information by an organization to conceal its abuse of the environment in order to present a positive public image.* In assessing corporate, institutional, and government behavior, it pays to probe deeply and to determine the extent to which perception is aligned with reality. Efforts to do so are gaining traction, and metrics have been developed for quantifying a company's policies and its impact on the environment (McGinn, 2009). The Union of Concerned Scientists examined statements made and actions taken on climate change by twenty-eight U.S. companies (UCSUSA, 2012). While several companies were either uniformly obstructionist or constructive on climate issues, most chose to have it both ways, "making inconsistent statements across different venues." A company might express concern for climate change on its website and in its annual report, while misrepresenting the science in the same and/or other venues and working to weaken climate policy by supporting organizations that lobby against global warming solutions.

Some more words of caution, this time for environmentalists who are inclined to misunderstand, if not discount, the views and needs of the business community. To do so is both a tactical and a strategic mistake. Like it or not, business is an important stakeholder, one with responsibilities to its shareholders and employees, as well as to the greater community. Those businesses approaching the problem of warming and climate in an honest and informed manner should be viewed as important collaborators in seeking solutions.

It is also important to remain respectful of the operation of free markets. Quoting from Hawken (2005, p. 7), "No system has so revolutionized ordinary expectations of human life – (and) enlarged the range of human choice – as democratic capitalism." However, while advocating on behalf of free market principles, he cautions against adopting a fundamentalist view of unrestrained capitalism. As an article of faith, "free-market purists believe that their system works so perfectly that even without an overarching vision, the market will attain the best social and environmental outcome." Rooted more in belief than in logic, it is an *ism* that rejects any evidence to the contrary, however compelling. In its worst manifestations, it is fixated on growth – removing all barriers, seeking the lowest cost of doing business, and leaving little room for social values, including environmental stewardship. Marshalling whatever resources are needed to sustain growth and increase "shareholder value," any intervention, including government regulation, is resisted, nonrenewable resources are commoditized, and environmental costs are externalized, endangering prospects for future generations.

8.9 Summary

Anthropogenic climate change is not a problem bounded by geography. It is neither local nor national; it is global. All humans contribute to GHG emissions, some more than others, and all are affected by climate change. If we're all in this together, it makes sense to address the problem collectively, working through our political institutions. That view was adopted in 1988 when the United Nations established the Intergovernmental Panel on Climate Change (IPCC) and convened its member countries to formulate the Framework Convention on Climate Change (FCCC).

The IPCC was charged with assessing and periodically updating the science of global warming and climate change, along with mechanisms for mitigating and adapting to change. The FCCC was formulated as a nonbinding international agreement to stabilize GHG concentrations at levels that would preclude disruption of the climate system, with signatories, termed the Conference of Parties (COP), left to work out the details. On the surface, it was a good start, but the devil would be in the details. The centerpiece achievement was the Kyoto Protocol (KP), which called on industrialized nations to reduce their GHG emissions to 95% of 1990 levels by 2012, while exempting developing nations from reductions. Although the KP was ratified in 2004, the United States – then the world's largest emitter – was not a signatory. And although a signatory, China – soon to surpass the United States as the largest emitter – was exempted from its provisions.

There have certainly been lessons learned from the KP, but if it were to be graded on the extent to which it achieved its goals, it would receive no better than a C. It lacked a plan for the future role of developing nations or for differentiating between various gradations of the term – for example, between China and Bangladesh. Because it also relied heavily on a cap-and-trade system that was far too generous with allowances and offsets, it was unable to price emissions at a level that had a material impact. Global emissions continued to increase, driven largely by robust economic growth in developing nations.

To date, subsequent efforts made by the COP to move from the KP to a workable and effective international agreement have been impeded by priorities given to national economic interests and a persistent divide between the views of developed/richer nations on one hand and developing/poorer nations on the other. Efforts have essentially been stalled since the Copenhagen meeting of 2009. If a meaningful successor is to be crowned, agreement will have to be reached among the world's largest GHG emitters – China, the United States, India, Russia, Japan, and the EU, which collectively account for 70 percent of the global total – and with China and the United States playing leading roles. Commitments made by the EU, United States, and China in 2014 provide some optimism that a substantive international agreement will be reached at the 2015 FCCC meeting in Paris. However, there are also reasons to be skeptical.

Skepticism stems from the existence of disparate interests, needs, and governance across the world's nations and absence of a universal mindset to the politics of climate change. In the run-up to Paris, the "idea is for each country to cut emissions at a level that it can realistically achieve, but in keeping with domestic political and economic constraints" (Davenport, 2014d). Will the EU, United States, and China be able to deliver on their commitments? To what extent will their commitments induce other nations to follow suit? Will other developed nations that depend on fossil fuels for export and/or domestic consumption consider meaningful reductions? Will developing nations that rely on fossil fuels commit to realizing peak emissions in an appropriate time frame? India, the world's third-largest GHG emitter, has indicated it will continue to use its large coal reserves to sustain economic growth. And whatever agreement is reached in 2015, it would not be enacted until 2020. Even if commitments are made and honored by all nations, will they simply be too little and too late?

Jamieson (2014, p. 201) provides a sense of the frustration that exists with the state of global cooperation on climate change:

Rather than a global deal rooted in a conception of global justice, climate policy for the foreseeable future will largely reflect the motley collection of policies and practices adopted by particular countries. There will be climate-relevant action, but it will be different in different countries and it will be pursued under different descriptions and with different objectives.... Some countries will do a lot and others will do little. In some countries there will be a great deal of subnational variation, while other countries will nationalize and even to some extent internationalize their policies. These policies, in different proportions depending on the country, will reflect a mix of self-interest and ethical ideals constructed in different ways in different countries.

He is hardly optimistic about prospects for bending the emissions trajectory downward.

To this point we have focused on competing economic and environmental interests as the principal obstacle to reducing GHG emissions. But there are other impediments, and in the next two chapters, we'll consider the influence of ideology, human behavior, and ethics.

CHAPTER 9

Dissenting opinions: the great hoax

Whether as questions or comments following a presentation on climate change or in the course of casual conversation, I am often confronted by strongly held views on the subject. The tone is typically collegial and open to dialog, but it can get tense, if not hostile. In the United States, the issue is ideologically polarized, and it doesn't take much for discussions to become emotional. I've had enough of these interactions to prompt me to wonder, given the knowledge base on warming and climate change: What differentiates those who summarily dismiss the issue from those who believe it merits serious attention?

9.1 The political-corporate axis

Early in the history of concerns for global warming, influential elements of the U.S. business community viewed the issue as a threat and were determined to suppress it by whatever means, including the use of political leverage. Establishment of the IPCC in 1988 was followed almost immediately by formation of the Global Climate Coalition (GCC), an industrial consortium of major GHG emitters seeking to question the science of climate change by supporting organizations such as the Competitive Enterprise Institute (CEI) and the American Petroleum Institute (API) to act as attack dogs. Numerous lobbying and public relations efforts were launched to cast doubt on evidence supporting global warming, even when engineers and scientists within the participating companies were affirming the contribution of GHG emissions to warming. Although the GCC disbanded in 2002, vestiges continued to challenge the authenticity of climate change. Casting doubt was the operative strategy, however specious the arguments, and became a tour de force for challenging the science of climate change in the U.S. Congress.

As described in Section 8.6, the strategy has been successfully used to kill legislative measures to curb GHG emissions. In the Senate, opposition has been led by James Inhofe of Oklahoma, an influential member of the Environment and Public Works Committee who views global warming as a hoax, perpetuated by alarmists (Carey, 2006a). In his pre-2007 role as chair of the committee, Senator Inhofe did not hesitate to go after those who thought differently, demanding tax and membership records from state and local air pollution and control groups, deriding the Reverend Richard Cizik – a dedicated Reaganite and vice president of governmental affairs for the National Association of Evangelicals, but an advocate for reduced GHG emissions – as "a liberal wolf in sheep's clothing," and referring to Dr. James Hansen, a prominent NASA climatologist who had researched the issue since the mid-1980s, as an "alarmist."

Because Senator Inhofe represents a state with large oil and gas interests, his opposition to anything that would curb the use of fossil fuels is understandable. But there's more. It would not be unfair or incorrect to classify Senator Inhofe as ultraconservative and to say that his views typify those on the far right of the political spectrum. In 2010, those views, whether actually shared or not, were professed by all Senate Republicans, even those who a few years earlier had viewed global warming as a serious issue. In 2011, all but one of a large field of Republican presidential candidates dismissed global warming. What could explain this unanimity on such a complex issue? Could moderate and conservative Republicans be cowed by the far right so much that they dared not admit to some plausibility behind the science of climate change?

In the U.S. House of Representatives, Senator Inhofe's use of the bully pulpit was matched by Representative Joe Barton's demands for detailed documentation of global warming studies conducted by three scientists on the hockey stick phenomenon (Section 4.1), as well as for a list of all grants made in the area of climate science by the National Science Foundation over a ten-year period (Monastersky, 2005; Regalado, 2005). As then chair of the House Committee on Energy and Commerce, Mr. Barton certainly had a legitimate stake in the global warming debate, but ironically, he had yet to hold a hearing on the subject and strongly resisted any efforts to launch related legislation from his committee. Mr. Barton's actions may have been motivated by his past employment in the oil and gas industry and by large contributions made by the industry to his campaigns. But like Senator Inhofe, his views stem from a political orientation that is sharply tilted to the right.

For its part, the Bush White House was not above attempts to dilute conclusions of the scientific community, within which evidence was growing to support the premise of global warming due to GHG emissions. For example, Philip A. Cooney, a former lobbyist for the API on climate issues and one of many government officials with close ties to fossil fuel industries, joined the Bush administration in 2001 as Chief of Staff for the White House Council on Environmental Quality. With the API's long history of opposition to any restrictions on GHG emissions, one might wonder about a conflict of interest. True to form, Cooney, with no scientific background, took it upon himself to remove or alter conclusions in the text of climate research reports published by government scientists (Revkin, 2005). The administration also discouraged government scientists from publicly sharing their findings and used political appointees to ensure predetermined policy outcomes (NYT, 2004, 2006b). Interference by the Bush administration with the science of climate change is documented in a report by the Committee on Oversight and Government Reform of the House of Representatives (House of Representatives, 2007).

Two science historians (Oreskes and Conway, 2010) paint a graphic picture of how a handful of scientists and conservative think tanks and politicians, abetted by pseudo-institutes and segments of the media, manipulate evidence to promote special interests. Whether the issue involves tobacco, DDT, acid rain, CFCs and stratospheric ozone, or climate change, the tactics are the same: *exaggerate uncertainties and discount harmful effects on humans and the environment*. And, regardless of the issue, the same cast of characters kept appearing, from scientists such as Fred Seitz and Fred Singer, to news outlets such as the *Wall Street Journal* and the *Washington Times*, to organizations such as the CEI. The playbook is the same: *cast doubt on the evidence, however compelling, and dismiss it as "junk science," while providing disproportionate attention to the doubters.*[1]

Oreskes and Conway use the saga of linking tobacco to human health as a case in point. Months after publication of a 1953 report connecting cancer in mice to cigarette tars, a Tobacco Industry Research Committee was formed by tobacco companies to cast doubt on the veracity of the report. In 1979, seeking to enhance its credibility, the industry employed Mr. Seitz as a consultant to add layers of doubt on the growing number of studies linking cigarettes to cancer. Although Mr. Seitz was no lightweight, having previously served as president of the U.S. National Academy of Sciences, he was a physicist by training and not particularly well equipped to judge the work of life scientists such as biochemists and pulmonary physiologists. Nevertheless, aided by Mr. Singer, another physicist, he was effective in casting doubt on any finding that linked smoking to human health. Relentless in its efforts, in 1993, the industry published a primer, *Bad Science: A Resource Book* (LTDL, 1993), which could be used by others to reinforce its case. The underlying messages were laden with suspicions that the science was corrupted by a political agenda and that any laws formulated to curb smoking would be an infringement of individual liberties that could presage other governmental controls on personal freedom. The strategy succeeded, and it would be almost fifty years since publication of the first studies that the industry would be brought to heal.

Now transition from 1953 to 1995, when the first IPCC report was released. Include politicians from states heavily invested in the production of fossil fuels; replace the tobacco industry with the oil, gas, and coal industries; and add talk radio, cable news, and the blogosphere to media addressing climate change. Then follow the same tactics: *cast doubt on the science (and the scientists) and delay, if not squash, any effort to implement remedial action.*

Whatever the specific health or environmental hazard, Oreskes and Conway highlight the fact that it was a small group of scientists of deeply conservative beliefs and closely linked to corporate interests who carried the banner of doubt and denial. A common set of values underpins their resistance: a belief in unfettered markets and disdain for government intervention; indifference to environmental degradation and the health and welfare of those impacted by degradation; and a preference for short-term benefits to special interests over long-term benefits to society at large. Similar views have been expressed by Pooley (2010a) and Klein (2014).

Pooley examines the futile attempts of Congress and the White House to deal with the issue, as well as roles played by the fossil fuel industry, lobbyists, organizations such as the CEI and Heartland Institute, and segments of the media. For this group the playbook is the same: spend millions on advertising and public relations to promote the use of fossil fuels while casting doubt on the veracity of anthropogenic climate change. Klein attributes the inability to curb GHG emissions to ideologies that embrace unregulated markets and render environmental interests subservient to economic interests. Books authored by Hansen (2009) and Schneider (2009) also examine efforts to distort the record on behalf of short-term political and economic interests.[2]

9.2 Science or pseudoscience – credible or conspiratorial?

Manipulation of the science of climate change has become a favorite tactic of those seeking to dismiss the issue. A case in point is the book entitled *State of Fear* by Michael Crichton (2004), a widely read novelist who for years has engrossed readers with fictional narratives at the intersection of science and suspense. In his book Crichton casts a group of environmentalists and scientists as fabricators of global warming threats and plotting to create deadly weather episodes that validate the threats. The plot proceeds with a scenario of scientific absurdities such as attribution of a deadly tsunami to global warming. Nevertheless, the book was treated seriously by those debunking global warming, for whom Crichton became a willing resource.

But, what about those who reject global warming? Don't they provide an element of credibility? In examining how science is used to discredit concerns for global warming, Kaufman (2009) focused on a former spokesperson (Mr. Marc Morano) for the Senate's Environment and Public Works Committee chaired by Mr. Inhofe. Mr. Morano's views are promulgated on his Web site, climatedepot.com, which is funded by the nonprofit Committee for a Constructive Tomorrow, an advocate for free markets that receives much of its support from individuals and foundations associated with conservative causes. Kaufman concluded that such dissenters are proficient at integrating "diffuse pieces of scientific research … into a political battering ram," misrepresenting "the work of legitimate scientists," and compiling lists of alleged scientists who dispute the science behind global warming. The alleged scientists include weather forecasters, who may or may not be meteorologists, and others with little knowledge of climate science. One example is a TV weatherman who has no degrees in meteorology and, when pressed, admitted that he doubts a human connection to climate change because "it completely takes God out of the picture." It would be difficult to discount an analogy between efforts to dismiss global warming by spinning the science of climate change on the one hand and using creationism or intelligent design to dismiss the theory of evolution on the other.[3]

In 1998, a Petition Project denying human responsibility for global warming was launched by Fred Seitz and subsequently reissued by the Oregon Institute of Science. The petition states, "There is no convincing evidence that human release of carbon dioxide, methane, or other greenhouse gases is causing or will, in the foreseeable future, cause catastrophic heating of the Earth's atmosphere and disruption of the Earth's climate." By 2007, the project had accumulated more than 31,000 signatures, many with links to the scientific community (Unruh, 2008). Quoting the founder of the Petition Project, Unruh states, without elaboration, that the Kyoto Protocol "would harm the environment, hinder the advance of

science and technology, and damage the health and welfare of mankind."
He further states:

> The inalienable rights to life, liberty, and the pursuit of happiness include
> the right of access to life-giving and life-enhancing technology. This is
> especially true of access to the most basic of all technologies: energy.
> These human rights have been extensively and wrongly abridged. During
> the past two generations in the U.S., a system of high taxation, extensive
> regulation and ubiquitous litigation has arisen that prevents the accumu-
> lation of sufficient capital and the exercise of sufficient freedom to build
> and preserve needed modern technology.

Translation: Americans should be unencumbered in their efforts to exploit
fossil fuels.

It is hard to treat the foregoing views as anything other than an ideological rant. Reducing carbon emissions and advancing carbon-free energy
technologies will harm the environment, hinder innovation, and degrade
human health and welfare? The issue is addressed by waving the flag and
invoking the Declaration of Independence, as well as a variant of American
manifest destiny that entitles us to exploit fossil fuels to the fullest. But,
there's more. Unruh states that "the issue [climate change] has nothing
to do with energy itself, but everything to do with power, control and
money, which the United Nations is seeking" and that "the U.N. [is] violat-
ing human rights in its campaign to ban energy research, exploration and
development." And there you have it, that old saw of the far right – the UN
as bogeyman.

I've interacted with scientists and engineers, including PhDs, who deny
the existence of anthropogenic climate change, and I'll use one anecdote
to frame conclusions drawn from these interactions. It was at a dinner
party, and the subject was raised by a colleague who is a distinguished
engineering professor and well respected in a specialty removed from the
climate sciences. After patiently listening to him unequivocally and emo-
tionally dismiss the issue, I asked some questions about the underlying
science, which he was unable to answer. I then asked whether he had
examined the issue in depth, and he honestly replied that he had not.
How then had he come to feel so strongly about the matter? Well, politically he swings hard to the right and without discernment is willing to be a
torch bearer for the party line. Regrettably, at least for me, this experience
is not atypical. Those whose views range from doubt to derisive dismissal
typically have done little to consider the issue in depth and are strongly
influenced by political ideology.

The perception that climate denial had become firmly entrenched in U.S. right-wing ideology was underscored during the run-up to the 2010 midterm congressional elections. All but one of the twenty Republican candidates with a serious chance of winning a Senate seat rejected the scientific consensus for anthropogenic climate change, carrying the torch for an array of conservative radio and television talk show hosts, organizations funded by the fossil fuel industry, and a burgeoning political movement called the Tea Party (Broder, 2010). Yes, some might say, but what about the views of a world-renowned scientist such as Freeman Dyson? Because this question was often put to me a few years ago, I'll provide more than a cursory response.

Dyson is a physicist at Princeton's Institute of Advanced Study. It would be an understatement to call him brilliant, and he is certainly not a dyed-in-the-wool conservative. But he is somewhat of a curmudgeon, who has been consistently skeptical of big science, such as the Star Wars missile defense program and the superconducting, supercollider (Dawidoff, 2009). His criticism of global warming proponents is rooted in their use of climate models, which he believes to be laden with too many uncertainties and to lack a sufficient observational (experimental) data base to predict the future reliably. The criticism is fair and reasonable. Despite efforts to unravel complexities of the carbon cycle and to refine climate models, uncertainties remain in their ability to predict paleo-climatic conditions and future climate trends. But he overreaches with other comments.

Dyson points to potential benefits of increasing atmospheric CO_2 concentration, such as increasing plant growth and mitigating effects of the next ice age. Yes, plant growth is enhanced by increasing CO_2, but only until other growth factors such as temperature, water, and nutrients become limiting. And, of course, there is no growth if productive soils are permanently flooded or the salinity of irrigation water becomes too high due to encroaching seas. As for circumventing the next ice age, we know that the effects of warming occur on time scales much smaller than those associated with glaciations. Dyson also views problems of ocean acidification as overblown, despite their potential to destroy important food chains. When asked by his wife, "how far do you allow the oceans to rise before you say, this is no good," his response was "when I see clear evidence." Of course, by then it would be too late to remedy the cause through mitigation, and perhaps even the effects through adaptation. If CO_2 concentrations become too high, he suggests they could be reduced "by the mass cultivation of specially bred carbon-eating trees."

Dyson admits he hasn't studied the subject in depth and doesn't "claim to be an expert on climate change." Yet, dissenters were quick to exploit his comments to advance their case. Without having done his homework, his skepticism was inappropriate, and it's regrettable that he chose to play it fast and loose with an issue of serious consequence. But can the same criticism be directed at scientists developing the knowledge base underpinning anthropogenic contributions to climate change?

The credibility of the climate science community was shaken in late 2009, when the website of the Climate Research Unit (CRU) of the University of East Anglia in the United Kingdom was breached and access gained to thousands of e-mail messages. A mission of the CRU is to gather, assess and use proxy data such as tree rings and ice cores to reconstruct the Earth's temperature history over thousands of years. Although the messages revealed the very real difficulties and frustrations of the reconstruction process, many were laced with arrogance, pettiness, dismissal of contrary views, and statements that could be interpreted as attempts to manipulate and conceal data. Collectively, they cast suspicion on the validity of the research used to support the case for climate change, as well as the transparency with which it is shared with the public.

Material related to the work of Dr. Michael Mann in reconstructing data that yielded the hockey stick (Section 4.1) received special attention. Proxy data can be limited and incomplete, and using it to reconstruct past temperatures is difficult and uncertain. Mann used the thickness of pine tree rings, which increases with increasing summer temperatures, to infer temperatures over the last millennium. But because the data differed from direct temperature measurements over the last decades of the twentieth century and, in fact, indicated a decline in temperature, they were deemed unreliable and dismissed for this period. In so doing, Dr. Mann used a mathematical procedure to merge the instrument record with past proxy data and concluded that approximately 1,000 years of nearly constant global surface temperatures was followed by a sharp twentieth-century spike. In some of the e-mail messages, the procedure was referred to as a *trick*, creating the perception that Mann had manipulated the data to strengthen the case for anthropogenic warming. However, the term is commonly used in the scientific community to describe clever (and valid) approaches to resolving problems. Although it is appropriate to question Dr. Mann's choice of statistical analysis methods, he was subsequently absolved of any research misconduct in investigations conducted by his home institution – the Pennsylvania State University – and by the National Science Foundation.[4]

An investigation of the CRU e-mails conducted under the auspices of the Science and Technology Committee of the British House of Commons concluded that appropriate methods were used to compile the proxy data and there was no intent to misrepresent the results. The panel conducting the investigation confirmed CRU's results by examining original data sources. Further confirmation is provided by Figures 4.5 and 4.6, where the CRU results agree well with those of other studies. The panel also reviewed the entire exchange of e-mails and found no evidence of wrongdoing. It did, however, advise the CRU to more fully engage statisticians to ensure use of the latest and most effective procedures for interpreting the data. A subsequent investigation conducted by a separate and independent British panel also exonerated members of the CRU, citing no evidence of dishonesty or inappropriate research methods (Russell et al., 2010). Its report did, however, chastise the CRU for a lack of openness in ignoring outside requests for detailed data and data reduction procedures and for not elaborating on the "trick" used to resolve the late twentieth-century discrepancies in proxy and instrument records. But, in the final analysis, multiple investigations affirmed the integrity of the CRU and affiliated scientists, despite efforts of the media to suggest otherwise.

While CRU staff may not be guilty of scientific misconduct, they can be fairly criticized for unprofessional behavior and for circling the wagons in the face of relentless attacks by global warming skeptics. A sine qua non of publicly funded research is that results and procedures must be shared with all interested parties. The most egregious behavior of the CRU was its unwillingness to share raw data and data-processing procedures with skeptics and critics. In this sense, by underscoring the importance of transparency, the e-mail revelation could be viewed as a positive outcome.

By the end of 2009, a second shoe had dropped, this time casting doubt on the credibility of the IPCC itself and specifically contents of its 2007 report. In the report it was suggested that the Himalayan glaciers could completely melt by 2035, when in fact it could take centuries, and that large economic losses associated with extreme weather events were directly linked to climate change. Both suggestions reflected sloppy scholarship and inadequate internal review processes. However, lest we be inclined to completely dismiss the IPCC and its reports, it's important to remember that they involve three Working Groups dealing with: (I) the scientific basis of warming and climate change; (II) the impact of climate change and adaptation measures; and (III) measures that could be taken to mitigate warming and climate change. The work of Group II is subject to the largest uncertainties, making it important for its members to avoid speculation and exercise caution in their recommendations. But whatever mistakes were made in the reports, they pale by comparison to the overwhelming body of information pointing to the reality of human-induced warming and climate change. And if the IPCC is to be faulted, it would be more for understating than for overstating the effects of warming.

In the United States, the attacks on the science of global warming, and seemingly on science itself, were unrelenting and in some cases took the form of conspiracy theories. As in Crichton's book, scientists around the world were thought to be fabricating the case for warming and climate change and accused of only wanting to maintain a steady flow of research dollars. All the while, scientific evidence pointing to anthropogenic warming was growing, and science organizations around the world, including the U.S. National Academies of Science and Engineering, were affirming the reality of global warming.

9.3 The human element

So just what drives rejection of anthropogenic global warming and climate change? Is it rooted in science, economics, and/or political ideology? Or are there other factors? And, whatever the factors, what is the range of human response to the issue?

Brand (2009) used four categories to classify possible responses. At one extreme are the denialists, who are vocal, political, and adamant that it's all a hoax, if not a conspiracy, to bring down America. They dismiss any and all results that say otherwise. At the other extreme are the calamatists, who play it fast and loose with data, dramatize the consequences of warming, and trivialize the difficulties of reducing GHG emissions. In the middle are the skeptics and the warners. Skeptics examine climate data and methodologies, looking for inherent weaknesses and inappropriate conclusions; warners examine the impact of increasing GHG concentrations and the possible effects on economic, environmental, and social systems.

Another taxonomy of perspectives falls under the rubric of what Leiserowitz (2010) and colleagues call Global Warming's Six Americas. Moving from one extreme to another, they include those who are alarmed, concerned, or simply cautious about the implications of climate change to those who are disengaged, doubtful, or dismissive. One conclusion of their study is that people are inclined to mistakenly jump on recent weather events to justify preconceived views about climate change. A winter storm that immobilized much of the Northeast in February 2010 (an El Niño year) provided a colorful example, with one dismissive U.S. senator using the event to erect an igloo in Washington, DC and post a sign encouraging motorists to "honk if you love global warming" (Zeller, 2010). Conditions were reversed in 2011 (a La Niña year), when much of the U.S. experienced extreme weather consistent with predictions of the climate models.

From the late 1980s to the present, debate on global warming in the United States has been highly polarized and dominated by extremes. An apt analogy in the world of fluid mechanics would be of two intense, counter-rotating vortices into which adjoining fluid cannot resist being drawn. So too for global warming, where even clear and independent thinkers can be swept into one camp or the other. In view of the risks and potential consequences of increasing GHG concentrations on the one hand and the costs of mitigating emissions on the other, as recently as a few years ago the preferred approach would seem to have been one of adopting the intermediate skeptic/warner or concerned/cautious mindsets delineated by Brand and Leiserowitz. Such approaches are underscored by Pielke (2007), who calls for honest brokers who can bridge the gap between competing issues to find sensible solutions. Pielke's point is well taken, but with evolving knowledge and events, we have passed the time for which skepticism and caution are appropriate responses.

Hoffman (2012) views climate change from the lens of cultural warfare. If there is a scientific consensus, he asks, why not a social consensus? The answer: because opinions are strongly influenced by ideology and increasingly by political affiliation. He cites a survey by McCright and Dunlap (2011), which found that from 2001 through 2010, the percentage of Republicans and conservatives believing that global warming effects had already begun decreased from about 50% to 30%, while increasing from roughly 67% to 74% for Democrats and liberals. If acceptance of the scientific consensus is increasingly seen as a liberal view, what's a good Republican to do? Regrettably, climate change has joined other issues, such as gun control, immigration, and health care, for which constructive dialog is impeded by deep partisan divisions. Hoffman recognizes the powerful economic and political interests blocking action to mitigate global warming, but drawing an analogy to the eventual weakening of similar interests on behalf of tobacco, he is optimistic about an eventual merger of the scientific and social consensus on climate change. I'm inclined to be less sanguine about the prospect.[5]

With a mismatch between the scientific and public consensus on climate change, it's tempting to think, *if only the public grasped the science, it would be concerned about climate change.* As many of my students have suggested, all that's needed is an effective public education campaign.

However, the facts suggest otherwise. In a study of public perceptions of the risks associated with climate change, Kahan et al. (2012) found that those with the highest level of scientific and technical literacy had the least concerns. Instead, their views were more strongly influenced by cultural values, be they political, economic, or religious. The operative term is *cultural cognition*. We've considered how these values are manifested politically and economically; in the next chapter we'll examine the impact of religious beliefs. As a foreword, consider one tenet of Christianity, namely that nature is God's gift to humankind. There are two interpretations of how to use the gift. One is to be good stewards by protecting nature and using its bounty in a sustainable fashion. Another is to consider God's gift as inviolate; nature exists to serve man, and man is incapable of degrading what God has created.

There is yet another factor that influences how people feel about global warming, one that's linked to a universal trait of human behavior. Gertner (2009) addressed the question of how decisions are made on environmental issues characterized by uncertainty. With respect to climate change, how does one deal with trade-offs? Should sacrifices be made in the short term to reduce the risk of adverse effects in the long term? Gertner identified two mechanisms for responding to the trade-offs, one analytical and the other emotional. The analytical response assesses costs and benefits associated with mitigating risk. On the surface, it would seem to be the proper response. However, driven by a "dislike for delayed benefits," Gertner speaks of a tendency to discount long-term consequences, thereby undermining the willingness to make changes that would reduce the effects of climate change.

The emotional response correlates with the extent to which fear and alarm are felt. If one has experienced the effects of persistent drought or the increased frequency of destructive storms, this response mechanism would prompt a call for decisive action. But because such experiences are still remote from the daily lives of most people, they move to the bottom of our hierarchy of fears, behind matters such as economic security and opportunity, personal safety, health care, political stability, or even the declining fortunes of our favorite sports team. In an annual poll of American policy priorities (Pew, 2013b), large majorities put the economy and job creation atop the list, followed by concerns for the federal budget deficit, terrorism, the Social Security system, and education. At the bottom of the list? Global warming! And what issue had the widest partisan gap between Democrats and Republicans? Protecting the environment! Opinions were likely influenced by lingering uncertainties in the scale, timing, and consequences of global warming. In this century, will temperatures rise by 2°C or 6°C from preindustrial levels? What exactly will be the effect of exceeding 2°C? Will the consequences be significant, and would the costs of avoidance have been less than those of adaptation?

Underlying complexities and uncertainties make it easier to dismiss climate change, particularly when self-interests make it beneficial to do so. If mitigation measures that substitute carbon-free sources of energy for fossil fuels – regulatory or otherwise – adversely affect my interests – pecuniary or ideological – I will selectively accept results that cast doubt on the seriousness of the problem. When I can no longer cling to these results, I will advocate adaptation while still resisting mitigation. And when adaptation is no longer practical, I will advocate geoengineering while continuing to eschew mitigation.

If we view the consequences of climate change to be distant and the risks long term, why change our behavior? As Americans, we are fond of supersizing our homes and vehicles, thereby increasing demands on the energy needed for heating, cooling, and transportation. We have also shown a preference for suburban living, with little interest in mass transit, adding to the use of energy for transportation. These tendencies have been enabled by ample supplies of low-cost fossil fuels. But even if supplies were to dwindle and costs were to rise, prompting greater interest in energy efficiency, responses would be restrained by inherent asset inertia. Replacement of oversized vehicles and homes with smaller, more efficient options and the development of modern mass transit systems would occur over time frames ranging from a few to many decades and would require large capital investments.

Why abstain from short-term gratification? Humans are inclined to respond best to a problem when it reaches crisis proportions. Then the threat is neither imminent nor uncertain; it's here and now. Of course, by then, any response to global warming could fall in the too-little-too-late category.

The human response to global warming is akin to the environmental time scales and inertia discussed in Appendix D. But instead of time delays due to the physical processes governing climate change, the delays are linked to human behavior and the response of socioeconomic systems. Not unlike society's predisposition to high levels of personal and government debt, human tendencies are likely to favor living for today and letting tomorrow take care of itself. Decades can pass from the time a threat is perceived to the time that solutions are implemented. For global warming the propensity to defer action is bolstered by environmental time scales and uncertainties in our ability to predict the future, both of which are powerful agents for complacency.

9.4 An analogy

As an engineer, I am drawn to analogies that exist between an appropriate response to climate change on the one hand and, on the other, the failure of complex engineered systems such as a nuclear power plant, a large electric power grid, an air traffic control system, or the space shuttle. Those responsible for the design and operation of such systems are driven by the need for high reliability and safety, including sensitivity to unexpected events and precursors to failure. Every failure is preceded by signals, however weak, before it becomes a failure, and engineers learn, sometimes painfully, that *weak signals should not be accompanied by weak responses*. The significance of this admonishment is underscored by well-publicized failures, such as those of the Challenger and Columbia space shuttles. To have prevented such failures, early warning signs should have been recognized and should not have been rationalized on the basis of prior success. *By responding strongly to signs of potential failure, failure can be averted.*

Returning to the global climate system, how should existing signals be interpreted? Do they portend serious consequences, if not to us then to future generations? Or, as humans conditioned to dealing with problems after they reach crisis proportions, should we continue with business-as-usual? There are simply too many indicators that anthropogenic warming and climate change are real and that consequences are too severe to risk ignoring the problem. The signals are no longer weak, and they are getting stronger.

In Section 2.2 it was noted that for the past twelve millennia humans have lived in the Holocene, a time of relative climate stability at the trailing edge of a one-million-year Quaternary period characterized by transitions into and out of ice ages. The Holocene has been very good to humankind, but at the dawn of the new millennium, some would say that we've transitioned to a new era termed the Anthropocene (Economist, 2011b; Science, 2011). Yes, we've arrived at last! By virtue of our growing population and activities, we can now affect our planet. We can alter the carbon, nitrogen, and water cycles; we can deplete ozone over Antarctica; we can contribute to species extinction; and we can erode terrestrial and oceanic biospheres. We are doing all of this, and with continued population and economic growth we can do more. Our environment has given very generously, but we have reached a threshold where its gifts can no longer be taken for granted.

9.5 Summary

For more than three decades, a confluence of powerful forces has success-fully thwarted efforts to restrain GHG emissions in the United States. The game plan parallels that used to delay action on other agents found to be harmful to human health and/or the natural environment: cast doubt on the science and give disproportionate attention to the doubters.

What are these powerful forces? Begin with a multi-trillion-dollar fossil fuel industry opposed to erosion of its markets. Enormous capital invest-ments have been made in power plants fueled by fossil fuels and in facili-ties for recovering, transporting, and processing the fuels. These facilities will not be decommissioned until realization of appropriate returns on investment. Add well-funded institutes dedicated to free-market principles and opposed to interference in the form of environmental regulations. Include the influence on public opinion of right-leaning newspapers, talk radio, cable news, and the blogosphere. And employ an army of well-funded and well-connected lobbyists to influence the political process. Aided by the existence of sharp political divisions and growing resistance to government infringement on individual liberties, these forces have done much to shape public opinion and impede miti-gation measures.

But there are also behavioral obstacles to addressing climate change. A preference for immediate over delayed benefits is not uniquely American; it is a human trait. To many, climate change is an abstraction whose con-sequences are neither evident nor imminent. And, to varying degrees, all of us can fall victim to confirmation bias, an inclination to treat informa-tion selectively – more inclined to accept what supports our *isms* and less inclined to accept information that challenges them. In an ever more complex world with exponential growth in access to information, a natu-ral response is to filter the information, accepting only what confirms our beliefs and to look at things as we'd like them to be.

Yet, the scientific consensus on the dangers posed by global warming and climate change is strong and well grounded. Within the scientific community there is overwhelming consensus on the anthropogenic origins of climate change. Yes, there's much we can still learn, but what we know to date points to existence of a serious problem that must be addressed. This view is shared by scientists across the world. The issue is controversial, but politically, not scientifically. The way forward must be guided by pragma-tism, not ideology or emotion. The longer we wait, the worse it will be for those who follow.

CHAPTER 10

The ethics of climate change

By now, you would probably agree that the issue of climate change is wrapped in science, technology, economics, and politics. But is there yet another dimension? Many would say "yes," maintaining the existence of a *moral imperative*. To address climate change from such a perspective, we'll examine some of the philosophical pillars of ethics, as well as foundations of ethical behavior derived from religious traditions. From both philosophical and theological perspectives, how do ethical considerations inform the debate on climate change?

Ethics involves reflection on human behavior and how to channel it in appropriate ways. A central question involves life and how it should be lived. Another involves the nature of *good* and standards by which an action is judged good or not. Such questions have been addressed by philosophers for millennia in attempts to delineate between right and wrong. But in applying these standards, difficulties often arise because many issues are multifaceted, complex, and nuanced.

10.1 Ethical dimensions of climate change

Technology has allowed humans to conquer space and time. Modern transportation systems provide movement of goods and people from one location to any other; modern communication systems enable ideas and knowledge to flow almost instantaneously across the world. Globalization has had an enormous impact on raising living standards throughout the world. But underpinning it all has been rising energy consumption, particularly from fossil fuels, and attendant environmental degradation. Some environmental damage is local, such as mining coal by mountaintop removal, or regional, such as acid rain. And most degradation is manifested over relatively short time scales, from immediate to months or years. But climate change is global and manifested over decades to millennia. Today's GHG emissions have consequences for all, anywhere on Earth, and for the unborn as well as the living. When I burn one gallon of gasoline, I discharge almost 9 kg-CO_2 to the atmosphere, putting the entire planet at greater risk to the effects of climate change. When the U.S. transportation sector consumes 215 billion gallons of fuel, as it did in 2012 (Davis et al., 2014), it adds about 2 Gt-CO_2 to the atmosphere, with Americans enjoying the benefits of consumption while calling upon the world to share the burdens.

For GHG emissions there are spatial and temporal separations of cause and effect. Spatial separation relates to the fact that richer nations have derived the greatest benefits and are the least vulnerable, while poorer nations have derived the least benefits and are the most vulnerable. Globally, benefits and vulnerabilities are asymmetrical. Countries that historically contributed disproportionately to emissions by building their economies using abundant and low-cost fossil fuels are wealthier for having done so and by virtue of their wealth are better able to adapt to the effects of climate change. In contrast, poorer countries that have contributed far less to legacy emissions are least able to adapt to climate change. This spatial separation of cause and effect adds a social justice implication to the dimensions of climate change.

The relevance of climate change to equity, fairness, and social justice is underscored in the Fifth Assessment Report of the IPCC (2014c), which emphasized the threat it poses to the food security of the world's poorest nations and its potential for exacerbating poverty and inequality in both developing and developed nations. The report also points to the increased frequency and intensity of extreme drought and flooding and their disproportionate effects on the world's poor. Recognizing that "water is life, but too much or too little of it can become a threat to life," Shiva (2002) equates "climate injustice" to "water injustice" and provides many examples of how the world's poorest have suffered from extreme weather events.

Garvey (2008, p. 67) associates justice and equality with a shared distribution of "goods, resources, burdens (and) benefits" – concepts that are well established in systems of civil and common law – while allowing for unequal distributions if "there are good reasons to the contrary." Justice can be served if collective benefits are derived from the goods and services produced by people and nations disproportionately using a limited resource and if associated burdens are shared. But justice is not served if the fruits of disproportionate use are self-serving and the burdens are borne by others.

From the time of the Industrial Revolution to the end of the twentieth century, Western nations contributed disproportionately to the world's annual GHG emissions. And when integrated over the last two centuries, Western nations contributed even more disproportionately to the atmosphere's store of GHGs. The ethical issue is therefore the following. Should wealthy nations that contributed so much to the atmosphere's inventory of GHGs make the greatest efforts to reduce emissions and the largest contributions to assisting those most affected by and least able to adapt to climate change?

If richer nations have benefited from past GHG emissions, do they have a moral responsibility to reduce their emissions, while ceding a larger portion of future emissions to developing nations? To some extent redistribution is already occurring, albeit for other than ethical reasons. Although OECD nations have disproportionately contributed to and benefited from GHG emissions, developing nations are rapidly increasing their emissions, with China now the world's largest contributor. Who bears the largest responsibility for reducing emissions – China which has yet to break the bonds of poverty for hundreds of millions of people, or a nation like the United States, which has achieved acceptable living standards for a large majority of its citizens and has far larger emissions per capita? Were a cap to be placed on global emissions, how would ethical considerations affect redistribution among the world's nations? To what extent would some nations be expected to reduce and others to increase their emissions?

A temporal separation of cause and effect is a matter of intergenerational ethics, of weighing the needs of future generations and the natural environment against the needs and wants of current generations. I and other Americans of my generation have been beneficiaries of living standards derived from industrialization enabled by fossil fuels. And we will likely live our remaining days without having to endure any of the more onerous effects of climate change. Can the same be said for our children and grandchildren and for those who follow?

Because of the inherent inertia of systems affected by GHG emissions (Section 3.5, Appendix D), adverse effects of past and current emissions are not borne as much by the emitters as by future generations. The long residence time of atmospheric GHGs and the long lag time associated with their full effect on warming and rising sea levels mean that, although the benefits of today's emissions are realized immediately, adverse consequences will be progressively experienced by future generations. In the words of Gardiner (2011, p. 198), the consequences are "substantially deferred" and the problem is "seriously back-loaded." The image is one of current and past generations of OECD nations as profligate consumers of fossil fuels, ergo wanton emitters of GHGs, leaving future generations with depleted fuel stocks and the adverse effects of climate change. By passing the burden to future generations, Gardiner (2011, p. 36) refers to the situation as a "tyranny of the contemporary."[1]

Yet another ethical issue relates to the effect of climate change on the natural environment. Are actions that eliminate other species and entire ecosystems in the biosphere morally justified? Is it right to have an exclusively anthropocentric perspective, one concerned only with the effect of climate change on humankind? Or should we extend our thinking to include its effect on all species?

In summary, the central ethical challenges posed by climate change deal with its impact on (1) the poor, (2) future generations, and (iii) the natural environment. Can ethical theories and religious traditions provide guidance in dealing with the problems of climate change?

10.2 Ethical theories and principles

Ethical theories lie in the realm of moral philosophy. They are systems (frameworks) of thought that address moral rights, duties, and behavior, and three such frameworks will be considered in terms of their ability to guide moral judgments on climate change. One theory is based on obligations termed categorical imperatives. In the context of climate change, an imperative could simply be that it's wrong to inflict the dangers of global warming on the poor, the unborn, and other species of the biosphere. A second theory, termed utilitarianism, couches ethical behavior in terms of outcomes. Actions should bring about the greatest good for the greatest number, where good implies outcomes such as happiness, pleasure, and satisfaction. The third theory, termed virtue or Aristotelian ethics, delineates character traits that work for the benefit of society and hence the greater good.

Imperatives

Categorical imperatives are rules that distinguish right from wrong. Termed deontological theory, an early proponent was Immanuel Kant who believed that activities can only be considered moral if pursued in a spirit of goodwill and a sense of duty. A spirit of goodwill is manifested by a desire to contribute to the well-being of others; a sense of duty is guided by imperatives tantamount to universal laws of morality. Codes of ethics espoused by professions such as medicine, engineering, and law are applications of deontological theory. They articulate professional responsibilities (duties) and delineate rules for their fulfillment, among which are the admonitions to do no harm and to follow the golden rule.

Applying categorical imperatives to a specific action involves considering its effect on others. Would the world be a better place? Would collective interests be well served? Imperatives are useful guides to determining the morality of many actions. How do we judge violent acts and theft? Do we expect an engineer to emphasize public safety in designing a product or a doctor to emphasize patients' well-being in diagnosing and prescribing therapeutic responses to an ailment? In such cases, categorical imperatives serve us well in determining our responsibilities to others. But can they help us assess the morality of anthropogenic contributions to global warming and climate change? Is the moral imperative one of ensuring sustainability and preserving natural capital for future generations? Or is it one of maintaining consumption and allowing for degradation of natural capital at levels sufficient to meet immediate needs and wants?

Drawing from Kant, Gardiner (2011, pp. 307–9) interprets moral corruption as an attempt to undermine moral claims by casting doubt on their validity and making them "better suited to our wishes and inclinations." It is a rationalization process by which moral requirements are diminished by introducing competing claims. Gardiner notes that such claims allow one to ignore inconvenient moral demands while retaining some semblance of morality. There is no paucity of examples.

When Rachel Carson (1962) published her book condemning the environmental threats posed by DDT, those with a vested interest in continued use pointed to the thousands of deaths that would be caused by its removal. In response to regulation of criteria pollutants from coal-fired power plants, those interested in minimal regulation can point to the impact of higher electricity costs on the poor. The same can be said for the costs of curtailing GHG emissions, to which proponents of fossil fuels would add the ability to remove hundreds of millions of people from poverty through use of the fuels in developing nations. On this basis, those advocating on behalf of fossil fuels can claim that continued use does more good than harm and point to Kant and deontological theory for moral justification. Using categorical imperatives to claim the moral high ground for measures that curtail GHG emissions is a slippery slope.

Consequences

Teleological theory, variations of which are termed utilitarianism and consequentialism, focuses on the consequences of human action. As articulated by Jeremy Bentham and John Stuart Mill (Sweet, 2008), it seeks to maximize utility or, in Bentham's words, to achieve the "greatest amount of happiness for the greatest number of people." Happiness, or pleasure relative to pain, is associated with goodness, and moral behavior corresponds to acting in ways that foster the greatest good for the greatest number of people. Everyone matters, rich and poor, young and old, productive and unproductive. But how does one measure outcomes on a happiness meter, and how does one concurrently maximize both the level of good and the number of people impacted at this level? And, under the guise of utilitarianism, is a nefarious act moral if it produces a good outcome? Does the end always justify the means?

If categorical imperatives provide a convenient ethical framework for professions such as medicine and engineering, utilitarianism is a convenient framework for economists. In Section 7.2 we discussed the concept of a Pigovian tax, one used to restore a balance between interests of the individual and society. In Section 7.2 we also discussed the Stern Review, which used economic analysis to make the case for "strong, early action" to reduce global warming. In economic terms, the greatest good for the greatest number could be achieved by imposing taxes that discourage the use of fossil fuels and by investing in mitigation measures.

But there is a counterargument, also based on economic analysis. Efforts to curb GHG emissions may be too costly relative to future benefits, and the greatest good would be achieved by focusing on wealth accumulation through economic growth. By ignoring costly mitigation measures and emphasizing economic growth, many more people in poorer, developing nations would achieve improved living standards, while all nations would achieve levels of economic growth that would better enable them to adapt to climate change. How can two economic analyses lead to such different conclusions concerning the morally correct approach? The answer has to do with what one chooses to assume and/or ignore in the analysis.

Criticism of the Stern Review focused on choice of the discount rate used in the analysis. Use of a low discount rate leads to the conclusion that the economic costs of deferring action would well exceed those of prompt and aggressive action. In contrast, a large discount rate leads to the conclusion that greater benefits are achieved by choosing deferred adaptation over immediate mitigation. But where does that leave us if ethical judgments are based on the value of an economic parameter for which there is little agreement among economists? By ignoring mitigation, wouldn't we be externalizing the costs of climate change, in this case at the expense of future generations – in effect kicking the can down the road?

Gardiner (2011, p. 269) is correct when he states that "discounting buries the important issues about our relation to the future in a single 'all-purpose' rate, and so undermines good ethical analysis." Utilitarianism is essentially an exercise in cost-benefit analysis, and Gardiner makes a strong case against using it to predict the consequences of climate change, pointing to its inability to deal with many relevant uncertainties – economic and scientific. Can we reasonably estimate the effects of climate change? How does one assess costs associated with extreme weather events and drought? How does one monetize the effect of receding glaciers, the loss of fertile soil and ocean acidification on food production? What are the social and economic costs of mass migrations due to dislocated populations? How can one be sure that degradation of natural resources won't leave future generations less able to deal with the problems? And, in determining costs and benefits for future generations, how far out in time does one go?

An ethical framework commonly invoked by global environmental movements as a call to sustainability is the Seventh Generation Principle. How will today's policies and actions affect those who come seven generations later? Is 175 years good enough? The principle deals unambiguously with the intergenerational component of environmental ethics, giving high priority to the well-being of those who follow. As implied by the title of his book, Hansen (2009) embraces the principle and challenges the premise that, in economic terms, the impact of climate change on future generations can be deeply discounted.

Summarizing, how well can one measure the benefits accrued from using fossil fuels against the costs inflicted by climate change, including effects on the natural environment and future generations? In short, not very well! The uncertainties are simply too large to make cost-benefit analysis anything more than an academic exercise. Like categorical imperatives, utilitarianism is unable to provide us with a compelling ethical approach to climate change.

Values and virtues

Value refers to the worth or importance of something and in economic terms can be quantified in monetary units, as in the cost of a car or a service such as medical or legal assistance. But one can also speak in terms of social and human values.

Sayre (2010, pp. 262–4) defines a social value as "something viewed in a favorable light within the society in which it operates" and that provides "a guide of behavior" approved within the society. Social values are ideals collectively embraced by most, if not all, members of a group. They differ from personal values in the sense that they are operative within the larger society. Although personal and social values can overlap, there is room for variance between the values of an individual and the society of which he is a part. Examples of social values operative in many societies are liberty, honesty, education, pleasure, and what Sayre terms the *consumer values* of comfort, convenience, and acquisition. A personal value can also be a virtue, that is, a character trait exemplifying moral behavior. Examples include Plato's four cardinal virtues of fortitude or courage (the ability to manage fear and uncertainty), wisdom or prudence (an ability to select appropriate courses of action), temperance or moderation (the ability to exercise self-restraint), and justice or fairness (the ability to recognize and the will to render the rights of others).

Sayre (chapter 16) draws a line between social values harmful to the environment and personal values that are environmentally benign. Focusing on consumer values, which pervade developing as well as developed nations, he cites many examples of adverse environmental effects, including contributions to GHG emissions and climate change. Worldwide, consider the billions of air-conditioning and heating systems currently used to maintain comfortable temperatures and the convenience afforded by 700 million automobiles. Consider also the marketing and social pressures that impel people to acquire ever more goods – needed or not. With sustained economic growth in developing nations, levels of consumption may well triple by 2050. Comfort, convenience, and acquisition become pervasive social values as societies become more affluent.

The antithesis of consumer values is what Sayre (chapter 17) terms the *survival values* of moderation, simplicity, and contentment. Moderation and simplicity are used in the context of eschewing overindulgence and adopting a simpler lifestyle. Contentment is used in the context of understanding what brings satisfaction and resisting the endless barrage of marketing devices designed to stimulate wants and convert perception thereof to needs. Self-reflection that critically examines the consequences of consumption on the lives of others and the natural environment can contribute to nurturing these values. In today's world, moderation, simplicity, and contentment are largely personal values that have not risen to the level of operative values in most societies. Were they to rise to this level, they would have a huge effect on curtailing GHG emissions. As social values, people would be expected to act accordingly. Moderation would be approved; excessive consumption and overindulgence would not.

Embracing the foregoing values does not imply a life devoid of pleasure, one necessarily marked by austerity or asceticism. But it does reject indiscriminate application of consumer values. For example, we don't have to renounce air-conditioning, but we can more critically examine how and when it's used. We don't have to increase cooling and heating loads by supersizing our domiciles, and we can prioritize fuel economy in our choice of vehicles while reducing vehicle miles traveled by availing ourselves of alternative modes of transportation. Unless consumer values are modulated by embedding heavy doses of moderation and simplicity, it is unlikely that humankind will be able to successfully mitigate and adapt to climate change. Philosophically, we have two choices.

Aristotle or Rand?

Aristotelian or virtue ethics – also termed Aretaic Theory – is rooted in ancient Greek philosophy and focuses on standards for achieving excellence (goodness) in human character and behavior. How can I become what I should be? Aretaic Theory speaks to being a better person by becoming more virtuous. Aristotle identified virtues of thought such as wisdom and empathy and virtues of character such as justice, temperance, prudence, generosity, and courage. Activities of a virtuous person are those that contribute to the common good, and the common good can only be achieved through expressions of virtue in all human endeavors.

In classical philosophy, the common good transcends what economists term collective goods such as roads and clean air, that is, "goods that cannot be enjoyed by one without being enjoyed by many" (Lewis, 2006). Because the common good implies a more comprehensive state of flourishing (eudaimonia) based on reason and personal virtue, the words can be viewed as too vague or theoretical to have practical value. Such a view misses a key point of Aristotelian ethics, namely the import of human values such as justice and prudence.

In the context of climate change, temperance and prudence have special relevance, the first implying moderation or self-restraint in action and the second implying careful management of resources with due consideration of the future. A prudent person is a thoughtful person, one who considers the consequences of her actions. Is it prudent to dismiss scientific evidence pointing to global warming and climate change due to GHG emissions? What would a prudent person do to reduce the risk of adverse effects? Would it be prudent to address the root cause by reducing GHG emissions, while concurrently increasing resilience to the effects of warming?

A temperate and prudent person might also be called a thrifty person, but not in a pejorative sense. In this context a thrifty person is the antithesis of a wastrel (one using resources foolishly and wastefully), or one given to avarice (excessive desire for wealth, also known as greed), or one given to profligate (extravagant and self-indulgent) behavior (Riley, 2004). Virtue is a state of mind that opts for actions that achieve balance in their outcomes, which are neither excessive nor deficient. The highest good is not achieved through excess, nor is it achieved through self-denial. Daly (1996, p. 36) captured the essence of this state of mind when balancing current needs with those of future generations, namely "that basic needs of the present should always take precedence over the *basic* needs of the future but that basic needs of the future should take precedence over the extravagant luxury of the present."

Drawing on the work of Daly and Cobb (1989, p. 138), Hawken (2005, pp. 58–9) uses Aristotelian ethics to distinguish between two forms of economic activity – chrematistics, which uses natural resources and man-made assets for short-term, personal gain; and oikonomia, which uses resources for the long-term benefit of a larger community. He associates chrematistics with economic growth – an increase in quantitative measures of the economy – and oikonomia with economic development. A developing economy seeks to continuously improve the efficiency of resource utilization and the quality of people's lives, while minimizing degradation of the environment and its natural resources. It is concerned with getting better more than with getting bigger. Since growth is in large part based on stimulating consumption, the difference between growth and development can be likened to that between wants and needs. Hawken views energy efficiency and conservation measures as development activities that clean our air, reduce global warming, and enhance energy security, all of which enhance people's lives, now and for future generations. Choosing between growth and development is a question of values. When the U.S. House of Representatives dismisses energy efficiency and conservation while encouraging unencumbered growth and the use of fossil fuels, it is clearly expressing its members' values, a choice of chrematistics over oikonomia.

But if Aristotelian ethics focuses on the common good, what about man as an individual? How are self-interests weighed against obligations to society? Enter Ayn Rand, a thought leader for many right-leaning politicians and businessmen.

Rand is a twentieth-century philosopher and a proponent of objectivism, a philosophy that views ethics as a code for human survival and sustenance, where "human" refers to the individual and not to humankind or humanity (Peikoff, 1993). At its core objectivism links reason to knowledge rooted in reality. Man's survival and sustenance depend on his mind, his ability to think and draw conclusions based on reality, and to act on those conclusions. The thought process must be disciplined and devoid of emotion, personal feelings, or compliance with cultural norms. It must in essence be a selfish act. Actions must be driven by rational self-interest and self-preservation and not by interest in the welfare of others. The objectivist does not succeed through fraud or brutality, but through dedication and hard work. But at the end of the day, his mind and actions are used in service to self and not to others. Politically, the philosophy is grounded in a libertarian view of the world. The individual should be free to act according to his own judgment and interests and should respect the right of others to do likewise. Economically, the philosophy is grounded in unrestrained free-market capitalism.

In her signature work (Rand, 1992), a novel about villainous socialism and heroic capitalism, Rand's views are succinctly articulated by her protagonists. In the words of Hank Rearden and Francisco d'Anconia, successful entrepreneurs, inventors, and industrialists who equated virtue to productive ability,

> "We haven't any spiritual goals or qualities. All we're after is material things. That's all we care for." (p. 87)
>
> "[T]here's nothing of importance in life – except how well you do your work. Nothing.... It's the only measure of human value. All the codes of ethics they'll try to ram down your throat are so much paper money put out by swindlers to fleece people of their virtues." (p. 100)
>
> "I work for nothing but my own profit ... I will not say the good of others was the purpose of my work – my own good was my purpose, and I despise the man who surrenders his." (p. 480)
>
> "[M]y highest moral purpose [is] to exercise the best of my effort and the fullest capacity of my mind in order to support and expand my life." (p. 560)

And in the words of Rand's superhero, John Galt, who rejects the idea that goodness and morality require denial of self-interest,

> "I swear by my life and my love of it that I will never live for the sake of another man, nor ask another man to live for mine." (p. 731) "[M]an – every man – is an end in himself, he exists for his own sake, and the

achievement of his own happiness is his highest moral purpose."
(p. 1014) "Discard the protective rags of that vice which you called a
virtue: humility – learn to value yourself, which means to fight for your
happiness – and when you learn that pride is the sum of all virtues, you
will learn to live like a man." (p. 1059)

Rand is a proponent of a morality tuned to fulfilling the needs and wants
of the individual through hard work and capitalism. Few would dispute
the virtue of hard work and the benefits of free markets, but how many
would embrace the notion that selfishness is a virtue? How do we feel about
the premise that "greed is good?" The statement is often attributed (incor-
rectly) to Adam Smith and is used in the belief that excessive acquisitive-
ness by some will benefit all through a "trickle-down" feature of efficient
markets.[2] The premise begs the question: Is it therefore ethical to behave
selfishly? Is greed a virtue?

The foregoing questions bear heavily on linkages between economics
and ethics (Piedra, 2006). As the agent of economic activity, man can
act in ways that are personally beneficial but detrimental to the larger
society. The Great Depression and the more recent Great Recession
are examples of what can happen when capitalism and selfishness run
amok. Do ethics and concern for the common good call for limits on
personal freedom and material accumulation? In an article critical of
post-1950s Americans, Andersen (2012) called to mind familiar words
in the Declaration of Independence – "Life, Liberty and the pursuit
of Happiness." Although they affirm the importance of personal freedom, they do not condone greed and selfishness. Quoting from Thomas
Jefferson (Washington, 1861), "Self-love, therefore, is no part of morality.
Indeed it is exactly its counterpart. It is the sole antagonist of virtue leading us constantly by our propensities to self-gratification in violation of
our moral duties to others." Jefferson embraced classical philosophy and
Aretaic Theory.

Aretaic Theory and objectivism are both rooted in values and virtues,
but with inherently divergent interpretations. In fact, Rand represents the
anti-Aristotle.[3] One views the highest good in terms of outcomes that are
neither excessive nor deficient, achieved neither through profligacy nor
self-denial. The other sees the highest good in terms of self-fulfillment
and personal happiness. One aspires to a balance between caring for personal and collective needs; the other emphasizes individual needs and
wants. How then might they view the ethical challenges posed by climate
change?

Rand would have little patience for using social justice or intergenerational ethics as a rationale for limiting GHG emissions. And if emissions had to be limited, she would award the lion's share to the most productive members of society who would be free to use the fruits of their productivity for self-fulfillment. Personal freedom would trump social responsibility. Aristotle would approve of emissions required to address the basic needs of the living but done with temperance and prudence. He would have little tolerance for emissions attributable to over-consumption and self-indulgence.

The Precautionary Principle

In addition to well-established ethical theories, there are numerous ethical principles. The Precautionary Principle states that if an action has potentially harmful effects, as to human health or the environment, measures should be taken to reduce the threat, even if linkages between cause and effect are not fully understood. Even without scientific consensus, if there is potential harm, the burden of proving otherwise rests with those advocating the action. The principle implies a preference for measures that reduce the risk of adverse effects. When combined with cost-benefit analyses, it has been used in developing international agreements such as the Montreal Protocol. It has also been adopted as a public policy tool by the European Union.

The Precautionary Principle underscores the need to curtail harmful effects of human activity for both current and future generations and to do so even when knowledge of the effects is uncertain. Garvey (2008, p. 111) addresses this issue forcefully when he says:

> A part of the argument against action based on scientific uncertainty … amounts to a kind of gambling with the lives of people elsewhere on the planet now or in the future – betting that we can keep our comfy lives in the wager. There is a similar viciousness in the thought that the cost of mitigation and adaptation should be a reason for doing little or nothing…. The recklessness shows up again in the wishful thinking underpinning the hope for a technological quick fix … a bet that we can continue with our lives as they are in the hope that something unknown or untested might make everything all right in the end.

The Precautionary Principle makes it difficult to rationalize a do-nothing approach – to continue with business as usual. When could the world no longer say it was unaware of global warming and its consequences? By the late nineteenth century, chemists such as John Tyndall and Svante Arrhenius had elucidated the atmosphere's greenhouse effect and identified carbon dioxide as a greenhouse gas. By the 1960s, anthropogenic CO_2 emissions were recognized as a source of global warming, and David Keeling had begun his CO_2 measurements atop Mauna Loa. Through the 1980s, society can perhaps be excused for not taking the problem seriously, but following the first IPCC report of 1990 and the subsequent accumulation of scientific evidence, there is no longer an excuse for ignoring the warning signs.

The Precautionary Principle can be used to justify aggressive measures for reducing GHG emissions to curb the effects of climate change. But at what cost? The Montreal Protocol was preceded by strong evidence that CFCs were depleting the stratosphere of its ultraviolet radiation shield; options for replacing the CFCs were available; and the costs of taking prompt and effective action were reasonable. But options and costs are more problematic for climate change, and it's not surprising that collective efforts to deal with the problem, such as the Kyoto Protocol, have been far less successful than the Montreal Protocol. Application of the principle is encumbered by the same issues that thwart use of categorical imperatives or utilitarianism as ethical frameworks for mitigating GHG emissions.

The bottom line

There is clearly a good deal of variability in the foregoing theories and principles and ample room for a range of interpretations. At the end of the day, are the ethical judgments of an individual simply expressions of his or her social and cultural norms (cultural relativism)? Or are the judgments a manifestation of personal views (*emotivism*)? If all moral judgments are expressions of personal feelings, there are no absolutes. Personal choice and autonomy are paramount.

So how do ethical theories and principles guide our thinking on climate change? Even if absolute standards and values are embraced, they can clearly vary from one person to another, from community to community, and from nation to nation. Does the good of the collective trump the good of individual autonomy and the pursuit of personal sustenance? Does the well-being of the living outweigh concerns for future generations? Is it better to master the environment than to preserve it? Before addressing these questions, let's add one more dimension to the discussion. What do the world's religions say about climate change?

10.3 Religious traditions

It would be a mistake to discount the role faith communities can play in the debate and the call for action on climate change. Religious foundations of ethics are built on values derived from traditions linked to divine revelation. If the world's poor are to bear the greatest burdens of climate change, is it not God's will to protect the powerless? As articulated in the Book of Genesis, is it not God's will to also care for all creation and by extension to protect the environment on which all life depends? Since it is what I know best, I'll start with positions taken by the Roman Catholic Church.

The Roman Catholic tradition follows from the Old and New Testaments of the Bible and from Catholic Social Teachings, which are rooted in papal statements, pronouncements of world councils and synods, and conferences of bishops. These teachings attach primacy to the human condition – particularly for the more vulnerable members of society – and the common good. The Church's perception of goodness is linked to views of morality delineated first by Aristotle and subsequently in the teachings of St. Thomas Aquinas. In the Catholic tradition, God's will is not transparent, but is better understood through reason and experience, which includes the fruits of scientific endeavors.[4]

The Catholic view of Creation (nature) is that its bounty should be used for the benefit of all humankind, but with good stewardship and options preserved for future generations. Pope Paul VI in an Apostolic Letter (Octogesima Adveniens, 1971) bore witness to social injustices manifested by environmental degradation when he wrote of "the dramatic and unexpected consequences of human activity. Man is suddenly becoming aware that by an ill-considered exploitation of nature he risks destroying it and becoming in his turn the victim of this degradation ... thus creating an environment for tomorrow which may well be intolerable. This is a wide-ranging social problem which concerns the entire human family."

In an Encyclical Letter (Redemptor Hominis, 1979), Pope John Paul II initiated what would become a sustained concern for man's use of and impact on the environment when he wrote that "it was the Creator's will that man should communicate with nature as an intelligent and noble 'master' and 'guardian,' and not as a heedless 'exploiter' and 'destroyer'." In a subsequent Apostolic Blessing (SollcitudoRei Socialis, 1987) devoted to the ramifications of economic development, he described the natural world as God's gift to the entire human race, and unbridled consumerism, as well as disparities in the fruits of development, as abuses of the gift.

The relationship of humankind to creation merits special attention. In Evangelium Vitae (1995), Pope John Paul II refers to Genesis 1:28, where in reference to humankind, "God blessed them, and God said to them, Be fruitful and multiply, and fill the earth and subdue it; and have dominion over the fish of the sea and over the birds of the air and over every living thing that moves on earth." The quote could be considered an invitation to exploit the Earth's resources in the name of human development. But drawing from Genesis 2:15 in which man is called to care for God's gift of creation, Pope John Paul II reminds us of God's satisfaction and pleasure with creation and that "man has a specific responsibility towards the environment in which he lives ... ranging from the preservation of natural habitats of the different species and animals and of other life forms to human ecology ... which finds in the bible clear and strong ethical direction."

In his World Day of Peace Message (WDPM, 1990), Pope John Paul II shared his concerns for what he viewed to be an ecological crisis. Although he only briefly mentioned the greenhouse effect, his words bear on the Church's view of climate change. Using the Old and New Testaments to clarify "the relationship between human activity and the whole of creation," he states: "Faced with widespread destruction of the environment, people everywhere are coming to understand that we cannot continue to use the goods of the earth as we have in the past." Little more than 200 years ago – a mere blip in the 4-billion-year history of the Earth – the ability of humans to alter the environment was limited and local. But with exponential growth in human population and consumption, that's no longer the case. In his message the "ecological crisis" is described as a "moral problem," for which "many ethical values" are deemed "particularly relevant." Consider the following: (1) "[W]e cannot interfere in one area of the ecosystem without paying due attention both to the consequences of such interference in other areas and to the wellbeing of future generations." (2) "While in some cases the damage already done may well be irreversible, in many other cases it can still be halted. It is necessary, however, that the entire human community – individuals, States and international bodies – take seriously the responsibility that is theirs." (3) "Often ... economic interests take priority over the good of individuals and even entire peoples. In these cases, pollution or environmental destruction is the result of an unnatural or reductionist vision which at times leads to a genuine contempt for man." (4) "Modern society will find no solution to the ecological problem unless it takes a serious look at its life style. In many parts of the world society is given to instant gratification and consumerism while remaining indifferent to the damage which these cause. As I have already stated, the seriousness of the ecological issue lays bare the depth of man's moral crisis.... Simplicity, moderation and discipline, as well as a spirit of sacrifice, must become a part of everyday life, lest all suffer the negative consequences of the careless habits of a few."

Responsibilities to future generations, the need for collective responsibility, violation of the common good by economic interests, and the impact of overconsumption are clearly articulated in the foregoing statements, as are references to Aristotelian virtues. To circumvent misinterpretation of its economic views, it should be noted that the Church views socialism as an abject failure in meeting human needs. But it also has concerns for the extent to which capitalism has historically fostered imbalances between rich and poor nations and between the rich and poor within all nations. While it supports private ownership and organization of capital, it does so with expectations of a social contract that includes responsibility for the common good.

Concerns for consumerism were also expressed in a 1991 Pastoral Statement of the U.S. Conference of Catholic Bishops (USCCB, 1991), which included the comment: "Consumption in developed nations remains the single greatest source of global environmental destruction." Fast forward to 2015, and the statement can be modified. With globalization in full bloom, the words "consumption in developed nations" can be replaced by "global consumption." How much humankind consumes and the sources of energy used to enable consumption drive GHG emissions.

The USCCB has since been unequivocal in its view that climate change is a serious environmental problem with significant moral dimensions. Consider the following excerpts from a 2001 Pastoral Statement (USCCB, 2001): (1) "Global climate is by its very nature a part of the planetary commons.[5] The earth's atmosphere encompasses all people, creatures and habitats." (2) "In facing climate change, what we already know requires a response; it cannot be fully dismissed.... [E]ven in a situation with less than full certainty, where the consequences of not acting are serious ... if enough evidence indicates that the present course of action could jeopardize humankind's wellbeing, prudence dictates taking mitigating or preventive action. This responsibility weighs more heavily upon those with the power to act because the threats are often greatest for those who lack similar power, namely, vulnerable poor populations, as well as future generations.... [T]he impact of prudent actions taken today can potentially improve the situation over time, avoiding more sweeping action in the future." (3) "All nations share the responsibility to address the problem of global climate change. But historically the industrial economies have been responsible for the highest emissions of greenhouse gases ... [and] significant wealth, technological sophistication, and entrepreneurial creativity give these nations a greater capacity to find useful responses to this problem. To avoid greater impact, energy resource adjustments must be made both in the policies of richer countries and in the development paths of poorer ones." The foregoing views have been underscored in more recent pronouncements of the USCCB.

Such views on creation, the environment, and climate change extend well beyond Roman Catholic traditions. They are embraced by Islam (Mahasneh, 2003), Judaism (COEJL, 2007), and other Christian faiths, as well as by Buddhism and Hinduism. Consider the following representative pronouncements (Brock, 2012):

American Baptist Church (Resolution on Global Warming): "Global warming affects hunger, access to clean water, environmental stewardship, health and peace. Addressing global warming will make it more possible for all to live the life of possibility that God intends."

Buddhism (His Holiness the 14th Dalai Lama): "Right now our greatest responsibility is to undo the damage done by the introduction of fossil carbon dioxide to the atmosphere and climate system during the rise of human civilization."

Eastern Orthodox Church (Bartholomew I, Ecumenical Patriarch): "Climate change constitutes a matter of social and economic justice ... it is a profoundly moral and spiritual problem."

Hinduism (Convocation of Spiritual Leaders): "We cannot continue to destroy nature without also destroying ourselves. The dire problems besetting our world – war, disease, poverty and hunger – will all be magnified many fold by the predicted impacts of climate change."

Islam (Sheikh Ali Gomaa, Grand Mufti of Egypt): "Pollution and global warming pose an even greater threat than war... the fight to preserve the environment could be the most positive way of bringing humanity together."

Judaism (Jewish Community Priorities for Climate and Energy Policy): "Minimizing climate change requires us to learn how to live within the ecological limits of the Earth so that we will not compromise the ecological or economic security of those who come after us."

Religious activism on behalf of restraining climate change is also promoted by organizations such as the World Council of Churches, the

National Council of Churches of Christ, and the National Association of
Evangelicals (NAE).

The foregoing pronouncements suggest that the world's religions are
united in the view that climate change represents a threat to humankind
and that dealing with the issue is a moral responsibility. There is, however,
an outlier. When it comes to climate change, the U.S. evangelical movement does not speak with one voice.

In late 2005, the NAE, which represents approximately 45,000 churches
and 30 million congregants, circulated a draft of a policy statement that
encouraged lawmakers to legislate controls on GHG emissions (Janofsky,
2005). However, Senator James Inhofe, himself an evangelical, was quick
to object, saying that the NAE was being "led down a liberal path." Still, in
early 2006, eighty-six evangelical leaders issued a statement calling for federal legislation mandating economy-wide reductions in carbon emissions
through the use of cost-effective, market-based mechanisms (Goodstein,
2006; Kintisch, 2006). Although not a signatory to the statement, it was
supported by the NAE's chief lobbyist and vice president of governmental affairs, the Reverend Richard Cizik. With statements to the effect that
global hunger and global warming are inescapably linked and that global
warming is a *spiritual issue,* moral dimensions of the matter were being
thrust front and center (Breslau and Brant, 2006). Thus began a schism
within the evangelical movement that continues to fester (EEnews, 2013;
Jose, 2013).

The opposing evangelical view can be found in a declaration com-
posed by the Cornwall Alliance (2011). Quoting from the Alliance's
declaration,

> "Earth and its ecosystems – created by God's intelligent design and infi-
> nite power and sustained by His faithful providence are robust, resilient,
> self-regulating, and self-correcting, admirably suited for human flour-
> ishing, and displaying His glory. Earth's climate system is no exception.
> Recent global warming is one of many natural cycles of warming and
> cooling in geologic history."

> "[M]andatory reductions in carbon dioxide and other greenhouse gas
> emissions, achievable mainly by greatly reduced use of fossil fuels, will
> greatly increase the price of energy and harm economies." And, "... such
> policies will harm the poor more than others because the poor spend a
> higher percentage of their income on energy and desperately need eco-
> nomic growth to rise out of poverty and overcome its miseries."

The Alliance denies that "Earth and its ecosystems are the fragile and
unstable products of chance, and particularly that Earth's climate system is

vulnerable to dangerous alteration because of minuscule changes in atmospheric chemistry."

This anthropocentric view is based on an interpretation of Genesis 1:28 in which God blessed man and woman and said to them: "Be fruitful and multiply, and fill the earth, and subdue it; and rule over the fish of the sea and over the birds of the sky and over every living thing that moves on the earth." But Bergant (2010) challenges this interpretation. Drawing from the Book of Job and the Wisdom of Solomon, she argues that biblical societies were fully aware of their dependence on the natural world and the need for harmony with all creation. She also notes that many of today's biblical interpretations were formed in the nineteenth century, when new technologies were greatly amplifying the extent to which the natural environment could be exploited and the notion of creation was expanded to include the many benefits derived from human ingenuity. The observation prompts her to question whether the anthropocentric interpretation of Genesis was imposed by relatively recent readers instead of implied by the biblical author.

Schaefer (2010) extends the concept of sin to all of God's creation and the notion of poor and vulnerable to the natural world. It is not just the most vulnerable members of human society that will be impacted by climate change but also the most vulnerable species and ecosystems. Moreover, sinfulness is amplified when it is manifested by "patterns of human behavior that become institutionalized." Pfeill (2006) speaks of social sin when people "succumb to a kind of moral blindness whereby they participate in their societal institutions or systems without realizing that their actions, both of commission and omission, contribute to structures of sin." The characterization fits those of us who collectively ignore and/or contribute to the consequences of climate change   alteration ?

The Cornwall Alliance views nature as an inviolable design of God, one that can be used for the advancement of humankind but cannot be altered by human activities. Other religious groups share the view that nature originates from God, but believe that it can be altered by human activities and must not be altered to the detriment of any humans, both the living and the yet to be born. Does dominion imply domination or stewardship? Is God's gift so robust that it is truly left unscarred by human activity?

On the surface, it might seem that the foregoing views are sincerely held by religious leaders with different belief systems. But there's more to it. In Section 9.1 we considered a political-corporate axis that has mobilized to discredit climate science and scientists. A similar group of individuals and organizations is behind the Cornwall Alliance (Romm, 2010). In an ideal world, one might think that faith nurtures values, which combined with empirical evidence guide public policy. In this case, the opposite seems to apply, with political ideology underpinning faith and providing a basis for dismissing empirical evidence.[6]

In a comprehensive study of religion in American politics, Putnam and Campbell (2010) found that a growing number of Americans adjust their religious beliefs to match their politics. A survey conducted by Baylor University (2011) also revealed that roughly 20% of Americans view the invisible hand of free markets to be the hand of God, and they object to government intrusion. They are therefore likely to seek religious affiliations that affirm this belief. By extension, governmental constraints on GHG emissions that impede economic activity are violations of their religious beliefs. Religious views can be and, in fact, are modified to fit economic and political ideologies.

So where do we go from here? The fact that the ethical and religious implications of climate change allow for different perspectives does not mean that these implications are irrelevant. We all have a system of values shaped by ethical and/or religious considerations. It behooves us to take a few moments to examine climate change in terms of these values.

10.4 A role for ethics

If humankind has a moral obligation to act now – by adopting mitigation and adaptation measures that reduce the amount of global warming and its adverse effects – how should it proceed? How should sacrifices and costs be distributed among nations? To what extent should nations that have contributed disproportionately to the atmosphere's current store of GHGs contribute disproportionately to reducing their emissions and to advancing adaptation measures? And to what extent should these countries compensate poorer nations for losses suffered from climate change and the costs of adaptation?

There is clearly a historical factor to assigning moral responsibility for reducing GHG emissions. If one counts emissions since the Industrial Revolution, developed nations have put the lion's share of today's GHGs into the atmosphere – while growing their economies – and it can be argued that they have a moral responsibility for dealing with the consequences. Through the 1990s and 2000s, it is this argument that shaped the distinction between richer/developed nations and poorer/developing nations, with the first group viewed as agents of global warming and the second as its victims. This distinction was embodied in the Kyoto Protocol, which established emission reduction goals for developed (Annex I) nations while imposing no constraints on developing (Annex II) nations. But the argument weakens if emissions are counted from the time climate change was generally recognized as a problem. If 1990 is chosen as the appropriate time, since then, cumulative emissions of developed and developing nations have been comparable, calling into question whether it is any longer appropriate to separate the two groups in charting the way forward. Moreover, holding developed nations accountable for past emissions is not without risk. If atonement leads to impairment of their economies, the global collaboration needed to deal with the problem would be impaired, and efforts to impose accountability for past emissions would be counterproductive. And the hard reality is that developed nations lack the fiscal discretion to redirect large resources for this purpose. Beyond economic impairment imposed by the great recession of the early twenty-first century, they face huge demands for investments in education, health care, and infrastructure, as well as obligations to aging populations (Myers and Kulish, 2013).

Focusing on the future, Garvey (2008, pp. 81, 126–35, 139) examines two approaches to the problem – one termed *shared emissions* and the other *comparable burdens*. To address fairness and justice criteria, one could argue that emissions should be apportioned equally among all humankind. To understand the implications of such an argument, consider circa 2011 global emissions of 33.9 billion tonnes of CO_2 (33.9 Gt-CO_2). Since 2001, emissions had increased at an average annual rate of 2.7% (Olivier et al., 2012) despite trends that could have bent the curve downward. Factors contributing to reduced emissions included recessionary effects in OECD nations, displacement of coal by natural gas for power generation in the United States, and increased adoption of energy efficiency measures and renewable energy worldwide. However, the reductions were more than offset by the increased emissions of developing nations. For 2011 emissions from China and India had increased 9% and 6%, respectively, from the preceding year and accounted for one-third of the global total, equivalent to the cumulative emissions of all OECD nations. It could be said that the trends align with a social justice agenda that calls for reduced emissions by developed nations and increased emissions by developing nations. But the net effect is still an increase in global emissions, contrary to the need for stabilizing and subsequently reducing emissions.

The top four contributors to circa 2011 CO_2 emissions were China (9.70 Gt), the United States (5.42 Gt), the twenty-seven nations of the EU (3.79 Gt), and India (1.97 Gt). On a per capita basis, the world's 7 billion people discharged an average of 4.8 tonnes per person, while per capita emissions in China, the United States, the EU, and India were 7.2, 17.3, 7.5, and 1.6 t-CO_2, respectively. Two features are note-worthy: the order-of-magnitude difference between the United States and India, and the fact that China's per capita emissions have essen-tially reached European levels. Since 1990, China's per capita emissions increased from 2.2 to 7.2 tonnes while decreasing from 9.2 to 7.5 tonnes in the EU and from 19.7 to 17.3 tonnes in the United States (Olivier et al., 2012).

What would it take for all nations to maintain per capita emissions at the global average of 4.8 t-CO_2? India could increase emissions by 200%, grow-ing its economy accordingly, as could many other developing nations. But China would have to reduce its emissions by 33%. For the United States it would mean a 72% reduction. And that's just to maintain total emissions at 33.9 Gt-CO_2 (9.2 Gt-C). Recall Pacala and Socolow's (2004) mitigation scenario of Section 6.5, which involved holding annual global emissions at 7 Gt-C (25.7 Gt-CO_2) for fifty years. To do so with a 2011 global population of about 7 billion, average per capita emissions would have to drop 23% to 1 t-C (3.67 t-CO_2). And assuming this level could be maintained through 2050, the Pacala and Socolow's scenario calls for a subsequent sharp drop in global emissions to about 2 Gt-C/yr (7.3 Gt-CO_2) or, assuming a fixed population of 7 billion, to about 1 t-CO_2 per person per year. But, of course, the population will change, with estimates of 9 billion people by 2050 requiring additional reductions in per capita emissions.

Consider for a moment what the foregoing numbers mean to you the reader. Assume you live in a developed/industrialized nation for which average per capita emissions are 10 t-CO_2/yr and your nation's energy port-folio is heavily weighted with fossil fuels. What lifestyle adjustments would you have to make to reduce your contribution by 90% – from 10 to 1 t-CO_2/yr? You can sort through the ways you use energy – space heating and cool-ing, transportation, all manner of electrical devices, manufactured goods whose production and transport consume energy, and so on – and you'll find that getting there borders on the impossible. Referring to the Human Development Index (HDI) introduced in Section 1.5, 1 t-CO_2/yr places you at the low end of the HDI scale, at about or below the level of many devel-oping nations. Your response is likely to be "yes, but emerging clean energy technologies will allow me to maintain my current lifestyle while signifi-cantly reducing my carbon footprint." To a degree, yes; but can technology do it all? And if so, how long will it take? Will lifestyle changes be needed to augment the role of technology? Hold these questions! We'll return to them in the next chapter.

While perhaps appealing to some, the foregoing model of *per capita equivalence* is no more than a straw man. It is neither practical nor ethical. It ignores many factors such as the size and climate of a nation, which influence energy requirements for transportation and heating/cooling. It also ignores energy-intensive aspects of a nation's productivity, particularly its contribution to the world's supply of food and manufactured goods. Without major advancements in energy production and storage technologies, as well as huge capital investments, it would be difficult, if not impossible, to achieve the foregoing reductions while maintaining acceptable living standards in developed nations and achieving such standards in developing nations. And, taking a page from Ayn Rand's *Atlas Shrugged*, there will always be variations in human aspirations, ambition, and accomplishments – within a nation and between nations – that will affect per capita emissions. Some people will always work harder and smarter than others and produce more for the collective good. The shared emissions model is untenable.

If emissions are referenced to a nation's GDP instead of its population, a different picture emerges. Based on purchasing power parity (PPP) and 2010 U.S. dollars, a 2011 gross world product (GWP) of $77.2 trillion (EPI, 2012) yields a global average emission of 0.44 kg-CO_2 per dollar of economic output. On this basis, OECD nations compare favorably, with below-average emissions of 0.40 kg-CO_2/$ for the United States and slightly less than 0.30 kg-CO_2/$ for the EU. In contrast, China and India had above-average 2011 emissions of 0.84 and 0.48 kg-CO_2/$, respectively. But again, the numbers don't tell the whole story, and they certainly don't suggest that equalizing emissions based on conventional measures of economic output would address moral concerns. China and India have used domestic coal reserves to sustain economic growth and, like many developing nations, have used low-wage labor and energy-intensive manufacturing activities as engines of growth. Hundreds of millions of people have achieved acceptable living standards, but hundreds of millions still live in poverty. For the foreseeable future, developing nations will need larger-than-average emissions per unit of economic output to sustain development.

This brings us to yet another measure of a nation's emissions, one that on moral grounds casts developing nations in an even more favorable light. Through globalization, much of the world's production has been exported to developing nations and with it related GHG emissions. Porter (2013) raises the following question: "Are emissions the responsibility of the countries that made (the products) or the countries for whom the products were made?" If emissions were referenced to a nation's consumption, those of the United States and the EU would increase by about 8% and 33%, respectively, while China's would drop by 20%. But for developing nations, the issue is a double-edged sword. While using consumption as a reference elevates the moral high ground for developing nations, measures to adjust for consumption would likely involve a carbon tax on imports, reducing their competitive advantage and impairing the economic growth of developing nations. For this reason, developing nations are not inclined to trumpet consumption-based accounting.

Garvey's "comparable burdens" approach, which calls for all parties to contribute according to their ability, combines elements of fairness and pragmatism and provides the best path forward. In absolute terms, wealthier nations would contribute more, but all nations would contribute in recognition of a shared understanding of the problem and in the spirit of shared sacrifice for the common good. Rich nations have a moral responsibility to take the lead in reducing emissions and adapting to global warming. In the face of growing evidence concerning climate change, arguments for not doing so are becoming increasingly indefensible, straddling a line between wishful thinking and recklessness. Inaction is tantamount to rolling the dice in the belief that whatever happens, we can deal with it later. At its root, it is driven by little more than a desire to maximize self-fulfillment and accumulation in the here-and-now.

Wealthier nations can lead by wringing emissions out of every process and activity, by aggressively implementing efficiency and conservation measures, decarbonizing the energy supply system, and, where feasible, by implementing carbon capture and sequestration. In the context of conservation, Garvey distinguishes between *subsistence emissions* – what's necessary to achieve an acceptable quality of life – and *luxury emissions* associated with overconsumption. Wealthy nations can also lead by developing robust adaptation measures that increase the resilience of infrastructure, food and water supplies, and biodiversity to climate change. Leadership and moral responsibility also imply assistance to poorer nations by sharing technology and best practices for mitigation and adaptation. But while necessary, the best efforts of wealthier nations are not sufficient.

Increasingly, the distinction between richer/developed nations and poorer/developing nations is being blurred by rapid economic development and increasing GHG emissions in the latter group. Today, nations such as China and India, as well as many in the Southern Hemisphere, are very much a part of the problem and should be active contributors to the solution. Yes, on the basis of their emissions since the IR, developed nations have a moral obligation to provide leadership in mitigating and adapting to climate change and to assisting poorer nations with losses incurred by the effects of climate change. But there will soon be a significant shift in culpability. Within a decade, developing nations will have contributed more than half of post-IR emissions, making them a significant contributor to the problem and imposing a shared responsibility for the effects of climate change (Myers and Kulish, 2013). They must therefore be aggressive in reducing emissions per economic output and developing a conservation mindset. It is regrettable but not surprising that, as wealth is accumulated in these nations, beneficiaries are quick to adopt habits of overconsumption and excess.

Finally, some words of caution for those inclined to view geoengineering as an appropriate remedy to global warming should consequences eventually overwhelm adaptation mechanisms. At that point, some might argue moral justification for manipulating the environment on a global scale to counter the effects of warming. But what of unintended consequences such as ocean acidification if the solution involved seeding the atmosphere with sulfates, or crop failures due to changing weather patterns? And what about the restoration of warming deferred by geoengineering measures if termination becomes necessary? And wouldn't getting to the point of implementing such measures be an indictment of moral failure by not having pursued mitigation and adaptation measures more aggressively? If geoengineering is deployed as a safety valve, current generations will clearly be guilty of having transferred all risks and consequences of global warming to future generations. Whether having to deal with the consequences of warming and/or geoengineering, it would be their problem, not ours – the ultimate abdication of moral responsibility. *If our solution make the result even worse?*

10.5 Summary

The ethical implications of climate change have been couched in terms of an asymmetry between benefits and burdens. Nations that have contributed disproportionately to global warming through their GHG emissions have benefited from their emissions, becoming among the world's richest and best able to adapt to the effects of climate change. Conversely, nations that have contributed far less to global emissions remain poor and among the most vulnerable to the effects of climate change. The asymmetry is also intergenerational. While the living and their ancestors have benefited from GHG emissions, the burden of dealing with adverse effects is left for future generations. Climate change is a social justice issue, as well as one of intergenerational ethics. Its largest impact will be felt by the most vulnerable members of society and the unborn.

In seeking guidance from ethical theories and religious traditions, there is ample room to couch morality in whatever terms reinforce one's inclinations, making it difficult to agree on what is good. How do we balance responsibilities to self against responsibilities to others, especially the poor and underserved? How do we weigh the needs of the present against those of the future, of the living against those of the unborn? Admonitions to follow *imperatives* or to *do no harm* are subject to conflicting interpretations, as are judgments guided by personal values. Although the world's religions overwhelmingly view climate change as a threat to humankind and dealing with the issue as a moral responsibility, scripture can be interpreted differently. It's not that we've become amoral; it's simply that we're inclined to adjust our moral standards and religious beliefs according to circumstances and predispositions.

But if one accepts the premise that climate change is a real and significant threat to humankind, one must acknowledge a moral responsibility for dealing with the problem. Although it must be a shared responsibility among all people and nations, it is one for which leadership and a larger burden must be assumed by those who have disproportionately contributed to and benefited from GHG emissions. They can do so by aggressively implementing energy efficiency and conservation measures, by accelerating the transition from fossil to noncarbon forms of energy, by proactively developing adaptation measures, and by assisting poorer nations that have the fewest resources to deal with the effects of climate change and would suffer the largest losses.

On the grounds that moral action is not conditional, Garvey (2008, p. 108) believes the United States was wrong to reject the Kyoto Protocol because it placed no restrictions on developing nations. Moral action by one party is not conditional on whether it is taken by others. The United States was in a position to exert moral leadership and it opted out. Had it not done so, some would argue that its efforts to reduce emissions would have been rendered inconsequential by the increased emissions of developing nations. However, to whatever extent the United States could have reduced its emissions, it would have lessened global warming and more importantly it would have provided an exemplar and best practices for others to follow.

Garvey (2008, p. 156) also takes the United States to task for its large per capita GHG emissions and views as morally indefensible its failure to provide leadership in addressing the problem, particularly since it is equipped with the economic and intellectual resources to do so and has such large and easy targets associated with its luxury emissions. He is quite blunt about the matter when he states that "we should change our comfortable lives, tighten our belts and cut back on our easy lives of high-energy expenditure." He is referring to an aspect of the American way of life tied to overconsumption and frivolous uses of energy, which brings the discussion back to values. As individuals and collectively, do American values align more with those of Aristotle or Ayn Rand? By embracing Aristotelian virtues, the United States would be setting the right example for poorer nations in demonstrating that a high quality of life can be achieved without frivolous consumption.

Aristotelian virtues provide a hospitable oasis, a place of intersection for religious beliefs – faith in the divine – on the one hand and secular modernity – faith in science, technology and social progress – on the other.

CHAPTER 11

A way forward

"Climate change is occurring, is very likely caused primarily by human activities, and poses significant risks to humans and the environment." This statement was made in a report prepared by the National Research Council (NRC, 2011), which is an agent of the U.S. National Academies of Science and Engineering and the Institute of Medicine. The view is shared by governmental science agencies across the world, as well as by an overwhelming majority of the global scientific community. We know that atmospheric GHG concentrations are rising, that the gases contribute to warming, and that warming is enhanced by feedback effects. We also know that atmosphere and ocean temperatures are rising, along with sea levels, as Arctic sea ice and glaciers continue to melt. The trends are due to the use of fossil fuels, and we know they will continue, *even if atmospheric GHG concentrations were to remain at today's levels*. And we are aware of the consequences, such as the impact of rising sea levels on low-lying coastal regions and the increased frequency of extreme weather events.

Yes, there are gaps in the scientific knowledge base. They've been addressed in previous chapters and are acknowledged in the NRC report. But the science underpinning a human impact on climate change is strong and getting stronger with each new study. This part of the debate is all but over. Most uncertainties are associated with partitioning of energy between the Earth's atmosphere and oceans and with understanding the consequences of climate change. Where and to what extent will the frequency and intensity of major storms increase; where will chronic drought and desertification be most pronounced; where and to what extent will ecosystems, biodiversity and agricultural production be adversely affected; and what will be the scope of attendant diasporas? It is not acceptable to focus on these uncertainties and to simply call for more research without addressing root causes of global warming and preparing to deal with the adverse effects of climate change. Because the potential consequences of global warming are profound, the prudent and morally responsible approach is to take action now. The NRC report states unequivocally that "uncertainty is not a reason for inaction."

The time for decisive action is now. The risks of doing otherwise – to the global economy, the environment, and the well-being of human kind – are real and large. But before discussing measures that should be taken, it is useful to interject a dose of reality by identifying two unlikely outcomes.

11.1 Economic realities and unlikely outcomes

Climate change is unlikely to be satisfactorily mitigated by a well-organized, collaborative, and effective global effort. It is troublesome to make this statement. After all, climate change is a global problem requiring global cooperation in achieving a global solution. Conditions have passed the point where the burden of action can fall exclusively on the developed nations. All nations must find ways to contribute. But is that a realistic expectation? Special interests abound, and in many nations perceptions of a serious threat are overshadowed by more pressing domestic issues. And, for all manner of economic and political issues, the world is becoming less manageable, making an integrated global approach to climate change much more problematic.

Thus far, achieving global consensus and cooperation has been elusive and is by no means imminent. The commitment made by developed (Annex I) nations under the Kyoto Protocol expired in 2012, and apart from the EU, many of the original signatories demurred from committing to a second round. Developing (Annex II) nations would welcome an extension of the Protocol, since they were absolved from commitments to reduce emissions and, through carbon offset programs, were the principal beneficiaries of emission-trading schemes. But with growing concerns over the effectiveness of offsets and at best modest progress toward the Cancún promise of providing climate change aid to developing nations, they are unlikely to derive significant benefits from any post-Kyoto agreement. Yet, significant progress toward reducing global emissions cannot be made without their participation.

In 2006, China's CO_2 emissions surpassed those of the United States, and in 2013, its emissions represented about 30% of the global total. In fact, China accounted for two-thirds of the global increase since 2000 (Economist, 2013b), and it can no longer fall back on the claim that its per capita emissions are well below those of Western nations (Olivier et al.,

2011). Although 25% of its emissions are attributable to its huge manufacturing base, much is tied to factors such as extensive use of coal and growing energy consumption by its affluent class. China is very much a part of the problem, and among developing nations it is not alone. Like China, India has a rapidly growing middle class, is well endowed with coal reserves, is adamant that the reserves will be used to foster economic growth, and is aggressively constructing new coal-fired power plants. For developing nations, economic growth will remain a high priority for decades.

That said, developing nations are hardly oblivious or indifferent to the consequences of climate change. In China, rising sea levels and storm surges will impact 80 million people living at sea level, and drought and desertification are already affecting agricultural production (Economist, 2013b). Aware of its vulnerability, China is determined to reduce its carbon intensity by increasing the efficiency with which it produces and uses energy and by leading the world in deploying renewable sources of energy (solar, wind, and hydro) and nuclear power. It is also experimenting with regional emission-trading schemes and is considering consequences of implementing a national cap-and-trade system. But with a population of approximately 1.3 billion and a per capita GDP of approximately PPP $7,500, it still has a long way to go in moving hundreds of millions of people from poverty to acceptable living standards. Through at least 2025, its emissions will continue to rise, with coal providing much of the energy for base load power. Where trade-offs must be made, economic considerations will trump environmental concerns, as they will in India and other Asian, African, and South American nations. Collectively, the world's developing nations have been experiencing rapid economic growth of 4% to 10% per annum, with accompanying growth in energy consumption, particularly from fossil fuels. Barring a severe and sustained global recession, they will likely maintain at least a modest growth trajectory, and by 2030 it's estimated that non-OECD emissions will be more than twice those of OECD nations (EIA, 2013b).

Pielke (2010) takes scientists, environmentalists, and politicians to task for failing to deal with the economic realities of carbon mitigation options. One of his points is that when economic and environmental trade-offs are juxtaposed in public discourse, it is the economic argument that prevails. Many developing nations are fully aware of the implications of climate change and their own vulnerability. But their longer-term environmental concerns are trumped by more immediate needs for economic growth, particularly in forms that improve living standards, employment opportunities, and political stability. What would it take for these nations to commit to a plan for bending their emission trajectories? A lot! OECD nations would have to contribute financial and technical support to the developing nations and, as a measure of good faith, would have to commit to large reductions in their own emissions. Both measures are unlikely, particularly at a time when many OECD nations are experiencing economic stasis.

So what can be expected from the 2015 COP meeting in Paris and beyond? Based on past experiences and existing economic and political conditions, it's hard to be sanguine about prospects for an agreement with binding and measurable commitments that call for shared sacrifices and make a substantive difference.

It is highly unlikely that the goal of reducing global emissions to 20% of 1990 levels by 2050 will be achieved. Since this is the approximate target associated with keeping atmospheric CO_2 concentrations below 450 ppm and warming below 2°C, it is again troublesome to make such a statement. But consider the trends. Emissions increased by 45% from 1990 to 2010 (Olivier et al., 2011). For China and India, they increased by 257% and 180%, respectively, underscoring the increased demand for energy that accompanies rapid economic growth and the need to meet most of the demand with fossil fuels.

From 2010 to 2040, global emissions from fossil fuels are projected to increase at an annual rate of 1.3% from approximately 31 Gt-CO_2 to 45 Gt-CO_2 (EIA, 2013b), driven to a large degree by economic growth in developing nations.[1] Yet another driver will be the rapid growth in oil and gas production occurring in the Americas, Africa, and possibly the Arctic. Advances in drilling technologies are providing access to large amounts of oil and gas embedded in shale rock formations and in deep ocean waters, while production of oil from the tar sands of Canada continues to increase. Even in the face of growing global demand, these developments will moderate fuel costs, sustaining high levels of consumption and emissions. It is difficult to see how the atmospheric concentration of carbon dioxide can be stabilized at 550 ppm, much less 450 ppm. And if business-as-usual prevails, reaching 650 ppm by the end of the century is within the realm of possibility.

Moving forward, the foregoing discussion suggests two conditions. First, whatever progress is made toward curtailing emissions, it is unlikely to be driven by binding international agreements. Instead, it is more likely to be made by uncoordinated or loosely coordinated efforts of individual nations, groups of nations such as the EU, nongovernmental organizations, corporations, and individuals, each according to its values, priorities, and limitations. Second, since global warming is likely to exceed 2°C, adaptation measures must be implemented and assistance provided to those most likely to be impacted by but least able to deal with the effects of climate change.

The following sections provide recommendations of what should be done to reduce the threat of climate change to manageable levels. We begin with the role of technology.

11.2 Elements of an action plan: mitigation and adaptation

Mitigation

Mitigation is not an option; it is a necessity, a cornerstone of efforts to deal with global warming. Mitigation measures have strong roots in technologies associated with energy production, conversion, storage, transmission, and utilization, and to reduce GHG emissions, these technologies must be advanced.

Mitigation measures were considered in Chapter 6, beginning with energy efficiency. Simply put, energy efficiency must be driven into every facet of human activity. Increasing efficiency translates to reduced energy demand, and to reduced emissions if demand is met by fossil fuels. Production, conversion, storage, and transmission are supply-side considerations. By increasing the efficiency of related processes, energy consumed by the processes, and correspondingly carbon emissions, are reduced. Utilization is a demand-side consideration. An increase in the end-use efficiency reduces the amount of primary energy that must be devoted to the task. The possibilities are endless: building heating, cooling, and lighting systems; ground, air, and sea transportation systems; manufacturing processes; all manner of appliances and electronic devices; and more.

Because quick and positive returns often accompany investments in more efficient technologies and processes, efficiency can be a win-win proposition, beneficial to both the environment and the bottom line. Efficiency measures are integral to energy policy across the world – more so in some nations than others – with opportunities for improvement in all nations. By itself, however, efficiency is not enough. A mitigation agenda must include decarbonization of energy sources, particularly for power production and transportation.

Of the three fossil fuels, coal has the largest ratio of carbon-to-hydrogen atoms (C/H) and the largest carbon intensity (the amount of carbon released per unit of energy released by burning the fuel). In 2011, coal provided about 40% of the world's electricity and accounted for 45% of its CO_2

emissions (EIA, 2013c). Moreover, the contributions of coal to power generation and CO_2 emissions are projected to increase, with coal-rich developing nations such as China and India feeding most of the growth (EIA, 2013b). Reducing emissions from coal-fired power plants must be integral to any comprehensive mitigation strategy.

Apart from increasing the efficiency of coal-fired power plants, carbon capture and sequestration (CCS) provides the only other option for reducing their emissions. In Section 6.4 a good deal of attention was devoted to CCS technologies, and the assessments weren't encouraging. Currently, all options are expensive and impose a large penalty on the plant's thermal efficiency. If CCS is to become a viable mitigation measure, there would have to be a major breakthrough in technology, one that significantly reduces capital and operating costs. Without CCS, coal-fired power plants are financially competitive with other sources of electricity; with CCS, they are not. The reality is that currently, the likelihood of using CCS to meet expectations for significantly reducing carbon emissions is low. Given the fact that the world has huge coal reserves, which many nations deem critical for economic development, this conclusion has serious implications. Currently, global strategies for stabilizing atmospheric CO_2 below 450 ppm allow for the use of coal, but with emissions curtailed by extensive implementation of CCS.

No more than a decade ago, I would often say that "it's not a matter of *whether* coal would be used, but *how* it would be used," believing that CCS would allow for continued use without GHG emissions. Today, having far less confidence in the role of CCS, I would not make this statement. While some nations may be able to wean themselves from coal by transitioning to other energy sources, developing nations will be less able to do so and will continue to burn coal without CCS. Yet, because coal contributes disproportionately to global power production and GHG emissions, doing nothing to reduce its impact is not an option.[2]

One option for decarbonizing power production involves transitioning from coal to natural gas, which has the lowest C/H ratio and the lowest carbon intensity. As discussed in Section 6.2, it would reduce carbon emissions per kWh_e of electrical energy by a factor of two. Globally, reserves of natural gas are sufficient to sustain such a transition, while maintaining a balance between supply and demand that moderates volatility in the price of gas. However, there are some caveats. A significant fraction of the world's reserves consists of nonconventional (tight) gas accessible by horizontal drilling and hydraulic fracturing, and leakage could negate the benefits of replacing coal by gas. In addition, water supplies could be contaminated by drilling and production operations. The technology exists to address both issues, but in both construction and maintenance, developers and system operators must shoulder the responsibilities and costs of getting it right, and regulators must hold them accountable.

Assuming that concerns for leakage and water contamination can be alleviated, substitution of natural gas for coal can contribute to reducing CO_2 emissions and should be part of a twenty-first-century mitigation agenda. But even for natural gas, emissions remain large without CCS, and a pathway to acceptably low levels can only be provided by carbon-free sources of energy.

Although nations should develop their renewable energy portfolios according to indigenous resources – solar, wind, hydro, geothermal, wave, tidal, and/or biomass – two sources merit special consideration. For more than a decade, solar and wind energy have experienced annual growth of 25–50%. This growth rate must be more than maintained; it must be increased. But for power generation, expansion of renewable energy to meet much more than 20% of a nation's needs depends on the ability to manage intermittency effects and to transmit the energy from regions of high production to high demand.

A better balance between production and demand can be achieved by developing cost-effective and robust batteries capable of storing and discharging electrical energy ranging from comparatively small amounts (10 to 100 kW_e for distributed solar) to utility-scale amounts (10 MW_e or more for solar and wind farms). Battery technology must be a high-priority target for research and development. Achieving appropriate balances between generation and demand will also require expansion of the grid and endowing it with greater intelligence to optimize transmission and distribution of electricity generated by large (central) and small (distributed) sources. An ambitious but attainable goal is one of supplying 50% of the world's circa-2050 electricity from renewable sources. Getting there will require significant private- and public-sector investments in storage, transmission, and distribution, as well as generation. But even if renewables achieve this goal and continue to grow thereafter, there will still be need for dispatchable (base-load) power. Unlike coal and natural gas, nuclear energy provides a carbon-free source of electricity that should be exploited.

In spite of three well-publicized events – Three-Mile Island, Chernobyl, and Fukushima – nuclear power has enjoyed a commendable safety record. For more than sixty years, high standards maintained by nations such as France, the United Kingdom, and the United States have averted adverse effects on human health and safety and have done so largely with vintage 1960s and 1970s plant designs. Next-generation designs are endowed with improved safety features, and there's no reason to believe that new plants, as well as existing plants, can't be hardened to withstand the most extreme acts of nature or terrorism.

Nuclear energy can make a significant contribution to large-scale base-load power without contributing to GHG emissions and should play a prominent role in generating electricity for nations capable of managing fuel cycles and maintaining stringent safety, security, and emergency response standards. Regulatory agencies within each nation must be independent of plant operators and empowered to take appropriate action when deficiencies are identified, and all nations must be open to periodic review by international teams of experts. Power plants constructed over the next twenty-five years can contribute significantly to reducing twenty-first-century GHG emissions and help bridge the gap to achieving a twenty-second-century noncarbon, nonnuclear, renewable energy future.

In the transportation sector, there is certainly much that can be done to reduce GHG emissions. The internal combustion engine and hybrid vehicle technologies will be with us for years, and there remains considerable potential for improving fuel economy. But if transportation-related GHG emissions are to be reduced to acceptable levels, the market share of vehicles powered solely by electricity and biofuels must increase over time. If EVs are to become the predominant LDV, better batteries – lighter, cheaper, with larger charging rate and storage capacity – must be developed along with the charging infrastructure. Of course, movement to EVs is only carbon free if the electricity is generated by carbon-free sources of energy. Biofuels must also play an important role, particularly for aircraft, and may become the fuel of choice for heavy-duty vehicles. But, here again, technology improvements are needed to convert nonedible forms of biomass to biofuels in a cost-effective manner. To accelerate the transition to carbon-free modes of transportation, the importance of advancing engine, battery, and fuel technologies through research and development cannot be overstated.

Although technology and associated research and development play significant roles in advancing the foregoing mitigation measures, there is another measure for which technology is not a factor. Existing terrestrial carbon sinks must be maintained and augmented by reforestation and afforestation, with special incentives provided to developing nations for which the sinks are an important source of income.

Pathways to deep decarbonization

The foregoing recommendations parallel those of a collaborative effort involving fifteen major GHG emitting nations (Australia, Brazil, Canada, China, France, Germany, India, Indonesia, Japan, Mexico, Russia, South Africa, South Korea, the United Kingdom, and the United States) (SDSN, 2014). Each nation was charged with identifying aggressive pathways that could reduce its per capita emissions to 1.6 t-CO_{2eq} by 2050 – the *shared emissions* approach of Section 10.4 – which is roughly the amount needed to prevent the atmosphere from warming by more than 2°C. Although the pathways were influenced by circumstances unique to each nation, there were three common threads: (1) increasing energy efficiency for all end-use sectors, from transportation and buildings to industrial processes; (2) decarbonization of electricity generation, with renewables and nuclear energy meeting a much larger share of demand and CCS applied to generators fired by fossil fuels; and (3) replacing fossil fuels with low-carbon electricity and sustainable biofuels and hydrogen for transportation, heating/and cooling, and industrial processes. While most nations proposed pathways that would meet the target, some were unable to do so, and the average endpoint for the fifteen nations was 2.3 t-CO_{2eq}.

If the world's major emitters can identify mitigation pathways for getting close to the desired outcome by mid-century, what's to prevent it from happening? For one, the pathways are technology-driven, and it's assumed that technologies currently under development – such as CCS, cellulosic biofuels, batteries, and hydrogen fuel cells – will be economically deployable on large scales by 2025. Perhaps, but the assumption is made without assessments of related costs and amounts to be borne by private and public sectors. Even with aggressive energy efficiency measures and decarbonization of the energy supply, carbon emissions are projected to increase by 2% a year from 2010 to 2040 in non-OECD nations and at best to remain flat in OECD nations (Porter, 2014). In non-OECD nations such as China and India, the effects of rapid economic growth more than counter decarbonization and energy efficiency measures. That brings us to adaptation.

Adaptation

Although there are many measures that can be taken to reduce GHG emissions, it is unlikely that implementation will be sufficient to reduce atmospheric concentrations to acceptable levels. Global warming will continue, and with manifestations of climate change becoming more probable, there will be a growing need for adaptation measures. Mitigation provides more time to implement adaptation measures and reduces related costs, but it will not eliminate the need for adaptation.

Since there will be regional variations in the effects of climate change, from extreme weather events in coastal regions to prolonged drought and heat waves in central continental regions, decisions concerning adaptation are best left to local and regional governments, with national governments and nongovernmental organizations providing scientific/technical assessments of vulnerability, guidance on adaptation options, and financial assistance.

Examples of adaptation measures include improved early warning systems, hardening of infrastructure, particularly in coastal regions, disincentives for construction in low-lying, flood-prone regions, improved water management in drought-prone areas, development of drought- and salt-tolerant agricultural grains, and procedures for dealing with relocation of large human populations. Adaptation should be integrated with new development, and to leverage potential financial losses from the effects of climate change, new financial tools should be developed with costs tightly coupled to anticipated risks. The financial tools should be market-driven, and governments – local and national – should not assume costs resulting from undue risk in private development activities. But without an aggressive mitigation strategy, adaptation will become increasingly more difficult and in danger of being overwhelmed by climate change. Then what?

Beware of those who advocate geoengineering as a final solution and couch their views in terms of the problem-solving potential of human ingenuity. As discussed in Section 6.8, the solution is rife with risks and unintended consequences, and it does nothing to deal with the root cause of the problem. Atmospheric GHG concentrations would continue to rise, and if geoengineering measures must be terminated, the full effects of warming would be restored. At its core geoengineering is nothing more than an abdication of moral responsibility by current and past generations and a transfer of the risks and consequences of global warming to future generations. Mitigation and adaptation must be the pillars of climate policy.

11.3 Elements of an action plan: public policy

Whether implemented through international agreements or independently by national and/or local governments, public policy provides a powerful arsenal for accelerating implementation of mitigation and adaptation measures. Carbon emissions can be reduced by regulating efficiency and fuel economy standards for appliances and vehicles, establishing renewable portfolio standards for electric utilities, and placing caps on large emitters such as power plants and refineries. Public policy can also be exercised through financial incentives such as investment and production tax credits or feed-in tariffs to encourage production of carbon-free energy.

In view of the strides made in reducing the costs of solar and wind energy, it could be argued that subsidies such as investment and production tax credits are no longer needed. Such arguments miss the mark. Without subsidies, growth would be slowed and would be insufficient to increase the relative contribution of renewables to the world's total energy portfolio. The reality is that, for decades to come, global energy demand will continue to increase in concert with population and average energy consumption per capita. Without special measures to accelerate growth in renewables, the world's energy mix in 2050 would likely look much as it does today, which would mean a significant increase in the use of fossil fuels and atmospheric CO_2. Regulations and financial incentives make a difference and should be continued if not amplified.

Due to inherent assumptions and large uncertainties, efforts to predict long-term economic consequences of climate policy decisions are problematic (Section 7.2). However, economic considerations are critical over the shorter timelines associated with investments in energy technologies. How much should capital markets invest in the development and implementation of renewable and nuclear energy vis-à-vis the exploration and production of fossil fuels? How should electric utilities, automakers, other businesses, and individuals channel their investments? Investments in clean energy have certainly been increasing. But circa-2014, many investors across the world remained confident that good returns could still be obtained by directing hundreds of billions of dollars into fossil fuels, as in new oil, gas, and coal reserves, transportation systems for moving fuels to market, and new gas- and coal-fired power plants. The world is awash in fossil fuels, and such investments are made in the belief that the global demand for energy will continue to grow and that fossil fuels will continue to meet a large fraction of the demand. What would change the calculus of such decisions? What would it take to amplify investments in renewable energy at the expense of fossil fuels? The consensus view of economists points to another policy option, one that prices carbon emissions at levels sufficient to alter behavior by increasing the cost of using fossil fuels.

A cap-and-trade system can be used to price GHG emissions according to market-based principles of supply and demand. But experiences with the EU's ETS reveal the difficulties associated with setting a cap. If it is too high, the cost of emissions is too low and emissions remain high. If the cap is too low, economic impairment engenders corporate, political, and public resistance. Consider a proposal by the U.S. EPA to cap emissions from new coal-fired power plants at 500 kg-CO_2 per MWh_e (Wines, 2013a). The cap is less than the 725 kg anticipated for advanced, yet-to-be commercialized USCPC plants and well below the 790 kg associated with circa-2013, state-of-the-art SCPC plants. Achieving a 500 kg limit would require costly CCS measures, which could be a death knell for construction of new coal-fired plants. In contrast to the 500 kg cap favored by environmental groups, an association representing publicly owned utilities – with the support of politicians from coal-producing states and states in which coal-fired plants provide low utility rates – recommended a limit of no less than 860 kg, one that could be met without CCS and would sustain coal's role as a major contributor to GHG emissions. For cap-and-trade, bridging the gap between environmental, economic, and political interests remains a challenge.

What most economists favor is the simplicity of a carbon tax. The tax could start small and increase over time, concurrently sending appropriate signals to the market and giving it time to adapt. Consider an initial tax of $10 per tonne of CO_2 for power plant emissions and an escalator of 5% per year. For nominal emissions of 1 t-CO_2 per MWh_e, $10 per tonne of CO_2 would add 1 cent per kilowatt-hour to the cost of electricity in the first year and about 1.6 cents after ten years. For most consumers in developed nations, the added cost would be manageable and could be negated by efficiency and conservation measures. For utilities, it would provide greater certainty in planning for new generating capacity. A two-tiered system could be used to differentiate between developing and developed nations, with developing nations imposing a lower initial tax but a larger annual increase designed to achieve parity over several decades.

A multitiered approach could be taken for transportation fuels. For largely oil-rich countries having no tax and generously subsidizing fuel costs, taxation would be preceded by a gradual reduction in the subsidies. For developed nations such as Germany and the United Kingdom, fuel consumption is already curtailed by circa-2013 taxes of about $4.70 per gallon (approximately 60% of total fuel costs), and there is little need for additional taxes (Davis et al., 2014). But in developed nations with comparatively low taxes, such as the United States ($0.46 per gallon or about 14% of total fuel cost), a carbon tax could have a significant effect on consumption.

In view of circa-2016 and -2025 fuel economy standards imposed in the United States, it could be argued that a tax on transportation fuels is unnecessary. But it's questionable whether the standards will be met, and a carbon tax remains the best way to reduce consumption. In the United States, the federal tax has been held at 18.4¢ per gallon since 1993, and with state taxes the 2013 national average for the total fuel tax was about 46¢.[3] Since Americans have become accustomed to short-term swings of 50¢ or more per gallon, a carbon tax could be initiated in this amount and gradually increased. At 50¢ per gallon, it would correspond to about $55 per tonne of CO_2. An annual increase of 5% would increase the tax to $0.81 after ten years, still small compared to taxes in many developed nations but enough to influence behavior. Additional incentives could be provided by imposing an excise tax on the sale of vehicles failing to meet a threshold fuel economy standard, graduated to increase with decreasing fuel economy.

Putting a price on emissions carries an immediate economic penalty, and the argument is often made that it would impair economic development and job growth. Several counterarguments can be made. Carbon taxes could be made revenue neutral by reducing other taxes (payroll, income, and corporate), with priority given to reducing the burden on low-income segments of the population. Remaining revenues could still have positive economic benefits if used to accelerate development and implementation of emission reduction measures. By increasing incentives to use energy more efficiently and spurring investment in carbon-free sources of energy – effects of the tax and reinvestment, respectively – putting a price on emissions aids the economy by reducing energy costs and providing new engines of job creation.[4]

GHG emissions can be reduced by putting a price on all emissions, whatever the source. It would provide clear and strong signals; consumers would be incentivized to alter their behavior; and markets would respond accordingly, spawning innovation that reduces the cost of carbon-free energy, increases the efficiency of energy utilization, and creates jobs through evolution of new products, services, and infrastructure. Although equitable and meaningful implementation on a global scale is improbable at best, the policy would still provide an effective tool for reducing emissions in nations choosing to use it. If it places them at a competitive disadvantage to those choosing to impose no restraints on emissions – the so-called free-rider problem – a carbon tariff could be imposed on imports from recalcitrant nations.

11.4 Elements of an action plan: personal and social values

Technology and public policy are necessary elements for decarbonizing the world's energy portfolio. But in the language of mathematics, what is necessary may not be sufficient. Technology and policy are reducing the rate at which GHG emissions are increasing, and over time advancements in both sectors will bring the rate to zero – at which point emissions will have peaked – and emissions will begin to decrease as the rate becomes negative. By themselves, however, policy refinements and advancements in technology are not sufficient to drop per capita emissions below 2 t-CO_{2eq}. Another agent of change is needed.

The concept of *sustainable development* was introduced in Section 1.5. It involves judicial use of natural resources and challenges the notion that living standards increase with increasing levels of consumption – the idea that bigger and more are better. A distinction was also made between *economic growth* and *economic development*, the first aligning living standards with quantitative measures of growth and the second with enhancing the quality of life by increasing the efficiency of resource utilization and reducing environmental degradation – better rather than bigger. In Section 10.2, a distinction was made between *consumer values* of acquisition, convenience, and comfort and *survival values* that eschew overindulgence while meeting basic needs and seeking pleasure through moderation and simplicity – essentially embracing the Aristotelian virtues of temperance and prudence. In many cultures and nations, consumer values have become systemic *social values*. In Section 10.4, a distinction was made between luxury emissions associated with overconsumption and subsistence emissions associated with decent living standards.

The foregoing concepts are neither whimsical nor tertiary to climate change. In seeking solutions, technological and policy options must be accompanied by nothing less than a global sea change in cultural and behavioral norms, one that moderates human expectations and consumption habits. It begins with energy conservation, which must become a personal and societal imperative.

In the United States, many people have a conservation mindset, but it is by no means a social norm. Enabled by historically low costs of energy, many are given to waste and overconsumption in their homes, vehicles, and other lifestyle choices. Yet, with little or no effect on our well-being, we can drive smaller, more fuel-efficient vehicles, live in smaller homes, set thermostats higher during summer months and lower during winter months, and become advocates of alternative modes of transportation that provide acceptable substitutes for driving. The United States is certainly not alone among developed nations, but exemplars do exist. Contrast per capita energy consumption for Group 1 nations in Table 1.1 with results for Group 2. The comparison underscores the fact that a good standard of living can be maintained with much less energy consumption. Differences in energy consumption between Group 1 and 2 nations are in no small measure attributable to cultural differences. In wealthy nations such as Germany and Switzerland, conservation has long been a core social value, but not in nations such as the United States. What trajectory will be followed by developing nations? If, as it appears, they are embracing consumer values, it bodes ill for moderating energy consumption, curtailing the use of fossil fuels, and reducing GHG emissions.

The third leg of the stool – what is needed to establish necessary and sufficient conditions for addressing climate change – involves a change in cultural norms. It is no longer acceptable to value the fruits of energy consumption while turning a blind eye to the consequences of consumption. Globally, conservation must become a core social value. Aristotelian virtues of moderation and prudence must trump self-indulgence and overconsumption. Getting there involves challenges far greater than those associated with technology and policy. Changing human behavior and expectations is never easy, and in this case it would be made more difficult by significant resistance from special interests that would be adversely affected by the shift. It would require nothing less than political and social leadership on a global scale, along with persistent and effective use of all communication channels, from classrooms to social as well as traditional media.

Opposition to movement away from a culture of consumption is driven by two arguments, one having to do with its impact on human fulfillment and the other dealing with its impact on the economy. The first argument is specious. There are many ways to achieve happiness, fulfillment, and a sense of well-being in a lifestyle that is not centered on consumption. The second argument merits reflection. Would the new cultural norm *trash* the global economy? Not necessarily!

Businesses and jobs would certainly contract in many traditional manufacturing and energy sectors, but they would grow in other sectors. The automobile industry would not disappear; it would continue to evolve in response to changing energy sources and consumer preferences. Public transportation systems would become more ubiquitous and efficient, while urban and suburban communities would respond to the challenges of using energy more efficiently and becoming more resilient to the effects of climate change. New energy and service sectors would emerge, and advancements in information and communication technologies would attract investments that increased energy efficiency and productivity across a broad swath of industrial and service sectors. People would continue to innovate in response to new opportunities, contributing to economic development and job creation. During a transition over several decades, the global economy would experience significant structural changes, but it need not crash.

11.5 Epilogue

Historically, climate change has been a natural phenomenon that has cycled the Earth through glaciations and interglacials at intervals of 40,000 to 100,000 years, with the interglacial of the current Holocene providing relative climate tranquility for nearly 12,000 years. But over the last two centuries a new agent of climate change has emerged, one that is operative over a much shorter time scale. Principally through the use of fossil fuels, humankind is perturbing the global energy balance in ways that are warming the planet and disrupting human existence. The world is awash in fossil fuels; capital markets continue to invest heavily in the production and use of the fuels; people around the world continue to aspire to a standard of living that is energy intensive; and globally, energy demand continues to rise, with carbon-free sources of energy meeting less than 15% of the demand (BP, 2014).

Anthropogenic climate change is a serious problem. Confirmation of the problem is rooted in science, while solutions involve an array of competing technological, economic, political, and cultural issues. These factors and the existence of many stakeholders with competing interests make it a *wicked problem*, one for which definitive solutions have yet to emerge. Whether a satisfactory solution can be obtained depends on a number of factors. Technology and policy decisions determine the pace and the extent to which the world's energy portfolio transitions from its current emphasis on fossil fuels to carbon-free sources. They also determine the extent to which efficiency measures reduce the global demand for energy. But reconfiguring the energy mix and increasing efficiency will not be enough.

In a circa-2050 world inhabited by more than 9 billion people, with billions aspiring to higher living standards, expectations of the level of consumption needed to achieve those standards will strongly influence the demand for energy. If expectations lean toward high levels of consumption, it is difficult to see how requisite energy needs can be met without forsaking the need to significantly reduce GHG emissions. Reducing emissions will require a sharp departure from the current norm of associating living standards with levels of consumption. It will also require all hands on deck. As individuals, organizations, and governments, the entire global community must be part of the solution.

Will this story have a happy ending? Will the effects of global warming be manageable? Will the global economy survive? At this time, there are no answers to these questions. But what can be said is that, through its actions or inaction, humankind will determine its own destiny.

Units and conversion factors

Large numerical prefixes

Prefix	Abbreviation	Multiplier
kilo	k	10^3
mega	M	10^6
giga	G	10^9
tera	T	10^{12}
peta	P	10^{15}

Energy

1 quad	$= 10^{15}$ Btu	$= 1{,}055$ PJ	$= 293$ TWh	
1 TJ	$= 947.8$ Mbtu	$= 10^{-3}$ PJ	$= 0.278$ GWh	
1 GWh	$= 3{,}412$ Mbtu	$= 3{,}600$ GJ	$= 10^6$ kWh	
1 kWh	$= 3{,}412$ Btu	$= 3{,}600$ kJ		
1 Mtoe	$= 39.7$ TBtu	$= 41.87$ PJ	$= 11.63$ TWh	$= 7.33$ Mboe

Base units of energy in the English and International (SI) systems are the British thermal unit (Btu) and the joule (J). The kilowatt-hour represents energy as a product of the rate at which it is used (1 kW = 1,000 J/s) over 1 hour (1 h = 3,600 s). Energy can also be represented in terms of an equivalent amount of oil measured in millions of tonnes (Mtoe) or barrels (Mboe).

Mass

1 t (tonne)	$= 1{,}000$ kg	$= 2{,}204.6$ lb	$= 1.102$ short tons	$= 0.984$ long tons
1 Mt	$= 10^9$ kg	$= 10^{12}$ g	$= 1$ Tg	

The short ton (2,000 lb or 907 kg) is used in the United States and Canada and the long ton (2,240 lb or 1,016 kg) in the United Kingdom. To convert from GHG emissions in mass of carbon to mass of carbon dioxide, multiply by 3.67.

Volume[a]

1 m³	= 1,000 L	= 35.3 ft³	= 264.2 gal	= 6.29 bbl
1 gal	= 3.785 L	= 0.1337 ft³	= 0.003785 m³	= 0.0238 bbl
1 bbl	= 159 L	= 0.159 m³	= 42.0 gal	= 5.62 ft³
1 ppmv	= 1 part per million by volume			
1 ppbv	= 1 part per billion by volume			
1 pptv	= 1 part per trillion by volume			

[a] U.S. gallons throughout (1 U.S. gal = 0.832 imperial gal).

APPENDIX B

Fossil fuels

To appreciate the important role played by fossil fuels, consider the results of Table B.1, which provide U.S. oil, gas, and coal consumption for each of the four economic sectors. Virtually all nations depend strongly on fossil fuels for each sector, albeit with differences in the relative contribution of each fuel.

TABLE B.1. *Approximate U.S. 2011 fossil fuel consumption per sector. All values are in quads, and numbers in parentheses are percentages of total petroleum, gas, or coal consumption for each sector (EIA, 2014a)[a]*

	Oil[b]	Natural gas	Coal[c]
Residential/ Commercial	1.8 (5%)	7.9 (32%)	–
Industrial	8.1 (23%)	8.2 (33%)	1.6 (8%)
Transportation	25.0 (71%)	0.8 (3%)	–
Electric Power	0.4 (1%)	7.9 (32%)	18.1 (92%)
Total	35.3	24.8	19.7

[a] Nuclear and renewable forms of energy provided an additional 8.3 and 9.1 quads, respectively, to total U.S. consumption of primary energy. Renewable energy forms included hydropower, biofuels, wind, solar and geothermal energy. Nuclear energy was used exclusively for electric power, as was 54% of the renewable energy. Other sector uses of renewable energy were transportation (13%), industrial (25%) and residential/commercial (8%).

[b] Oil supplied 93% of the energy used for transportation, with natural gas and renewables providing 3% and 4%, respectively.

[c] Coal provided 46% of the primary energy for power production, with the balance supplied by nuclear energy (21%), natural gas (20%), renewable energy (13%) and petroleum (<1%). Since 2011, coal's share has dropped below 40% at the expense of increasing natural gas and renewables.

Apart from transportation, uses of natural gas are well balanced, with residential/commercial, industrial and utility applications accounting for roughly 32%, 33% and 32% of the total. In recent years U.S. gas production has been resurgent, largely from access to nonconventional (shale) gas enabled by horizontal drilling and hydraulic fracturing. Increased production, correspondingly low prices, and a more benign environmental impact have made natural gas an attractive fuel option, and its use is expected to increase in all four sectors – as a substitute for coal in generating electricity and for oil in each of the others sectors.

Although gas consumption is well balanced across three sectors, the same cannot be said for oil and coal. Although oil is used for residential/commercial heating and industrial process heat, fully 71% of circa-2011 consumption was for transportation (land, air, and sea). The importance of oil to transportation is underscored by the fact that it provided 93% of all primary energy consumed by the sector. Even with increased use of biofuels and electric vehicles, oil will remain the dominant transportation fuel for decades to come, globally as well as in the United States.

Coal also has a primary application, with 92% of 2011 consumption used to generate electricity. However, while power production will remain its principal function, coal's share of the primary energy used for this purpose has been in decline, dropping from 50% in 2008 to 46% in 2011 and 39% in 2013. The decline is attributable to environmental and economic factors, with coal ceding market share to natural gas and renewables. However, this trend is hardly universal, and in nations such as China and India, coal continues to fuel more than 80% of power generation. Because of its abundance and relatively low cost, coal has been and will remain a popular fuel for global power production.

Whatever the application, fossil fuels provide a potent and readily exploited source of chemical energy. The energy is associated with molecular bonding of the hydrogen and carbon atoms (H-C bonds) and is released (converted to thermal energy) when the fuel is oxidized in a combustion reaction. The chemical energy of a fuel, and hence the amount of thermal energy derived from the reaction, is termed the heat of combustion or, alternatively, the heating value (HV) of the fuel. A distinction is made between higher and lower heating values (HHV and LHV) according to whether water in the products of combustion condenses from the vapor to the liquid state or remains a vapor. Heating values are typically provided in units of energy per unit mass of fuel, such as kilojoules per kilogram (kJ/kg-fuel).

The heating value is one of several important properties of a fuel. Another is its carbon content (CC), which provides the mass of carbon

TABLE B.2. *Nominal higher heating value (HHV), carbon content (CC) and carbon intensity (CI) of selected fossil fuels and their derivatives*[a]

Fuel type	HHV (kJ/ kg-fuel)	CC (kg-C/ kg-fuel)	CI (kg-C/GJ)[b]
Crude oil	44,700	0.85	19.0
Gasoline	46,400	0.86	18.5
Kerosene	46,200	0.86	18.6
Diesel (#2 fuel)[c]	45,400	0.86	18.9
Natural gas	49,400	0.74	15.0
Coal			
Lignite	19,500	–	–
Sub-bituminous	23,000	0.48	20.9
Bituminous	32,400	0.67	20.7

[a] Adapted from DOT (2006), tables 6.11 and B.4; Rubin (2001), tables 5.1 and 12.6; and Tester et al. (2005), table 7.1. The table provides nominal values, with variations of only a few percent for oil and natural gas. Values for coal can vary by as much as 30%, according to the grade and the region from which the coal is mined.
[b] With molecular weights of 44.01 and 12.01 for carbon dioxide and carbon, respectively, 3.67 kg-CO_2 are associated with the production of each kg-C ($CO_2I = 3.67 \times CI$).
[c] Although the heating value of diesel fuel is slightly less than that of gasoline, its mass density (nominally 850 kg/m³) exceeds that of gasoline (nominally 750 kg/m³), providing it with approximately 11% more energy on a volumetric basis (146 MJ/gal versus 132 MJ/gal).

per unit mass of fuel (kg-C/kg-fuel). A third parameter, termed the carbon intensity (CI), is obtained by dividing the fuel's heating value (LHV or HHV) into its carbon content (CC).

$$CI(\text{kg-C/GJ-fuel}) = \frac{CC(\text{kg-C/kg-fuel})}{HV(\text{kJ/kg-fuel})} \times \left(10^6 \frac{kJ}{GJ}\right)$$

The carbon intensity provides the amount of carbon released relative to the amount of energy released when the fuel is burned. The corresponding CO_2 intensity (CO_2I) is larger by a factor of 3.67 (the ratio of molecular weights for CO_2 and C). In the interest of reducing CO_2 emissions, fuels of smaller CI are preferred. Of the three fossil fuels, coal and natural gas are characterized by the largest and smallest carbon intensities, respectively.

Representative heating values, carbon contents, and carbon intensities are provided in Table B.2. The carbon intensity of coal is about 40% larger than that of natural gas. However, in terms of using these fuels to generate electricity, the comparison is weighted even more favorably for natural gas. Nominal efficiencies for converting primary energy to electricity in coal-fired and combined-cycle natural gas power plants are 35% and 55%, respectively. Hence, when viewed in terms of carbon emissions per unit of power production, emissions for coal are approximately twice those for gas. Through the last decade, coal and gas supplied approximately 50% and 20%, respectively, of the primary energy used for U.S. power production. More recently, however, coal's share has been declining as the use of gas and renewables has increased.

The carbon intensity of a nation is the ratio of its emissions to its total energy consumption. Its value depends on the efficiency with which energy is used, as well as on the mix of fossil fuels and carbon-free sources of energy. Intensities are generally smaller for developed than for developing nations, and the trend for many nations has been one of declining intensities over time. The trend is attributable to an increase in the efficiency with which energy is used and a transition to primary energy sources of reduced or no carbon content.

While fossil fuels provide an enormous source of chemical energy, their use is not without detrimental effects on the environment and human health. Such effects are manifested in production, processing, and use of the fuels and by pollution of the atmosphere, land, and water bodies. Ideally, one would want fossil fuels to consist exclusively of energy-rich H-C bonds and, when burned, to experience complete combustion yielding only two products of combustion (POCs), namely H_2O and CO_2. However, if combustion is incomplete, the POCs can also include carbon monoxide (CO) and fine particles in the form of soot (C). Moreover, coal and oil can contain significant amounts of sulfur and nitrogen, which contribute to the formation of pollutants such as sulfur dioxide (SO_2) and oxides of nitrogen (NO_x or NOX), as well as toxic inorganic substances such as lead (Pb) and mercury (Hg).

In the United States, serious consideration of the environmental consequences of fossil fuel combustion took root with passage of the Clean Air Act (CAA) of 1963. The Act initiated studies to identify the principal sources of air pollution and to establish air quality standards that capped pollutant atmospheric concentrations at acceptable levels. The Environmental Protection Agency (EPA) was created to develop national standards, and its work culminated in passage of the landmark Clean Air Act of 1970, which regulated concentrations of what are termed criteria pollutants. The pollutants included CO, SO_2, NO_x, ground-level ozone (O_3), and particulate matter (PM). A sixth pollutant (Pb) was added in 1978 and a seventh (Hg) in 2005.

The 1970 CAA obligated the EPA to establish standards that ensured an adequate margin of safety in protecting public health. The standards had to be based solely on rigorous scientific studies, and the EPA could only include cost-benefit (economic) considerations in prescribing timelines for implementing the standards. The CAA also obligated the EPA to conduct a review of the standards for each pollutant every five years. As of 2013, the EPA had yet to establish emission standards for CO_2, although regulation of emissions from power plants was pending for 2015.

A point often made by those inclined to dismiss the significance of global warming is that CO_2 is not a pollutant, at least not in the sense that POCs such as NO_x and SO_2 are viewed as detrimental to the environment and human health. However, this view was rejected by the U.S. Supreme Court on April 1, 2007 in the case of *Massachusetts v. the Environmental Protection Agency*. Massachusetts and eleven other states sued the EPA over its refusal to regulate GHG emissions from the transportation sector. The Court's decision was marked by two key features. First, it ruled that because CO_2 emissions contribute to global warming and attendant climate change is detrimental to human welfare, CO_2 should be treated as a pollutant. The court also ruled that the EPA is empowered by its existing authority to regulate CO_2 emissions and, by extension, the use of fossil fuels.

Anthropogenic sources of natural gas and methane

In situ, natural gas is a mixture consisting largely of methane, CH_4 (70–90% by volume); smaller amounts of ethane, C_2H_6, propane, C_3H_8, butane, C_4H_{10}, pentane, C_5H_{12}, and/or hexane, C_6H_{14} (0–20%); carbon dioxide, CO_2 (0–8%); and trace amounts of other gases such as hydrogen sulfide, H_2S. As the pressure and temperature of the gas decrease during extraction from a well, pentane and hexane condense to their liquid states and separate from the gas. At the wellhead, CO_2 is removed, as is H_2S, which is both toxic and corrosive. The remaining gas can then be compressed and routed to a pipeline, or it can be liquefied at a temperature of $-162\,°C$ ($-260°F$) to increase its energy content per unit volume, facilitating transport as *liquefied natural gas* (LNG).

As late as 2006, conventional wisdom was that U.S. production of natural gas was in decline, with increasing imports needed to meet demand. But perceptions began to change with the discovery and exploitation of large gas-bearing shales, beginning with the Barnett Shale in Northeast Texas. Just two years later, with the discovery of the Marcellus Shale extending southwest from upstate New York through Appalachia, as well as the Haynesville (Louisiana), Fayetteville (Arkansas), Woodford-Arkoma (Arkansas and Oklahoma), and Bakken (North Dakota and Montana) shales, there was growing belief that the nation would soon be awash in natural gas and its reserves-to-production ratio would exceed 100 years (Krauss, 2008a, 2008b).

Conventional gas is extracted primarily from porous sandstone lying below impermeable cap rock. The large permeability of the sandstone makes it easy to extract the gas once a well has been drilled. That's not the case for unconventional gas found in coal seams, tight sands, and shale rock. The low permeability of these formations limits gas production by conventional means.

Shale is a soft, fine-grained rock formed from mud deposits in shallow seas some 400 million years ago. Although shale formations are buried up to 4,000 meters below ground and gas is trapped within the shale, advanced recovery techniques have made production economically viable. A well can be drilled to the requisite depth and extended horizontally for thousands of meters, thereby greatly expanding the size of a gas field accessible to a single well. In a process termed fracking, a pressurized slurry of water, sand, and chemicals can then be injected to hydraulically fracture the shale, enlarging preexisting cracks and creating permeable zones that enable natural gas to flow to the well. Horizontal drilling and fracking have increased the production of shale gas from virtually nil in the early 1990s to 30% of total U.S. production in 2011(EIA, 2013a, 2013b).

In the densely populated Dallas/Fort Worth area, horizontal drilling facilitated gas extraction from underlying portions of the Barnett Shale. In most cases, drilling and extraction were achieved without blighting the landscape and with financial benefits to property owners leasing rights to the gas below their land. By mid-2008, there were 7,500 wells in the Barnett Shale, covering approximately 5,000 square miles and producing approximately 40 billion cubic meters of gas annually (Jones, 2008). In 2008, production from the Barnett Shale alone accounted for 7% of total U.S. production. Another benefit of the Barnett and other shales is their proximity to existing pipelines, reducing the amount of additional infrastructure needed to get the gas to market.

Although U.S. 2008 proven reserves of natural gas were only 6.9 trillion cubic meters (Tcm) (BP, 2014), estimates commissioned by the U.S. gas industry placed the total amount of recoverable shale gas at approximately 24 Tcm (Kraus, 2008b). Including the contribution of other unproved forms of conventional and unconventional gas, the ultimate recoverable resource could be as large as 59 Tcm. Of the 52 Tcm in the unproved category, prospects for recovery were deemed to be probable or possible for nearly 70% and speculative for the remainder. It's no wonder there was growing optimism for sustaining high levels of production. But, like other forms of unconventional gas, developing shale gas is not without environmental issues.

There are three main environmental concerns: (1) that injection of the highly pressurized slurry could initiate earthquakes; (2) that the slurry could contaminate water supplies; and (3) that gas leaks into the atmosphere could exacerbate global warming. In the United States these concerns are exacerbated by ill-defined standards and whether regulation is to be the purview of local, state, or federal agencies.

Large amounts of water are needed to fracture the shale – from 2 million to 7 million gallons per well – and different chemicals are used to facilitate extraction (Mouawad and Krauss, 2009; Vidic et al., 2013). To enhance well production, the chemicals are used as antibacterial and scaling agents, flow friction reducers, proppants to keep fractures open, and surfactants to reduce surface tension. In all, about 750 chemicals and other additives are used, many of which are toxic, raising concerns for long-term contamination of groundwater and watersheds. Shale gas developers counter the concerns by emphasizing that drilling is to depths well below groundwater aquifers, which are separated from the shale by layers of dirt and rock, and that vertical well casings consist of several concentric steel pipes sandwiched between layers of cement (Vidic et al., 2013). However, attempts to alleviate concerns have not been helped by the reluctance of developers to divulge the nature of their chemicals, citing the need to protect their intellectual property and competitive advantage, nor by documented cases of water contamination at drilling sites for conventional gas in western states (Lustgarten, 2009). In some cases, where waste water must be returned to the surface, the water has contained unacceptable levels of toxins and radioactivity (Urbina, 2011).

Moving forward, the large potential of shale gas to meet future needs will likely spur continued development, but with greater attention paid to mitigating environmental effects. And development will not be limited to the United States. Shale gas deposits exist in many regions of the world, and by the end of 2009, exploratory wells were being drilled in seven European nations (Economist, 2009b), while nations such as China and Argentina were preparing to develop their significant resources. The potential of unconventional sources to satisfy future needs is huge, and globally, recoverable shale, coal-bed, and tight gas resources may be as large as 1,000 Tcm (Rotman, 2009).

In addition to being the principal constituent of conventional and unconventional sources of natural gas, large amounts of methane are also encapsulated by an ice lattice in what are termed clathrates – aka methane hydrates. Stable forms exist over a wide range of low temperatures and high pressures beneath Arctic permafrost and sea beds. They can also be found beneath deep waters off all continents, with most of the world's store believed to be in sea beds on outer continental shelves. Although clathrates typically exist in large horizontal zones at depths of only 450 to 600 meters, recovery without releasing methane to the atmosphere is not a trivial task.

A cubic meter of solid hydrate can contain up to 180 cubic meters of methane gas at standard temperature and pressure (Boswell, 2005), and globally the resource is estimated to contain up to 20 quadrillion (20×10^{15}) cubic meters (Boswell, 2009). Although a large uncertainty is attached to such estimates, even if they are overly optimistic by an order of magnitude, the resource could contain more energy than all of the world's other fossil fuels.

Not surprisingly, efforts are under way to locate major clathrate deposits and to develop production technologies. Production processes will depend on the sediment type and porosity and the influence of its permeability on flow in the deposit. Options include heating (thermal breakdown) of hydrates in permafrost, depressurization of sea deposits, and injection of fluids such as brines or alcohols to lower the freezing point (Boswell, 2009). However, any successful production technology must be able to control the separation of methane from the deposit, and there are many research questions yet to be answered. Ultimately, technical and economic viability will be determined by the existence of concentrated deposits in large fields of moderate to high permeability, as well as the environmental impact of large-scale production. Yet to be resolved is the impact of production on the stability of deep-water continental shelves and the impact of production on global warming, which would be accelerated by release of methane to the atmosphere.

The effect of a rapid increase in methane emissions could be twofold. In addition to the direct and immediate effect of adding large amounts of CH_4 to the atmosphere, there could be a long-term effect of increasing the atmospheric lifetime beyond the currently accepted value of twelve years (Table 3.1). The lifetime would increase if the rate at which CH_4 is converted to CO_2 and H_2O by reaction with the hydroxyl radical is overwhelmed by the rate at which CH_4 is discharged to the atmosphere.

Environmental time scales and inertia

Environmental effects can be differentiated in terms of *time scales* – also called time constants. Pursuant to an input that alters the equilibrium of a system, a time scale provides an approximate measure of how long it takes for the input to achieve a significant portion (e.g., 67%) of its final effect and hence to reach a new equilibrium. The larger the time scale, the larger the *inertia* of the system.

Consider the infamous fogs that plagued London from the mid-1700s to the mid-1900s. They were not a natural phenomenon, but were anthropogenic and caused by the use of coal for everything from space heating in homes and businesses to, in the twentieth century, generation of electric power. During cold and still winter days, the soot and sulfur-laden gases produced by burning coal would hover over the city, reducing visibility to near zero and inducing serious illness and death among those with respiratory problems. Created by several million sources of coal combustion, the last London fog in 1952 resulted in 4,000 deaths. The problem could no longer be ignored. In 1956, Parliament passed a clean air act prohibiting coal combustion in the homes and businesses of England's largest cities, and the problem, if not eliminated, was significantly alleviated within a few years. This example illustrates an environmental problem for which the causal agents were well understood, the scope was local, there was no uncertainty concerning the nature and seriousness of the consequences, and remedies could be quickly implemented.

The ozone hole in the Earth's stratosphere provides a more recent example, one that's global in scope and for which time scales associated with remediation are much longer. Stratospheric ozone absorbs much of the ultraviolet (UV) components of solar radiation. With depletion of the ozone, more of the UV reaches the Earth's surface, with adverse effects on human health and the environment. The effects could not be ignored, and chemical reactions with the chlorofluorocarbons (CFCs) used in vapor-compression refrigeration and air-conditioning systems were determined to be the culprit.

Recognition of the problem spurred global cooperation in obtaining a solution, highlighted in 1987 by the Montreal Protocol, which committed the world's industrial nations to 50% and 85% reductions in CFC production by 2005 and 2007, respectively. Implementation of the Protocol began the process of restoring atmospheric ozone, with complete restoration expected by the middle of the century. Like the London fog, causal agents were well understood and there was no uncertainty concerning the seriousness of the problem. In this case, however, time scales associated with remediation are measured in decades instead of years.

The foregoing problems were characterized by serious and widely acknowledged consequences and by solutions that could be implemented over time scales ranging from a few years to decades. And, just as important, the costs of remediation were manageable. London homes and businesses resorted to alternative fuels, while more benign fluids were developed and adopted for refrigeration and air-conditioning systems. But for global warming and climate change, things are different.

While the scientific principles underlying anthropogenic contributions to global warming are known, there is some uncertainty concerning effects of warming on the climate and, in turn, the effects of climate change on the environment and humankind. Issues are further clouded by uncertainties in the economic impact of climate change, as well as by the cost of mitigating measures. Since global warming is driven principally by the burning of fossil fuels, which also happen to drive the global economy, the economic implications of significant curtailment are problematic. The task of replacing fossil fuels with carbonless sources of energy is challenging, with barriers imposed by an array of technical, political, social, and economic considerations.

Time scales provide another distinction between global warming and traditional environmental problems. Effects of adding anthropogenic GHGs to the atmosphere include rising temperatures and sea levels. But how long does it take for the effects to fully manifest themselves? How long does it take for CO_2 released yesterday, today, or tomorrow to have its full effect? In fact, it can take a very long time.

As indicated in Sections 2.1 and 3.5, the Earth's climate is determined by a complex array of interactions between the atmosphere and entities such as the hydrosphere and cryosphere. While the time scale associated with atmospheric mixing of GHG emissions is less than a year, time scales associated with reductions in atmospheric GHG concentrations due to natural processes can range from decades to centuries. Time scales associated with response of the Earth's average surface temperature to increasing GHG concentrations are also measured in centuries, while those associated with changes to the hydrosphere and cryosphere are even longer. Time scales for rising levels of the Earth's oceans are influenced by their enormous capacity for storing thermal energy and by the resistance to transfer of energy from surface layers to deeper waters. Large energy requirements associated with the melting of ice sheets also affect time scales associated with their contribution to rising sea levels.

To provide some meat to the foregoing discussion, consider the results of a study performed for two hypothetical CO_2 stabilization scenarios – scenarios that consider the effect of different CO_2 emission trajectories on the time required to yield new (equilibrium) values of the atmosphere's CO_2 concentration and the Earth's average surface temperature and sea level (IPCC, 2001a). One scenario assumes that annual emissions peak at 11 Gt-C in 2025 and decay to less than 2 Gt by 2300. The other scenario assumes constant emissions of approximately 9 Gt-C from 2000 to 2300.

For the first scenario the atmospheric CO_2 concentration stabilizes at 550 ppm by the end of the century, but it would take another two centuries for rising temperatures to stabilize. Due to limitations on the rate at which energy is transferred within the oceans and the slow response of ice sheets to changing temperatures, sea levels would continue to rise for millennia. Outcomes are worse for the second scenario. If emissions remained constant for three centuries, atmospheric CO_2 concentrations would reach 800 ppm by 2300 and would continue rising well thereafter, as would temperatures and sea levels.

The principal takeaway from the foregoing results is that the effects of today's GHG emissions are not manifested immediately but gradually and over long timelines that can encompass millennia. Hence, adverse effects will be felt much less by those responsible for – and benefiting from – the emissions than by future generations. Another point to be made is that hypothetical emissions associated with the results are, if anything, overly optimistic. From trends revealed in Sections 3.3 and 3.4, actual emissions will exceed the hypothetical scenarios, likely by large amounts. Increasing emissions, as well as the time at which they peak, will increase values at which the atmospheric CO_2 concentration, temperature, and sea level ultimately stabilize, as well as the times required to achieve stabilization. Reducing emissions will require a significant departure from business as usual, which introduces yet another time scale.

Coal-fired power plants: operating conditions and costs of carbon capture and sequestration

Basic elements of a complete end-to-end CCS system include CO_2 capture and compression, transport from the source to a repository, and monitoring the status of the sequestered CO_2 over time. Each element contributes to the total cost of CCS. Capture and compression costs increase with decreasing concentration and pressure of CO_2 in the gas stream from which it is extracted, while transport costs increase with increasing distance from the source to the storage site. Costs of monitoring the sequestered CO_2 are relatively small.

The cost of carbon capture for coal-fired power plants depends on the type of power plant, its efficiency, and whether the carbon is captured by pre- or post-combustion processes. To understand these options, it's helpful to first examine the makeup of a conventional plant. Consider the system shown in Figure E.1.

Pulverized coal (PC) and air enter the boiler, where the coal is burned to produce high-temperature products of combustion (POC). As the POC pass over tubes within the boiler (the steam generator), heat is transferred to pressurized water flowing through the tubes, converting it to steam. The steam is then routed through a set of steam turbines, where work is done on the turbine blades, enabling the turbine shaft to do work on an electric generator. To reduce losses in transmission by the grid, the voltage is increased by a transformer before the electricity enters the grid. The turbine has multiple stages to accommodate reheating in the boiler and to maximize the efficiency with which work is done by the steam. After leaving the last turbine stage, the steam is converted to water in a condenser and passed through a series of feedwater heaters and pumps before returning to the boiler to repeat the process.

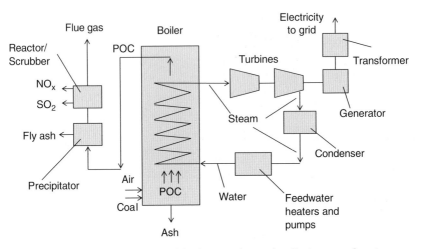

FIGURE E.1. Schematic of a coal-fired power plant with pollution control equipment.

The efficiency with which the chemical energy of the coal is converted to electrical energy depends strongly on the temperature and pressure of the steam emerging from the boiler. Most PC plants operate under what are termed subcritical conditions and have a nominal efficiency of 35%. The maximum pressure of the steam is less than the critical pressure of water (221 bar, or 218 times normal atmospheric pressure) and temperatures are less than approximately 550°C. However, newer supercritical pulverized coal (SCPC) plants operate above the critical pressure and at higher temperatures, with yet higher pressures and temperatures expected from evolution of a new generation of ultra supercritical pulverized coal (USCPC) plants. With the ability to operate at progressively higher pressures and temperatures, efficiencies will increase to 45% or more. The larger the efficiency, the smaller the amount of coal that must be burned to generate the electricity and the smaller the carbon emissions.

The power plant of Figure E.1 is equipped to remove pollutants from the POC. Comparatively large solids are removed from the bottom of the boiler as *ash*, while smaller particulates (fly ash) are removed by an electrostatic precipitator. Oxides of nitrogen (NO_x) are removed by selective reactions with other chemicals and sulfur dioxide (SO_2) by passage through a scrubber, where the POC interact with a lime emulsion. Although the system of Figure E.1 lacks provision for removing CO_2 from the POC, it could be retrofitted to remove CO_2 from the flue gas.

Just as the POC are scrubbed to remove SO_2, the flue gas can be routed through an absorption tower for removal of much of the CO_2

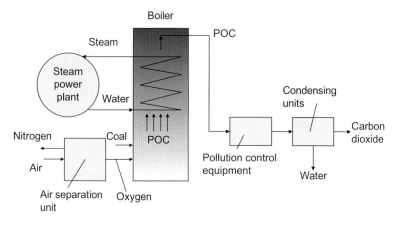

FIGURE E.2. Oxy-fuel combustion steam power plant with carbon capture.

(Socolow, 2005). The gas interacts with droplets of an absorbent such as amine (an ammonia-based compound), which preferentially absorbs CO_2 at low temperatures. The CO_2-rich absorbent leaving the tower is routed to a regenerator in which it is heated to release (capture) the CO_2. The cycle is completed by cooling the amine and returning it to the absorber. The process works, but energy requirements of the absorbent cycle are large, as are capital and operating expenditures. A significant fraction of the cost is attributable to the relatively low concentration of CO_2 in the POC – typically less than 15% by volume – which increases demands on all elements of the removal process. The cost of CCS can therefore be reduced by increasing the relative concentration of CO_2 in the POC.

Since the large amount of nitrogen introduced with the combustion air is the principal contributor to the small relative concentration of CO_2 in the POC, the concentration can be increased significantly by removing nitrogen from the air. Termed oxy-fuel combustion (OFC) and integrated gasification combined cycle (IGCC), two alternative coal-fired power plants are designed for that purpose.

Apart from two additional components and a significant modification to one component, OFC resembles the conventional coal-fired power plant of Figure E.1. As shown in Figure E.2, one addition is the upstream separation unit used to purge the air of nitrogen and thereby to supply the boiler with a nearly pure stream of oxygen. The major modification is to the boiler, which must be redesigned to deal with the higher temperatures resulting from oxy-fuel combustion. The other addition is a downstream cooling system to condense water vapor in the flue gas, leaving an enriched

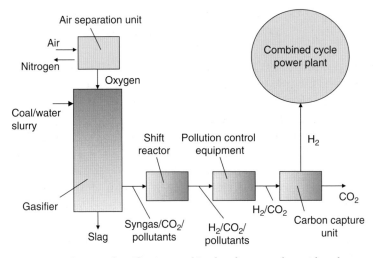

FIGURE E.3. Integrated gasification combined cycle power plant with carbon capture.

carbon dioxide stream that can be compressed and sequestered. Oxy-fuel combustion systems are in development, but have yet to be operated at scales of a hundred megawatts or more.

IGCC power plants represent a marriage of two well-established technologies, one involving gasification of the coal and the other a combined-cycle power plant to generate electricity. Like OFC, the IGCC process begins with separation of nitrogen from the air, but unlike OFC, the coal is gasified and not burned. As shown in Figure E.3, a coal-water slurry and high-purity oxygen enter a gasifier, where they react at very high temperatures and pressures to produce syngas (a mixture of hydrogen and carbon monoxide), carbon dioxide and pollutants such as acid (sulfur-based) gases, fine particulates and mercury. In what is termed a water-gas shift reaction, the carbon monoxide in the syngas can be converted to carbon dioxide and additional hydrogen, and measures can then be taken to remove the pollutants, leaving a mixture comprised largely of hydrogen and carbon dioxide. Because of its high concentration, the carbon dioxide can now be captured at lower cost and sequestered. The remaining hydrogen-rich gas is burned, and the POC are used to power a combined cycle comprised of integrated gas and steam turbines.

Costs of the foregoing CCS options have been assessed for a hypothetical 500 MW_e (net) power plant (MIT, 2007). The assessment considered air-blown pulverized coal-combustion plants operating under subcritical (PC), supercritical (SCPC) and ultra-supercritical (USCPC) conditions, as well as oxygen-blown supercritical oxy-fuel combustion (SCPC-OXY) and integrated gasification combined cycle (IGCC) systems. For all but the SCPC-OXY system, results were obtained for two cases, one without carbon capture and the other with 90% capture. The SCPC-OXY system is designed specifically to meet the needs of CCS and would not be suited for use without carbon capture. For the air-blown systems, carbon capture was presumed to involve a post-combustion amine absorption process. Since the SCPC-OXY system is only suitable with carbon capture, the cost of capture (COCC) is measured relative to an SCPC system without capture. Estimates for the oxy-fuel and IGCC systems presume n^{th} plant conditions, that is, conditions reflecting improvements and cost reductions associated with lessons learned from construction and operation of the first few plants.

Table E.1 reveals the significant costs incurred by implementing CCS. For each of the PC systems, there is a reduction of 9.2% in the efficiency of converting the chemical energy of the coal to electricity. For the SCPC system, the efficiency goes from 38.5% to 29.3%, with the difference attributed to parasitic power requirements of CCS. Hence, to generate a net electric power of 500 MW_e for transmission, more coal must be consumed, with increased CO_2 production. For the SCPC system, the rate of CO_2 production would increase from 415,000 to 545,000 kg-CO_2/h, and with carbon capture, the rate of carbon emissions (RCE) would decrease from 415,000 to 54,500 kg-CO_2/h. The carbon intensity (CI) would drop from 830 to 109 kg-CO_2/kWh_e. With or without carbon capture, emissions would be reduced by efficiency improvements associated with transitioning from subcritical PC to SCPC and USCPC conditions. Efficiencies for the oxygen-blown systems with carbon capture are intermediate to those of the SCPC and USCPC plants, as are values of RCE and CI.

Although total plant costs (TPC) with carbon capture are comparable for the five systems, ranging from $1,900/$kW_e$ for the oxygen-blown systems to approximately $2,150/$kW_e$ for the air-blown plants, incremental plant costs associated with carbon capture vary, ranging from a high of $950/$kW_e$ for the subcritical PC plant to a low of $460/$kW_e$ for the IGCC system.

Total plant costs contribute to the levelized cost of electricity (LCOE), as do fuel costs, nonfuel operating expenses, and costs associated with gas cleanup and CO_2 removal. With carbon capture, the LCOE is largest for the PC system and decreases with increasing efficiencies for SCPC and USCPC conditions. However, the lowest costs are associated with the oxygen-blown systems. Although the IGCC plant provides the highest LCOE without carbon capture ($51.3/$MWh_e$), it offers the lowest

TABLE E.1. *Impact of coal-fired power-generating technologies and carbon capture on efficiency (η), rate of CO_2 emissions (RCE), intensity of CO_2 emissions (CI), total plant cost (TPC), levelized cost of electricity (LCOE) and cost of carbon capture (COCC).[a] Adapted from The Future of Coal (MIT, 2007)*

	Air-Blown			Oxygen-Blown	
	PC	SCPC	USCPC	SCPC-OXY[b]	IGCC[b]
Efficiency, η (%)					
W/O	34.3	38.5	43.3	–	38.4
With	25.1	29.3	34.1	30.6	31.2
RCE (kg-CO_2/h)					
W/O	466,000	415,000	369,000	–	416,000
With	63,600	54,500	46,800	52,200	51,200
CI (kg-CO_2/MWh$_e$)					
W/O	931	830	738	–	832
With	127	109	94	104	102
TPC ($/kW$_e$)					
W/O	1,280	1,330	1,360	–	1,430
With	2,230	2,140	2,090	1,900	1,890
LCOE ($/MWh$_e$)					
W/O	48.4	47.8	46.9	–	51.3
With	81.6	76.9	73.4	69.8	65.2
COCC ($/t-$CO_2$)	48.2	40.4	34.8	30.3[c]	24.0

[a] Estimates correspond to a net power of 500 MW$_e$, use of Illinois #6 coal at $1.50/MBtu, a plant capacity factor of 0.85, 90% CO_2 capture, and 2005$.

[b] Estimates are for the nth plant.

[c] Relative to a baseline of SCPC without CO_2 capture.

cost ($65.2/MWh$_e$) with capture. At an estimated cost-of-carbon capture (COCC) of $24/t-$CO_2$, it also captures carbon at half the cost of a PC plant. If limits on CO_2 emissions are imposed, the COCC will become an important factor in choosing from available power plant options. However, before one of the lowest-cost carbon avoidance options (USCPC, SC-OXY or IGCC) can be chosen with confidence, it is essential that the technologies be demonstrated at commercial scales of 100 MW$_e$ or more.

Following capture, CO_2 is compressed and transported to a geological formation for injection and storage. Transport costs increase with increasing distance between the source and repository, while injection costs increase with increasing reservoir depth and the number of wells needed to facilitate injection. Once injected, additional costs are associated with monitoring the sequestered CO_2. Although the cumulative costs of transport, injection and monitoring depend on site selection, there is no paucity of attractive sites and nominal costs are approximately $5/t-$CO_2$.

The cost advantage historically enjoyed by coal relative to other means of power production had begun to erode by 2010, and from the foregoing estimates, it's clear that further (significant) erosion would occur if constraints are imposed on CO_2 emissions. At this point, however, cost estimates of CCS options are just that, estimates. Confidence in the estimates must await construction and operation of the first large (commercial) scale power plants. On the one hand, actual costs of CCS may exceed projections; on the other hand, innovation may yield technology advancements that reduce costs (TR, 2008).

In a separate study (WorleyParsons, 2009), the percentage increase in the cost of power generation due to CCS was estimated to be approximately 39%, 60% and 76% for IGCC, oxy-fuel and SCPC systems, respectively, with an additional increment of $11 to $14 per tonne for transport and storage of the CO_2. The estimates exceed those of Table E.1, and the study pointed to the hard realities of securing the capital investments needed for wide deployment of CCS. Three requirements were cited for securing the capital: carbon emissions would have to be appropriately monetized; national and internationally homogenized environmental laws and regulatory frameworks would have to be established; and government incentives would have to be provided to mitigate the effect of large costs. Under existing global economic and political conditions, challenges to meeting these objectives are daunting, and it's unlikely that capital markets will respond with enthusiasm.

Coal-fired power plants are considered to be point sources of CO_2 emissions. That is, unlike large numbers of widely dispersed low-level emitters, such as automobiles, the power plants emit large amounts of CO_2 at a specific location. Other point sources include gas-fired power plants, petrochemical refineries, and facilities for producing natural gas, ammonia, cement and steel.

Dooley et al. (2006) assessed net costs associated with implementing CCS for ten point source categories. The most attractive category was an ammonia-processing plant located within 10 miles of an oil field for which CO_2 could be used for enhanced oil recovery (EOR). A large CO_2 concentration in the gas stream and close proximity of the ammonia plant to the oil field combined to significantly reduce gross costs, while revenues derived from EOR yielded a negative net cost (a profit) of approximately $10 per tonne of CO_2. Costs are reduced any time CCS can be linked to enhanced oil or coal-bed methane recovery. However, the norm is more closely associated with sequestration in deep saline formations, and for the least attractive category – a gas-fired power plant located more than 50 miles from a deep saline reservoir – costs were projected to be $90 per tonne of CO_2. Cost estimates for coal-fired power plants located within 10 miles of an enhanced coal-bed methane recovery project and within 25 miles of a deep saline formation were estimated to be $35 and $50 per tonne, respectively. Development and deployment of new IGCC or oxy-combustion technologies would reduce these costs. Of the 500 largest point sources in the US, 95% are within 50 miles of a potential repository, and related costs of CO_2 transport and injection would not exceed $15 per tonne.

Notes

1 Energy, economics, and climate change

1   To specify an amount of energy or the rate at which it is used, it must be quantified in terms of appropriate units. A quad is commonly used to quantify the amount of energy that is annually consumed by nations, and a terawatt may be used to quantify the rate at which it is consumed. Many different units are used to quantify energy consumption, with factors for converting from one to another.

   Because the numbers associated with national or global energy consumption are huge, scientific notation is often used to represent specific values. Five kilowatts (5 kW) is 5,000 watts, or in scientific notation 5×10^3 watts. The notation is based on multiples of 10, with the superscript (in this case 3) indicating the number of zeros following the lead number (in this case 5). Five megawatts (5 MW) is five million (5,000,000) watts or 5×10^6 watts; five gigawatts (5 GW) is five billion (5,000,000,000) watts or 5×10^9 watts. You can appreciate the value of scientific notation. Who wants to write all those zeros? A terawatt (1 TW) is one trillion or 10^{12} watts, where inclusion of the lead number of 1 is unnecessary. A quad, or one quadrillion Btu (one million billion Btu) is 10^{15} Btu. Common numerical prefixes are: "k" for kilo (one thousand or 10^3), "M" for mega (one million or 10^6), "G" for giga (one billion or 10^9), and T for tera (one trillion or 10^{12}). Conversion factors for common units of energy, mass, and volume are provided in Appendix A.

2   Gross domestic product (GDP) represents the total market value of a nation's annual production of goods and services and includes the value added at each stage of production. For a system such as an automobile or an aircraft, GDP accounts for the value of each component that goes into the system, as well as the value of the final assembly. Purchasing power parity (PPP) accounts for differences in the cost of living, and hence the relative purchasing power, of different countries.

3   Member nations of the OECD are largely from the European Union and North America but also include countries from the Middle East (Israel and Turkey), Oceania (Australia and New Zealand), Asia (Japan and South Korea), and South America (Chile).

2 The earth's climate system

1   When it comes to weather, one aspect that merits special attention and could provide a clue to changing climate is the frequency of extreme weather

events, such as torrential rains and flooding, longer and more intense heat waves, prolonged droughts, and Category 4 or 5 hurricanes. The probability of an extreme event corresponds to the frequency with which it is expected to occur. A one in one hundred (1:100) year event would have a 1% probability of occurring in any given year. However, if such an event begins to occur with greater frequency, perhaps becoming a 1:20 phenomenon, it may be a harbinger of systemic and accelerated climate change.

2   The term *flux* is used to describe the *rate of energy transfer per unit surface area*, and it is commonly quantified in units of watts per square meter (W/m²). As applied to solar radiation at the outer edge of the Earth's atmosphere, the TSI represents the rate at which a surface of one square meter oriented perpendicular to the Sun's rays intercepts solar energy. For much of the Holocene, the TSI has had a nominal (average) value – called the *solar constant* – of 1,366 W/m². However, it is not precisely a constant since it varies by approximately 1.3 W/m² (~0.1%) over a typical sunspot cycle and annually due eccentricity in the Earth's orbit about the Sun (NAS, 2012).

3   An El Niño is characterized by higher-than-normal water temperatures in the tropical Pacific and an attendant disruption of the ocean-atmosphere circulation system that can significantly alter weather patterns. Effects can include excessive rainfall and flooding in the East Pacific, drought in the West Pacific, and winter temperatures in the continental United States that are cooler than normal in the Southwest and Southeast and warmer in north and central regions. In contrast, La Niña involves lower-than-normal temperatures in Pacific waters and winter temperatures that are cooler and warmer, respectively, in the Northwest and Southeast. Detailed information on the nature of ENSO events is provided by NOAA (2014c), along with periodic updates on current conditions.

3 Greenhouse gases

1   Through the latter half of the twentieth century, chlorofluorocarbons such as CFC-11 and CFC-21 were used extensively. But, in addition to being potent GHGs, they contributed to yet another environmental problem. When released to the atmosphere, the long-lived gases can ascend to the stratosphere where they absorb ultraviolet (UV) solar radiation and undergo photochemical reactions that release chlorine atoms. In turn, these atoms react with ozone (O_3) to reduce its concentration. Since O_3 has large absorption bands for UV radiation, the net effect is to increase transmission of this radiation to the Earth's surface, enhancing the risk of human skin cancer and damage to other life forms.

Concern for erosion of the stratospheric ozone layer was raised in the middle 1970s and resulted in a U.S. law prohibiting the use of CFCs as an aerosol propellant. However, it was not until the mid-1980s, when reports of a huge hole in the ozone layer over Antarctica became credible, that the issue gained world attention. An important outcome was approval of the Montreal Protocol of 1987, which was signed by thirty-six largely developed nations and provided exemptions for developing nations such as China and India. The protocol placed an immediate freeze on production of the five most widely used CFCs at 1986 levels, with subsequent reductions of 20% and 30% targeted by 1993 and 1998. However, with mounting evidence of further depletion of stratospheric ozone, the United States and the European Union agreed to an outright ban on production by 2000, providing an excellent example of cooperation and leadership among major industrialized nations of the world.

CFCs have since been replaced by chemicals that are less reactive or nonreactive with O_3, beginning with hydrochlorofluorocarbons (HCFCs) and more recently hydrofluorocarbons (HFCs). But these chemicals are potent GHGs, and with HFCs increasingly becoming the refrigerant of choice for cooling systems and demand for air conditioning and refrigeration growing in developing nations, they could become a more significant contributor to global warming.

Regulatory measures to reduce GHG emissions could include plans to replace HFCs by refrigerants that do not contribute to global warming or ozone depletion. If such plans prove unworkable within the framework of a comprehensive GHG treaty, an alternative regulatory pathway could involve use of the Montreal Protocol. However, all signatories would have to approve such an extension of the treaty, and it's problematic whether China and India, the world's largest producers of HFCs, would do so without incentives. Nevertheless, small steps are being taken. Beginning in 2011, the twenty-seven nations of the European Union embarked on a path to phase out HFCs in mobile air-conditioning systems, and in 2011 the U.S. Environmental Protection Agency announced its intention to withdraw approval of using HFC 134a in new automobiles.

2   Concentration measurements had been made for many years before 1958, and Beck (2008) combed the literature to gather the data and, in the process, became a *cause célèbre* for those wishing to dismiss an anthropogenic link to climate change. Beck accessed literature published between 1812 and 1961 and concluded that there were periods for which CO_2 concentrations were well above today's background levels. In a bar graph he showed levels ranging from approximately 550 to 800 ppm between 1928 and 1935, and in a plot of data for the Northern Hemisphere from 1812 to 1961, he showed values ranging from about 290 to 480 ppm. Beck's results also revealed a good deal of scatter and sharp variations over time, dropping, for example, from 480 ppm in 1825 to 380 and 310 ppm in 1830 and 1840, respectively, and increasing from 310 to 420 ppm between 1920 and 1945. If one accepted the data as a measure of the background concentration, it would follow that atmospheric concentrations are more strongly influenced by natural causes than by anthropogenic agents.

By his own admission, Beck focused on the concentration measurement methods and their inherent accuracy (and inaccuracy). However, he was remiss in failing to critically interpret the results. What were the atmospheric conditions associated with the measurements? It's doubtful that many, if any, could be considered background concentrations, and it's likely that all of the high readings were in locales of large CO_2 emissions. The large spikes and dips over relatively short periods are implausible. A discussion of the Earth's *carbon cycle* in Section 3.5 will show that for every 2.13 billion tonnes of carbon (2.13 Gt-C) discharged to the atmosphere, the average concentration of CO_2

increases by 1 ppm. Consider Beck's data indicating a drop of 170 ppm from 1825 to 1840 or an increase of 110 ppm from 1920 to 1945. It would follow from a carbon balance that 362 Gt-C would have been extracted from the atmosphere between 1825 and 1840 and 234 Gt-C added from 1920 to 1945. Such transfers are huge and would have required *un*natural phenomena that could not have gone unobserved. For comparison, consider that global carbon emissions were approximately 9 Gt-C in 2010. The amount of carbon withdrawn from the atmosphere between 1825 and 1840 would then have been forty times 2010 emissions, and the amount added from 1920 to 1945 would have been twenty-six times 2010 emissions.

Beck also questioned the validity of using air bubbles in ice cores to indirectly determine the atmospheric CO_2 concentration in preindustrial times. It's these measurements that led to determination of the nearly constant concentration of 280 ± 10 ppm shown in Figure 3.1a. There are clearly uncertainties in the measurement process, but actual concentrations are *not* likely to fall much outside a ±10 ppm range.

The bottom line is that the trend of increasing CO_2 concentrations revealed in Figures 3.1a,b and its anthropogenic origins are irrefutable. Beck's work is just one example of using flawed results to cast doubt on the science of global warming.

3   Use of a 100-year period to determine the GWP underestimates the contribution to warming by GHGs with much shorter lifetimes (Jackson, 2009; Shindell et al., 2009). For example, if based on a 20-year period, methane would have a GWP of 72, almost three times larger than its value for 100 years. Hence, over the shorter time frame, methane is a much more potent GHG.

4   Formulas for computing the radiative forcings of selected GHGs from knowledge of their atmospheric concentrations are provided in NOAA (2014b), along with periodically updated values for the GHGs of Figure 3.2.

5   In the Fifth Assessment Report of the IPCC, radiative forcings are reported in two ways. One is based on actual emissions and the contributions of all substances affected by the emissions; the other is based on the actual atmospheric concentrations of each substance. The first approach provides a more direct link to human activities, while the second, which was used in previous IPCC reports and is used by the NOAA, better represents the state of the atmosphere. The largest differences relate to *RF* values reported for CO_2 and CH_4, which are 1.68 and 0.97 W/m^2 for the first (emission-based) approach and 1.82 and 0.48 W/m^2 for the second (concentration-based) approach. The contributions of CO_2 and CH_4 are larger and smaller, respectively, for the second approach due to atmospheric chemical reactions involving CH_4. However, when all effects are considered, both approaches provide the same value of the total anthropogenic forcing. Results appearing in Table 3.1 and Figure 3.3 are based on the second approach.

6   This result corresponds to an effective radiative forcing, which accounts for rapid adjustments of the Earth's atmosphere and surface to the anthropogenic aerosols (IPCC, 2014a). Without the adjustments, the net radiative forcing has a slightly smaller negative value of −0.35±0.5 W/m^2, which is consistent with findings of −0.3±0.2 W/m^2 obtained by using a global aerosol model with satellite radiative transfer measurements (Myhre, 2009).

7   From a screening of approximately 400 methods of reducing methane and soot
     emissions, Shindell et al. (2012) identified 14 measures that, if implemented,
     would realize nearly 90% of the maximum possible reduction in their GWP
     and reduce their projected circa-2050 contribution to global warming by about
     0.5°C. For methane the measures focused on oil and gas production, gas
     transmission, coal mines, landfills, livestock manure, and rice paddies. For soot
     they included diesel engines, brick kilns, coke ovens, biomass (wood and dung)
     stoves, and the burning of agricultural wastes. All measures are well proven
     and their efficacy well documented. The study also went beyond mitigation of
     warming effects to include benefits to human health and agriculture.
          Atmospheric methane has a catalytic effect on the production of
     tropospheric ozone. In addition to augmenting the adverse effects of soot on
     cardiopulmonary diseases, ozone is deleterious to crop production. Were the
     recommended measures implemented, estimates of the annual reduction in
     premature deaths due to outdoor air pollution would range from 700,000 to
     4.7 million, and the annual yields of four staple crops (maize, rice, soy, and
     wheat) would increase by 30 million to 135 million tonnes. For methane, a
     cost-benefit analysis projected savings in the range from \$700 to \$5,000 per
     t-CH_4, well above abatement costs of less than \$250 per tonne.

8   Greenhouse gases discharged to the atmosphere comprise gross emissions, in
     contrast to *net* emissions that account for the effect of land use practices and
     net growth of a nation's forests. In the United States, the combined effect of
     land use changes and forestry provides a sink for atmospheric CO_2. That is,
     more CO_2 is extracted from the atmosphere to renew and sustain forest growth
     than is discharged to the atmosphere by deforestation and changes in land
     use. For 2012, gross U.S. GHG emissions of 6,526 Mt-CO_{2eq} were mitigated by
     extraction of 980 Mt-CO_2 from the atmosphere, yielding net GHG emissions
     of 5,546 Gt-CO_{2eq} (EPA, 2014a).
          Land use change and forestry have not always had a mitigating effect
     on U.S. GHG emissions. Through the 1800s and early 1900s, the effects of
     deforestation and land use change exceeded those of new forest growth,
     augmenting CO_2 emissions from other sources. But by the 1950s, the balance
     shifted in favor of forest growth for the United States and most industrialized
     nations of the Northern Hemisphere, while deforestation continued to provide
     a net source of atmospheric CO_2 in Africa, Latin America and Asia. Globally,
     tropical deforestation contributed approximately 10,600 Mt-CO_2 to annual
     emissions from 1990 to 2007, while assimilation of about 5,900 Mt from
     regrowth of tropical forests and 8,800 Mt from growth of the world's intact
     forests provided a net global sink of about 4,100 Mt-CO_2/yr (Pan et al., 2011).
     The significance of land use changes and forestry to the overall assessment
     of GHG emissions speaks to the need for accurate estimates of their effect.
     Related uncertainties can be reduced by using satellite observations, and
     results for forest canopies suggest that emissions for tropical forests are 50% to
     75% smaller than previously estimated (Harris et al., 2012).

9   In addition to contributing to global warming as a long-lived, tropospheric
     gas, N_2O is also transported to the stratosphere where it contributes to ozone
     depletion (Ravishankara et al., 2009). Unlike the CFCs, N_2O is unregulated
     and could become the largest ozone-depleting agent of the twenty-first century.

10   As the third-largest source of anthropogenic methane emissions, landfill gas
     contributed 18% of U.S. 2012 emissions, releasing about 103 Mt-CO$_{2eq}$ (EPA,
     2014a). However, there are opportunities to reduce emissions by collecting the
     gas and using it to generate electricity on-site or treating it and routing it to
     a pipeline for distribution to consumers. In 1994, the EPA created a Landfill
     Methane Outreach Program to encourage landfill operators to harness the
     gas, and by 2013 more than 600 energy projects were operating and another
     400 were identified as viable candidates for future development. Costs and
     financing of such projects are discussed by Pierson (2013).

11   The world has been on a trajectory of exponential population growth for
     centuries, moving from 1 billion people in 1800 to 2 billion in 1930, 4 billion
     in 1974, 6 billion in 1999, and 7 billion in 2011 (Bloom, 2011). Could Thomas
     Malthus have been right, albeit premature in his predictions? In 1798,
     Malthus warned of massive famine and environmental degradation due to
     overpopulation. More than two centuries later, it has yet to happen. But could
     it? In 2011, the UN projected a year-2050 global population between 8.1 and
     10.6 billion people, with a nominal estimate of 9.1 billion corresponding to an
     average fertility rate of 2 children per woman (UN, 2011). For 2100, estimates
     ranged from 6.2 to 15.8 billion people, with a nominal value of 10.1 billion.
     But estimates depend on fertility rates, and with rates not slowing as fast as
     expected in many developing nations, the UN has had to periodically revise
     its predictions. Two years later it projected nominal populations of 9.6 and
     10.9 billion in 2050 and 2100, respectively (UN, 2013). More recently, an
     analysis of UN data suggested that world population is unlikely to stabilize
     this century and assigned an 80% probability to reaching 9.6 to 12.3 billion by
     2100 (Gerland, et al., 2014). By 2050, India is likely to overtake China as the
     world's most populous nation, and Nigeria should surpass the United States,
     moving into third place. Although fertility rates have been declining, the UN
     may be underestimating the rate of decline and they remain large in some
     of the world's poorest nations expected to account for most of the world's
     twenty-first-century population growth (Norris, 2013).
         If population growth is concentrated in poorer nations and these nations
     experience little economic development, their impact on global GHG emissions
     will be less than that of developing nations able to transition large numbers of
     people into higher living standards. This point is underscored by the fact that per
     capita GHG emissions in the United States are almost twenty times the emissions
     of most African nations, where much of the population growth will occur. A critical
     issue is the extent to which economic growth will accompany population growth
     in developing nations and the impact on GHG emissions. In any case, stabilizing
     global population is a sine qua non of sustainable development.

12   To underscore difficulties in restraining the rise of GHG concentrations,
     consider the following unrealistic, but instructional, scenario. Davis et al.
     (2010) assumed that from 2010 to 2060, there would be no new construction of
     infrastructure that contributed to CO$_2$ emissions by burning fossil fuels. There
     would be no new coal-, gas-, or oil-fired power plants, no new transportation
     vehicles (road, off-highway, air) that consume fossil fuels, and no new
     residential, commercial, and industrial systems that rely on fossil fuels. The

authors then estimated emissions that would result from existing systems, as they were gradually decommissioned over the fifty-year period according to their typical lifetimes.

Termed committed emissions due to infrastructural inertia, the analysis yielded nominal cumulative emissions of 135 Gt-C, with lower and upper uncertainty bounds of 77 and 191 Gt-C. Adding 54 Gt-C due to emissions unrelated to the combustion of fossil fuels, the nominal circa-2060 CO_2 concentration was estimated to be approximately 405 ppm with lower and upper bounds of 380 and 425 ppm. Of course, with an actual circa-2013 CO_2 concentration of 395 ppm, the lower bound is moot. But even for the lower bound, warming would continue through 2060. The message is sobering. The constraints on emissions considered by Davis and colleagues would be draconian, and whatever is actually done to curb emissions would be far less drastic. The world is clearly continuing on a sharp trajectory of increasing atmospheric GHG concentrations.

13   A progression of deforestation into regions of high biomass density, such as the Amazon and Southeast Asia, is increasing CO_2 emissions and, in absolute terms, could increase deforestation from approximately 20% to 25% of the global total (Loarie et al., 2009). A significant driver is the demand for biofuels and the ability to meet demand by converting natural forests and grassland to cropland for fuels such as ethanol and biodiesel (Fargione et al., 2008; Searchinger et al., 2008). Depending on the nature of the biofuel, the agricultural feedstock, and the converted ecosystem, atmospheric CO_2 emissions due to the release of carbon stored in the soil and plant biomass of the ecosystem are estimated to be 17 to 420 times larger than the annual reduction in GHG emissions due to replacement of fossil fuels by crop-based biofuels. Stated differently, from 17 to 420 years of biofuel production would be needed to repay the carbon debt associated with transformation of the natural ecosystem (Fargione et al., 2008). Despite some uncertainty concerning the effect of warming on the large stores of carbon in the soil (Trumbore and Czimczik, 2008), there is growing evidence that increasing temperatures are increasing respiration rates of soil-based organic material, with an attendant increase in CO_2 emissions. *In situ* measurements performed for subsurface peat deposits in subartic regions reveal a 50–60% increase in respiration for a 1°C rise in temperature (Dorrepaal et al., 2009). Since peatlands contain approximately one-third of the world's soil-based organic carbon and more than half of the atmosphere's current carbon content, the phenomenon could provide a significant boost to global warming. Were the temperature of northern peatland soils to increase by 1°C, emissions could increase by up to 100 Mt-C/yr.

4 Global warming

1   The full glacial-to-interglacial transition provides an example of the extent to which regional differences in the temperature anomaly can occur. During the most recent transition, temperatures increased by 10°C in Antarctica but only 3°C in the equatorial Pacific (Skinner, 2012).

2   Silver (2012) provides a detailed discussion of short-term noise and long-term signals associated with global warming, including effects of the ENSO cycle and sulfur emissions. He notes that, even when decadal averages are taken, warming trends could only be discerned 75% of the time since 1900. He also concludes there is "about a 15% chance that there will be no net warming over a decade even if the global warming hypothesis is true because of the variability in climate."

3   A discussion of the effect of GHG emissions on global warming would not be complete without mentioning the concept of equilibrium climate sensitivity. The concept frames global warming in terms of the following question: How much will the steady-state or equilibrium value of the Earth's mean surface temperature increase due to a doubling of the atmospheric CO_2 concentration? In other words, what would be the ultimate temperature rise if the CO_2 concentration increased from its preindustrial value of about 280 ppm to 560 ppm, which is essentially the benchmark of 550 ppm for which adaptation would be difficult? Although Kerr (2004a) and Skinner (2012) report a nominal value of 3.0°C, uncertainties are large and the full range of estimates is from 1.5° to 11°C (Stainforth et al., 2005; Kerr, 2006a). Acknowledging the large uncertainties, the IPCC dropped its lower estimate from 2.0°C in AR4 to 1.5°C in AR5 and refrained from providing a best estimate (IPCC, 2014a). In AR5 equilibrium climate sensitivity was deemed likely (high confidence) to be in the range from 1.5 to 4.5°C.

   Two other measures of climate sensitivity are provided in AR5 (IPCC, 2014a). The transient climate sensitivity provides the amount by which the global mean temperature would increase when the CO_2 concentration has doubled and not years later after an equilibrium temperature is reached. Its *likely* value is in the range from 1.0°C to 2.5°C, below that of the equilibrium sensitivity. The other measure of sensitivity is the transient climate *response to cumulative carbon emissions*. It provides the amount by which the temperature would increase in response to cumulative CO_2 emissions of 1,000 Gt-C and is likely in the range from 0.8°C to 2.5°C.

   Finally, climate sensitivity can also be expressed as a ratio of the temperature change to the corresponding change in radiative forcing. For example, a forcing of 3.7 W/m² associated with a doubling of the atmospheric CO_2 concentration and a temperature rise of 3°C (Skinner, 2012) correspond to a climate sensitivity of approximately 0.80 °C/W/m². Ascribing increases in forcing and temperature of 6.6 ±1.5 W/m² and 5 ±1°C, respectively, to differences between full glacial and interglacial conditions, Hansen (2006) obtained a sensitivity of 0.75°C/W/m². These results imply that for each increase of 1 W/m² in the net radiative forcing, the Earth's mean temperature increases by 0.75° to 0.80°C. Correlations have been developed for determining the effect of increasing CO_2 (and other GHG) concentrations on the radiative forcing (NOAA, 2014b).

4   To better appreciate the positive feedback associated with increasing atmospheric water vapor, consider the effect of a hypothetical 10°C increase in the tropospheric temperature from 15°C (288 K) to 25°C (298 K). Assuming saturated conditions (a relative humidity of 100%), the partial pressure of water vapor at 15°C is 0.01705 bar, and for a standard atmospheric pressure of 1.01325

bar and ideal gas behavior, the corresponding mole fraction and concentration of the vapor are 0.0169 and 16,900 ppm, respectively. At a temperature of 25°C, saturated vapor has a partial pressure of 0.03169 bar, a mole fraction of 0.0313 and a concentration of 31,300 ppm. The change from 15°C to 25°C corresponds to a large and highly nonlinear effect. An increase of only 3.5% in the absolute temperature corresponds to an 85% increase in the amount of vapor. Of course, the troposphere is not saturated with water vapor, but so long as its average relative humidity is unaffected by (or increases with) rising temperatures, the same argument can be made.

Updated measurements suggest that atmospheric radiation models may have underestimated the positive feedback associated with increasing vapor concentrations. Ptashnik et al. (2011) have determined that absorption of radiation by water vapor in four *windows* on the wings of strong absorption bands between wavelengths of 1 to 4 μm is approximately ten times stronger than previous estimates, adding to the radiative forcing of the vapor.

5   Clement et al. (2009) reviewed circa 1953 to 2007 data for low-level clouds (below about 2 km) covering 3 million square kilometers of the Pacific Ocean, paying close attention to periods of significant warming and cooling of surface waters during the late 1970s and 1990s. With warming and cooling, cloud cover decreased and increased, respectively, pointing in both cases to existence of a positive feedback. With reduced cloud cover, more solar radiation makes it to the ocean and warming is enhanced (a positive feedback). Diminished cloud cover due to warming is attributed to increased emission of thermal radiation from the warmer ocean surface and a corresponding increase in absorption of this radiation by the clouds. Increased absorption enhances evaporation and hence thinning of the clouds. Using satellite data to determine radiation trapped by clouds over a ten-year period from 2000 to 2010, Dessler (2010) also concluded that warming due to GHG emissions enhances radiation absorption by clouds and provides a positive feedback. However, the contention has not gone unchallenged. Spencer and Braswell (2010, 2011) have suggested that, through their effect on the global radiation balance, the Earth's temperature is more strongly influenced by *natural* variations in cloud cover during El Niño and La Niña stages of ENSO. This premise has, in turn, been challenged by Dessler (2011), as well as by Trenberth et al. (2011) who supported Dessler's conclusions while pointing to errors used by Spencer and Braswell in their data analysis.

5 Consequences of global warming

1   The premise that basal lubrication by moulin flows accelerates movement of outlet glaciers has been challenged by Joughin et al. (2008). Their observations suggest a negligible contribution and instead a dominant contribution due to annual advancement and retreat of the glaciers' calving fronts.

2   Figure 5.1 and other results for the global-average sea level obscure two important features of changing sea levels. The first is that changes are not the same at all locations (Sallenger et al., 2012; Willis and Church, 2012). Several factors influence local conditions, such as differences in temperature and

salinity, variations in ocean circulation, and the effect of mass redistributions on gravitational forces and the Earth's shape. In concert with mass loss from the GIS and reduced circulation in the Atlantic Ocean, sea levels are projected to rise most rapidly along mid- to northern portions of the U.S. Atlantic coast.

The second feature deals with the fact that, even with persistent warming, other factors can cause sea levels to drop. The transition from the 2009/2010 El Niño to a strong 2010/2011 La Niña moved enough ocean water to the land to cause a temporary 5 mm drop in the mean sea level (Boening et al., 2012). Nevertheless, apart from the intermittent effect of such excursions, sea levels will continue to rise.

3   For a tropical storm to nucleate, the temperature of surface waters must be at least 26°C, and the atmosphere must be moist with little wind shear. Once nucleated, a storm intensifies by moist air rising from the sea beneath it and by the entrainment of surrounding air. Intensification continues as long as the storm moves over warm seas. When it reaches land and detaches from its energy source, it begins to dissipate, but not without retaining the potential to inflict significant damage.

4   Before Hurricane Sandy struck the U.S. Atlantic Coast, the NFIP was carrying a debt of $18 billion, on which interest was being paid to the U.S. Treasury Department (Lipton et al., 2012). Following Sandy, its debt was projected to increase to at least $25 billion. With net receipts from premiums of only $2.5 billion per year and the increased risk of flooding from extreme storms, the NFIP is unlikely to ever repay its debt, even with a large increase in premiums. To understand the scope of the problem, consider the fact that the total circa-2012 NFIP projected liability was $1.25 trillion, second only to Social Security (Kildow and Scorse, 2012). Of the total, $527 billion was tied to coastal property.

Examples of egregious NFIP payments abound. Lipton et al. (2012) cite two cases, one for which property valued at $183,000 was flooded fifteen times over a decade, costing the NFIP $1.47 million, and another for which more than $2 million was paid on a residence valued at $116,000. The poster child for egregious spending may well be Dauphin Island on the Gulf Coast of Alabama (Gillis and Barringer, 2012), which is frequently battered by large storms. Since 1988, property owners received $72.2 million from the NFIP, while having paid only $9.3 million in premiums. Additionally, since 1979, more than $80 million of federal money has been used to repair communal infrastructure, amounting to more than $60,000 per permanent resident.

The U.S. Congress should resist pressures from the real estate industry and property owners to maintain the NFIP. The program and its drain on taxpayer money should be abolished. And access to Stafford funds should be precluded for property development in areas deemed of high risk to disasters of natural and/or anthropogenic origins.

6 Mitigation, adaptation, and geoengineering

1   Conventional coal-fired power plants operate under subcritical conditions for which the pressure of the steam is less than the critical pressure of water

(220 bars) and the temperature is less than roughly 550°C. The efficiency with which these power plants convert the chemical energy of coal to electrical energy ranges from 33% to 36%, depending on system design, type of coal, and the temperature of the water used to extract waste heat from the power cycle.

Enabled by advanced materials and manufacturing processes, coal-fired power plants can also operate under supercritical (SC) or ultra-supercritical (USC) conditions. The efficiency of a contemporary SC unit operating at a nominal steam temperature and pressure of 565°C and 240 bars ranges from approximately 37% to 39%. For USC plants, temperatures and pressures are projected to approach 600°C and 320 bars, with efficiencies ranging from 40% to 43%. Long-term targets for USC systems call for steam temperatures and pressures up to 700 °C and 380 bars and efficiencies up to 50%. But getting there requires advancements in high-strength metal alloys and thermal management technologies for key power cycle (boiler and turbine) components (Powell and Morreale, 2008).

2   Although they are comprised largely of hydrogen-carbon (H-C) bonds, fossil fuels typically contain other atomic species. Neglecting the mineral components but including the common inorganic components (S, N, O), the chemical formula of a fossil fuel can be represented as $CH_xS_jN_kO_l$, where x, j, k, and l correspond to the number of hydrogen, sulfur, nitrogen, and oxygen atoms associated with a single carbon atom. Values of j, k, and l are typically less than 0.1 and vary with the type of fuel, decreasing from coal to oil to gas.

For oil and its derivatives such as gasoline and diesel fuel, the number of hydrogen atoms (x) is approximately 2.0. For natural gas and coal, x is approximately 3.7 and 0.8, respectively. Since the number of hydrogen atoms per carbon atom increases from coal to oil to gas, the ratio of carbon-to-hydrogen atoms (1/x) decreases to values of approximately 1.25, 0.5, and 0.3 for coal, oil, and natural gas, respectively. In units of kilograms of carbon per gigajoule of energy (kg-C/GJ), nominal values of the carbon intensity (CI) are 21, 19, and 15 for coal, oil, and gas, respectively.

3   Estimates of natural gas resources soared in the late 2000s, largely due to the successful application of technologies that extract gas from unconventional sources such as coal beds, shale, and sandstone. Circa-2011 estimates of recoverable unconventional gas were 75 trillion cubic meters (Tcm) in North America alone and 250 Tcm globally (Economist, 2012b). The estimate for North America is more than 100 times larger than its consumption in 2013 (BP, 2014a). Accounting for conventional as well as unconventional resources, the circa-2011 world total is estimated to be 752 Tcm (Economist, 2012b), almost 250 times global 2011 consumption of 3.2 Tcm. And resource estimates will likely increase over time.

However, a cautionary note must be added for the production of shale gas. Although shale gas is readily accessible in the United States, that may not be the case in other regions. In China, which has the world's largest resources and a large incentive to use gas in lieu of coal, much of the gas exists in deeper seams and in regions that are semiarid and earthquake prone – conditions that make extraction more problematic.

4   There's nothing new about natural gas vehicles. In 2012, about 110,000 were operating in the United States and about 13 million worldwide

(Economist, 2012b). In the United States, use was limited to buses and truck fleets with restricted routes and access to central refueling stations. But were filling stations more readily available along interstate highways, the long-haul trucking industry would be motivated to reap the economic benefits of transitioning to natural gas. And were filling stations provided in cities and towns, owners of light-duty vehicles might be quick to follow.

5   From 2010 through 2013, spot prices of U.S. Central Appalachian Coal varied from roughly $71 to $87 per tonne, while coal prices in Europe and Asia ranged from approximately $80 to $125 per tonne (BP, 2014a). With one tonne of coal corresponding to roughly 27.5 MBtu, the energy equivalent price of coal in the three regions ranged from roughly $2.60 to $4.50 per MBtu.

6   Hydraulic fracturing requires millions of gallons of water, typically drawn from a natural source and transported by trucks to the well site where it is stored in an artificial holding pond. Water from the pond is mixed with sand and chemicals before it is routed to high-pressure pumps and discharged to the well. Some geological formations allow for permanent sequestration of the fracking fluid thousands of meters below groundwater. But for other formations, the fluid – now laden with additional toxins and radioactive materials from the shale – must be returned to the surface. Even with reuse, which increases its radioactivity and the concentration of toxins, fluid must ultimately be transported from the site for treatment and disposal. If stored in retention ponds prior to treatment, the ponds must be designed for zero leakage and subsequent treatment must guarantee restoration of the water quality to environmentally acceptable levels. As an indicator of what can happen in the absence of regulation, some drillers and well operators are known to have taken the water to treatment plants ill-equipped to process the toxic and radioactive contents, resulting in the release of untreated water to the natural environment (NYT, 2011).

In a study on the future of natural gas (MIT, 2011), it was concluded that the foregoing problems are "challenging but manageable," and expectations are that the technologies can be implemented in an environmentally responsible manner. But for that to happen, shale gas companies must commit to the highest production standards, and strong state and federal regulations must be implemented to hold them accountable. The resource is simply too large and the opportunity to reduce carbon emissions too great to do otherwise.

As an interim measure or one that would complement an eventual regulatory framework, a group of energy companies, environmental organizations, and philanthropic foundations has established a Center for Sustainable Shale Development (CSSD, 2013). A set of fifteen performance standards has been developed – aimed at minimizing the risks of shale development to environmental safety and human health – along with a third-party certification process based on the standards. Participation of shale developers is optional, and it remains to be seen whether the procedures will be effective and whether fracking can be undertaken safely. There is no such thing as clean coal, and the jury is still out on whether we can speak of clean fracking.

7   In terms of mitigating global warming, there is an insidious downside to replacing coal by natural gas. From Section 3.1 we know that atmospheric

sulfates formed from SO_2 emissions due to coal combustion have a negative radiative forcing and hence a cooling effect. Wigley (2011) examined how a 50% reduction in the use of coal for power production from 2010 to 2050 would affect global warming. With replacement of the coal by natural gas, two mechanisms would contribute to increased warming, one having to do with reduced SO_2 emissions and the other with leakage of the gas. Leakage was varied from 0 to 10% of gas consumption, and even with no leakage, it was found that a reduction in the global mean temperature due to replacement of coal by natural gas would not begin until 2050. A reduction in the negative radiative forcing due to reduced atmospheric sulfate concentrations was enough to delay the benefits associated with reduced CO_2 emissions. For 10% leakage, the reduction in temperature would not kick in until 2140.

Draw your own conclusions, but two points must be made. First, the fact that SO_2 emissions reduce the solar input and thereby contribute to global cooling does not justify a business-as-usual approach. The cooling effect is temporary, and there are too many adverse effects of SO_2 emissions and of coal utilization in general. Second, in using natural gas, a near zero-tolerance approach should be taken to leakage.

8   A fuel cell (FC) is a device that converts chemical energy to electrical energy by means of electrochemical reactions. Like a battery, it includes an anode and a cathode and an intermediate electrolyte. However, unlike a battery, it is not a closed system. Reactants of a battery are contained within the battery and are depleted and restored during discharging and recharging, respectively. In contrast, a FC is an open system for which reactants, a fuel and an oxidizer, are continuously supplied from external sources. Unlike the metallic electrodes of a battery, the anode and cathode of a FC consist of the fuel and oxidizer, respectively, along with catalysts to facilitate the desired chemical reactions. Fuel cells developed for automotive applications use hydrogen as a fuel source.

9   An attractive option for sequestering CO_2 is to inject it into aging oil fields to enhance recovery of the remaining oil. Many sites exist that are well below the Earth's surface and well removed from sources of potable water. The CO_2 is pressurized to more than 73 atmospheres – the critical pressure of CO_2 – and pumped into the oil field as a supercritical liquid. Because the mass density of the liquid – about 500 kilograms per cubic meter – is well above that of the gaseous state, the storage capacity of the reservoir is increased significantly. In addition to increasing the reservoir pressure, the CO_2 has a viscosity comparable to that of a gas and reduces the surface tension of the oil. Collectively, these features reduce the resistance to oil flow, making it easier to extract it from the pores of geological formations and enabling enhanced oil recovery (EOR). At depths below approximately 800 meters, pressures naturally exceed the critical pressure of CO_2, thereby maintaining supercritical conditions.

In the United States, EOR is already achieved by annually routing more than 30 million tonnes of CO_2 per year through 3,200 kilometers of pipelines. One thirty-year project involves piping CO_2 330 kilometers from a North Dakota coal gasification plant to a Saskatchewan oil field, where it is pumped down a 1.5 kilometer pipe into oil wells lying beneath four layers of impermeable rock. The project will ultimately sequester 20 Mt-CO_2, while increasing oil extraction by approximately one-third. But this and similar projects use only a small fraction of the available capacity. For the lower forty-eight states alone, it's estimated that 24 billion barrels of oil accessible with circa-2012 EOR technology would provide storage capacity for 9 Gt-CO_2. Use of this capacity would be economically viable if the market price of crude oil remained above $85/bbl and the CO_2 could be provided at a cost of no more than $45/t (DOE, 2012). If oil fields that are not yet economically viable are included along with fields in Alaska and the Gulf of Mexico and next-generation EOR technology is considered, the potential rises to 137 billion barrels of oil and sequestration of 45 Gt-CO_2.

10   In addition to opportunities afforded by EOR, carbon dioxide can be stored in depleted natural gas reservoirs, deep coal seams, and deep geologic brine formations (Dooley et al., 2006; DOE, 2012). Criteria for safe storage include large reservoir porosity and permeability and an impenetrable layer of overlying cap rock. Including all possible repositories, global reservoir capacity is estimated to be about 11,000 Gt-CO_2 (Service, 2004; Dooley et al, 2006), enough to accommodate today's rate of point-source emissions for centuries.

If sequestration is to achieve the intended objective of reducing GHG emissions, there must be assurances that the CO_2 will remain in storage for millennia. So far the news is good. Unless the cap rock is fractured, leakage rates are negligible, a condition that's been confirmed by the more than thirty-year, largely problem-free history of using CO_2 for EOR (Benson and Friedman, 2014). Nevertheless, more must be done to monitor the fate of the CO_2 after it is sequestered.

By 2009, about a dozen sequestration projects were either in operation or planning stages (AAAS, 2009), mostly involving separation of CO_2 from natural gas production. The best-known and longest-standing Sleipner Project has been operating in the North Sea since 1996. Natural gas extracted from wells 2,500 m below the sea floor is routed to a platform where its CO_2 concentration is reduced from approximately 9% to 2%. At a rate of 1 Mt-CO_2 per year, the separated CO_2 is injected into porous sandstone 1 kilometer below the sea floor, where it is expected to remain for thousands of years. The economic motivation was provided by the desire to avoid a $55 tax per tonne of CO_2 imposed on energy producers by the government of Norway. The avoidance of approximately $55 million per annum in taxes more than pays for the $15 million cost of sequestration. A much larger project off the coast of Australia (the Gorgon project) was started in 2009 with plans to separate 3.3 Gt-CO_2 from natural gas fields and inject it into saltwater reservoirs 2,800 meters below ground.

11   Even with rapid growth in wind, solar, and possibly nuclear power, as well as increased use of natural gas, the world's growing appetite for electricity will sustain demand for coal-fired generation. The large costs associated with retrofitting existing plants to include CCS make this option highly unlikely, and most, if not all, plants will continue to discharge CO_2 throughout their days of service. And, for a decade or more, the construction of new power plants will likely precede the availability of well-tested CCS technologies. The plants could be endowed with capture-ready features to facilitate a future

retrofit for CCS, but not without incurring significant additional costs. Should limits and/or a price be placed on CO_2 emissions, utilities will lobby hard to grandfather existing plants, and with nominal lifetimes of forty years, CO_2 emissions would continue well into the future.

7 Public policy options

1   Although current goals call for increasing average fuel economy standards for LDVs to 34.5 mpg by 2016 and 54.5 mpg by 2025, progress has been slow with an average of only 24.1 mpg achieved in 2013 – an increase of 0.5 mpg from 2012 – and 24.2 mpg projected for 2014 (EPA, 2014c). Results may fall short of the 2016 goal, and meeting the 2025 goal is more than problematic.

8 The politics of global warming: a history lesson and future prospects

1   Through the latter half of the twentieth century, CFCs were used extensively as refrigerants, aerosols, and foaming agents. However, when released to the atmosphere, they are long-lived and can ascend to the stratosphere, where ultraviolet (UV) radiation drives photochemical reactions that release free chlorine atoms. These atoms react with ozone (O_3) to reduce its concentration. Since O_3 has large absorption bands for UV radiation, the net effect is to increase transmission of this radiation to the Earth's surface, enhancing the probability of human skin cancer and damage to other life forms.

     Concern for erosion of the stratospheric ozone layer was first raised in the middle 1970s and resulted in a U.S law prohibiting the use of CFCs as an aerosol propellant. However, it was not until the mid-1980s, when reports of a large hole in the ozone layer over Antarctica became credible, that the issue gained serious attention. An important outcome was approval of the Montreal Protocol, which was signed by thirty-six industrialized countries and provided a waiver to less-developed nations. The protocol placed an immediate freeze on production of the five most widely used CFCs at 1986 levels, with subsequent reductions of 20% and 30% targeted by 1993 and 1998, respectively. But with mounting evidence of further depletion of stratospheric ozone, the United States and the EU agreed to an outright ban on production by 2000 and replacement with non-ozone-depleting refrigerants such as HFCs.

     The Montreal Protocol provides an example of cooperation and leadership among industrialized nations of the world. If one is inclined to ask the question, *if for ozone depletion, why not for global warming,* one must consider the significant differences. Ozone depletion is due exclusively to anthropogenic activity, while climate change is associated with overlapping natural and environmental effects. Threats posed by the ozone hole are immediate, while those of climate change are more uncertain and manifested over longer time scales. And fossil fuels play a much more important role in the global economy than do CFCs. The economic impact of curtailing the consumption of CFCs was minor compared to what could follow such a rapid transition from fossil to carbon-free sources of energy.

If extended, the Montreal Protocol could also provide a powerful tool for curbing one component of GHG emissions. HFCs do not deplete stratospheric ozone, but they have large GWPs and their emissions are increasing more rapidly than any other GHG. Although the Montreal Protocol does not explicitly cover HFCs, it could be amended to include them and to thereby reduce or eliminate their contribution to global warming.

2   Prestowitz viewed the U.S. reaction to the KP as one of several unilateral decisions that fostered a world view of the United States as arrogant, wasteful, indifferent, and heavy-handed. Other unilateral decisions with geopolitical implications included abrogation of the anti–ballistic missile accord with Russia and refusal to participate in establishment of an international criminal court. But his strongest criticism was directed at America's stance on energy and environmental issues and its divergence from approaches adopted in other regions of the world, particularly Europe.

3   The Bush administration also advocated a Climate Change Science Program (CCSP) to obtain a better understanding of the climate system, with much of the $2 billion in annual funding allocated to NASA's Earth Orbiting System (EOS). Relying on numerous instrumented satellites, the system was designed to obtain data on a broad range of climate variables such as ocean and atmospheric temperatures, atmospheric aerosol concentrations and cloud cover, solar radiation incident upon and reflected by the Earth, and changes in land use and ice cover. Some argued that the need to mitigate warming was immediate and the CCSP was little more than a delaying tactic since several decades of data would be needed before the results would prove useful. Results obtained to date have in fact been useful, despite declines in federal support for Earth sciences and the CCSP.

4   As a signatory to the KP, Russia gained two immediate advantages (Kramer, 2005). First, with 1990 taken as the baseline year and its significant economic downturn in the 1990s, Russia's 2005 emissions were 43% less than those of 1990. Second, its industries were being driven by hugely inefficient energy systems that provided excellent opportunities for improvement at low cost. These opportunities could be exploited by Russia to further reduce its GHG emissions, while sustaining economic growth, or they could be used in a cap-and-trade system by other signatories unable to meet their KP requirements. By funding capital improvements in Russia that reduce GHG emissions, a nation could receive credit for the reduction. The first such transaction involved Denmark, which paid to upgrade two Russian power plants, reducing emissions by approximately 1.2 million tonnes of CO_2 per year and receiving 1.2 million credits.

5   By 2007, expectations of regional, if not national, caps on GHG emissions in the United States were stimulating interest in markets for trading allowances. Although the Chicago Climate Exchange (www.chicagoclimatex.com) was the first U.S. entity to offer such services, others were expected to follow (Dalton, 2007). The New York Mercantile Exchange (NYMEX), in particular,

would be well positioned to leverage its world-leading exchange for energy futures and its more recent venture into futures trading for SO_2 and NO_x emissions.

6   A centerpiece of efforts to reduce emissions from deforestation is an initiative termed REDD, an acronym for reducing emissions from deforestation and degradation in developing nations (Mongabay, 2011). The initiative focuses on tropical forests and seeks to reduce emissions by compensating countries for the loss of income that would accrue from deforestation. Compensation would depend on the cost assigned to emissions and the extent of the reductions, which would be monitored for verification.

7   Plans for realizing the EU target included a binding commitment to obtain 27% of its energy from renewable sources and a nonbinding commitment to increase energy efficiency by 27%. The commitments would not apply to each nation, but to the EU overall, raising questions about enforceability. And because they are referenced to 1990, a year when EU emissions were high due to heavy use of coal, they are not as ambitious as they might seem. Several questions are left unanswered. How will energy consumption be managed when the European economy begins to grow again? Will the former Soviet republics now in the EU be able to reduce their dependence on coal? How will the ETS be reconfigured? There is also a clause in the agreement that allows the EU to reconsider its targets if the Paris meeting does not yield a meaningful international agreement. Although the new EU target is intended to encourage comparable commitments by other nations before the Paris meeting, it is couched in broad strokes and could be considered aspirational rather than binding.

8   For China, large government and private investments as well as a huge and stable domestic market were propelling it to global leadership (Pooley, 2010b). In 2009, China's investment of $35 billion in clean energy more than doubled that of the United States, and in the second quarter of 2010 its investment of $11.5 billion exceeded that of the United States and Europe combined ($9.4B).

9   In the California cap-and-trade system an auction begins with companies or investors submitting confidential bids for a specified number of allowances. The highest bidder moves to the top of the line, the next highest bidder to second in line, and so on until all of the allotted permits have been claimed. All allowances are then sold to the successful bidders at a single price corresponding to the lowest successful bid (Doan, 2013).

10 By 2013, twenty-nine states had adopted renewable portfolio standards. However, pressed by intense lobbying and influenced by the low cost of natural gas, sixteen of the states were considering legislation that would reduce or repeal requirements (Martin, 2013). The pressure was applied by the same organizations that successfully impeded GHG mitigation measures in Congress – oil, gas, and coal producers, utilities, and advocacy groups such as the American Petroleum Institute and the Heartland Institute. To the extent this success is replicated at the state level, investments in renewable energy and energy efficiency would be adversely affected.

11 Consider the effect of taxing the carbon content of gasoline and coal. Assuming 2.4 kilograms of carbon per gallon of gasoline, a tax of $10 per tonne

of carbon would amount to \$0.024 per gallon and would do nothing to alter consumer behavior. But a tax of \$200 to \$400 per tonne (\$0.48 to \$0.96 per gallon) would get some attention. Assuming a nominal carbon intensity of 22.8 kilograms of carbon per gigajoule of coal and a thermal efficiency of 35% for a coal-fired power plant, the plant would release 0.00025 tonnes of carbon per kWh_e of generation. A tax of \$10 per tonne would add \$0.0025 or 0.25 cents per kWh_e to the cost of electricity, again hardly enough to alter the economics of production or habits of consumption. But at \$100 per tonne (2.5 cents per kWh_e), the tax would have a significant impact on the use of coal for power generation.

9 Dissenting opinions: the great hoax

1   One example illustrates how the *Wall Street Journal*'s editorial board manages the climate change debate. In 2010, a letter addressing the causes and effects of human-induced climate change was submitted to the publication as an opinion piece. The letter was signed by 255 members of the U.S. National Academy of Sciences, and its message was consistent with views expressed by science academies and societies around the world. The *WSJ* declined to publish the letter (Gleick, 2012), which by itself is not unusual, since numerous pieces are offered and space is limited. However, it subsequently chose to publish a piece signed by sixteen scientists dispelling the scientific basis for decarbonizing the world's economy (WSJ, 2012). The piece appropriately identified uncertainties in computer models and the temperature stasis of the early twenty-first century (Section 4.1) as reasons for questioning the science, but proceeded on that slippery slope of touting the benefits of increased CO_2 to plant growth while dismissing its role as a GHG. The piece also touted the questionable premise of achieving a high benefit-to-cost ratio by allowing GHG emissions to continue unabated. But the most egregious features of the piece were its aspersions on the scientific community, labeling its views alarmist and attributing them to the influence of money (research grants) and speciously paralleling its behavior to a Soviet-era incident. The 2010 letter rejected by the WSJ was subsequently published by the journal *Science* (Sills, 2010).

Arguments posed by denialists are undermined by tendencies to selectively choose, shade, and/or ignore the underlying facts. Expressions of strongly held beliefs are understandable, but not when they are shrouded in scientific distortions or motivated by financial support from those who have the most to gain from the use of fossil fuels.

2   Efforts to impede curbs on GHG emissions have another parallel in efforts to resist implementation of the Clean Air Act. Passed in 1970, the CAA authorized the EPA to regulate nearly 200 toxic atmospheric emissions and to establish automotive fuel economy (CAFE) standards. But for forty years, led by carmakers and oil and coal producers, measures were strongly resisted by the business sector. It was not until 2009 that meaningful CAFE standards were established.

3   Resources for countering claims made by climate deniers can be accessed at www.skepticalscience.com/argument.php and www.realclimate.org.

4   The allegations of research misconduct included (1) falsification of data; (2) concealment or deletion of e-mails, data, or other information; (3) misuse of privileged information; and (4) deviation from accepted research and scholarly practices (NSF, 2011). In each case, thorough examinations revealed no evidence of research misconduct.

5   By applying Bayes' theorem, Silver (2012) shows how one's prior beliefs can influence interpretation of new evidence contrary to those beliefs. Unless the evidence is overwhelming, prior beliefs retain significant weight, consistent with the principles of cultural cognition and confirmation bias.

10 The ethics of climate change

1   An interesting parallel to the generational ethics issue appeared in a *Wall Street Journal* article dealing with U.S. federal budget deficits (Wessel, 2004). Drawing on a book by Peterson (2004), the article portrayed the deficits as immoral. Quoting: "The real reason (to worry about the deficits) is that we are robbing our children and grandchildren. We are buying now and figuring they will pay later. We are enjoying the benefits of more government and spending and ignoring the fact that they will have to settle for fewer benefits and a slower growing economy as a result." I agree completely. But the *WSJ* has never thought to apply the generational ethics argument to climate change. In its desire to eliminate any and all obstacles to economic growth, it has instead been consistently dismissive of climate science and scientists. Generational ethics bears on climate change, as well as deficit spending.

2   The statement can be traced to the actor Michael Douglas, who in the 1987 movie *Wall Street* makes the following assertion: "The point is, ladies and gentlemen, that greed, for lack of a better word, is good. Greed is right, greed works."

3   Although their views are divergent, Rand was respectful of Aristotle, to whom she acknowledged her only philosophical debt. In a postscript to *Atlas Shrugged*, she states: "I must emphatically disagree with a great many parts of his philosophy – but his definition of the laws of logic and of the means of human knowledge is so great an achievement that his errors are irrelevant by comparison."

4   The Catholic Church's view of science is that, when done correctly, it can lead to a better understanding of God's will. Representative statements of the harmony between faith and science have been made by Pope Paul VI in Gaudium et Spes 36 (www.cin.org/v2modwor.html) and Pope John Paul II in *Science and Faith in the Search for Truth* (www.its.caltech.edu/~nmcenter/sci-cp/sci80111.html).

5   The notion of a planetary commons calls to mind *the tragedy of the commons* (Hardin, 1968). In his article, Hardin uses the example of herders grazing their livestock on a common pasture, each having the option to increase the size of his herd and use of the land. Acting independently, each herder can therefore

increase his income. But if too many act for personal gain, the pasture is eventually overgrazed to the detriment of all. The essence of the article is that individuals exploiting a common resource in their own interests can ultimately damage the resource with severe consequences for all. If instead the herders cooperate by distributing grazing rights in a sustainable manner, all share in the long-term benefits. For environmentalists, the commons represents any shared resource, such as bodies of water, forests, and, of course, the atmosphere.

Although an analogy between the atmosphere – also a finite resource – and a common pasture has limits, it does provide some insight to what the future may hold. By increasing its emissions, each nation acts on its own behalf to the detriment of all. But by cooperating in reducing emissions and stabilizing GHG concentrations at an acceptable level, all nations benefit.

The analogy breaks down over the consequences of overuse. For a common pasture, the herdsmen can mend their ways. They can cull and sell portions of their herd, allow a few years for regrowth of the pasture, and resume grazing in a sustainable manner. Not so for GHG emissions whose impact is felt for centuries. The different time scales point to short-term reversibility in one case and irreversibility, or at best long-term reversibility, in the other. The analogy also breaks down in contrasting herdsmen, who have common interests and values, to nations of highly disparate cultures and economic conditions. A local problem can be managed more effectively than a global one.

6   Different views of climate change are a subset of a larger issue splitting the evangelical community, that being the manner in which secular knowledge, including science, is integrated with biblical writings. On the one hand there are fundamentalists who adopt a literal interpretation of the bible and reject any scientific findings to the contrary. On the other there are those who are open to the progression of secular knowledge and integration with their faith.

11 A way forward

1   Equation 3.1 of Section 3.3 delineates the principal determinants of carbon dioxide emissions. Globally, population will continue to increase while energy intensity continues on a trajectory of gradual decline. Although the use of carbon-free sources of energy will increase, it is questionable whether the increase will be sufficient to match growth in energy demand. Fossil fuels will remain the dominant source of energy. Of the four terms on the right-hand side of the equation, the largest change will be in GDP per capita, driven by the pace of economic growth in developing nations. In 2010, the GDP per capita for China and India, respectively, was only 16% and 7% of that for the United States (Economist, 2011c).

Although large [annual] economic growth rates of up to 11% in developing economies are not likely to continue, economic growth will still exceed that of developed economies, ensuring growth in global carbon dioxide emissions. In 2010, gross world product (GWP) grew by 4.9% to $63.1 trillion, but with 6.6% and 2.9% coming from developing and developed economies, respectively (CIA, 2011). It would take an average annual growth of only 1.7% to double the GWP by 2050.

2   Apart from its carbon emissions, other benefits accrue from reducing the use of coal. Coal is a dirty fuel that degrades the environment and human health on many levels. From air pollution due to emissions of sulfur and nitrogen oxides, heavy metals, and particulates to ground and water pollution due to mountaintop removal and ash disposal, coal's deleterious effects are ubiquitous. It also takes a heavy toll on the health and lives of those who mine it. Benefits associated with reducing coal consumption accrue to any nation that relies heavily on the fuel, not the least of which are China and India whose air quality problems are among the world's worst.

3   A distinction should be made between the existing 18.4¢ federal tax and a new carbon tax. Revenues from the existing tax go to the federal Highway Trust Fund, which is used to maintain transportation infrastructure such as roads and bridges. For several years, the Trust has been underfunded, and the obvious remedy is to have those who use the infrastructure pay for needed repairs through an increase in the fuel tax. Although prospects for implementation by Congress are remote, a 50% increase in the tax for the Highway Trust Fund plus a 50¢ carbon tax would concurrently address infrastructure needs and reduce GHG emissions.

4   Aggressive implementation of mitigation and adaptation measures would not be a drag on the economy, as some argue, but would stimulate economic development and create jobs in industrial sectors dealing with energy efficiency, carbon-free energy technologies, and systems resiliency. In the fourth quarter of 2011 there were 3.4 million green jobs – defined as those that benefit the environment and/or conserve natural resources – in the United States, and 110,000 new jobs were created in 2012 (Bagnied, 2013). Other nations have experienced similar, if not greater, success. Nations that lead in the development of low-emission and energy-efficient technologies will compete successfully in large emerging markets. From lighting to appliances to heating and cooling systems to photovoltaics and biofuels, tomorrow's industrial leaders will come from those nations that invest heavily in developing and implementing related technologies. Reducing consumption of fossil fuels need not come at the expense of economic development, a point emphasized by Senators Bingaman and Murkowski (2011), who link the development of clean energy to stimulation of innovation, economic development, and job creation.

References

AAAS (2009). Carbon Sequestration. *Science* 325, 1644–5.

AMS (2014). Explaining Extreme Events of 2013 from a Climate Perspective. Special Supplement to *Bull. Amer. Meteor. Soc.*, 95(9) (www2.ametsoc.org/ams/index .cfm/publications/bulletin-of-the-american-meteorological-society-bams/ explaining-extreme-events-of-2013-from-a-climate-perspective).

APS (2011). Direct Air Capture of CO_2 with Chemicals. A Technology Assessment for the APS Panel on Public Affairs. *Amer. Phys. Soc.* (www.aps.org/policy/ reports/assessments/upload/dac2011.pdf).

Allen, D.T., et al. (2013). Measurements of Methane Emissions at Natural Gas Production Sites in the United States. *Proc. Natl. Acad. Sci.* 110(4), 17768–73 (www.pnas.org/cgi/doi/10.1073/pnas.1304880110).

Allenby, B.R. (2003). Engineering and Ethics for an Anthropogenic Planet. *Proceedings of the Workshop on Emerging Technologies and Ethical Issues*, October 14–15, National Academies Press, Washington, DC, pp. 9–28.

Alley, R.B. (2000). The Younger Dryas Cold Interval as Viewed from Greenland. *Quat. Sci. Rev.* 19, 213–26.

Alley, R.B., et al. (2005). Ice-Sheet and Sea-Level Changes. *Science* 310, 456–60.

Andersen, K. (2012). The Downside of Liberty. *The New York Times*, July 4, A23.

Anderson, D.M., et al. (2013). Global Warming in an Independent Record over the Past 130 Years. *Geophys. Res. Lett.* 40(1), 189–93.

Appenzeller, T., and D.R. Dimick (2004). Global Warming: Bulletins from a Warmer World. *National Geographic* 206(3), September, 12–75.

Aston, A., and B. Helm (2005). The Race Against Climate Change. *Business Week*, December 12, 59–66.

Baylor University (2011). The Values and Beliefs of the American Public, Wave III Baylor Religion Study (www.baylor.edu/content/services/document .php/153501.pdf).

Berkeley (2011). The Berkeley Earth Surface Temperature (BEST) Study (www .berkeleyearth.org). Accessed November 3, 2011.

BP (2014a). Statistical Review of World Energy 2014 (www.bp.com/statisticalreview).

(2014b). BP Energy Outlook 2035 (www.bp.com/en/global/corporate/press/ press-releases/energy-outlook-2035.html).

BPC (2011). Geoengineering: A National Strategic Plan for Research on the Potential Effectiveness, Feasibility, and Consequences of Climate Remediation Technologies. Bipartisan Policy Center (http://bipartisanpolicy.org/projects/task-force-geoengineering/about).

Baede, A.P.M., et al. (2001). The Climate System: An Overview. *Climate Change 2001: The Scientific Basis*. Third Assessment Report of the Intergovernmental Panel on Climate Change. Cambridge University Press, UK, 85–98.

Bagnied, O. (2013). Jobs in Renewable Energy and Energy Efficiency. Environmental Energy Institute (www.eesi.org/fact-sheet-jobs-renewable-energy-and-energy-efficiency-11-jun-2013).

Ball, J. (2003). Russia Will Back Kyoto If It Gets Payoff in Return. *The Wall Street Journal*, December 12, A3, A6.

   (2004a). As Kyoto Protocol Comes Alive, So Do Pollution-Permit Markets. *The Wall Street Journal*, November 8, A2.

   (2004b). AEP, Cinergy to Disclose Details On Ways to Cut Carbon Dioxide. *The Wall Street Journal*, February 19, A8.

   (2007a). In Climate Controversy, Industry Cedes Ground. *The Wall Street Journal*, January 23, A1, A17.

   (2007b). Fuel-Economy Rules May Face Tightening. *The Wall Street Journal*, December 16, A3.

   (2007c). Kyoto's Caps on Emissions Hit Snag in Marketplace. *The Wall Street Journal*, December 3, A1, A19.

Ball, J. and J.J. Fialka (2003). Bush Global Warming Plan Draws Heat. *The Wall Street Journal*, February 11, A4.

Ball, J. and A. Regaldo (2004). Cinergy Backs U.S. Emissions Cap. *The Wall Street Journal* (http://online.wsj.com/news/articles/SB110193701170688416).

Balmaseda, M.A., et al. (2013). Distinctive Climate Signals in Reanalysis of Global Ocean Heat Content. *Geophys. Res. Lett.* 40(9), 1754–9.

Barbier, E.B. (2014). A Global Strategy for Protecting Vulnerable Coastal Populations. *Science* 345, 1250–1.

Barnett, T.P., et al. (2008). Human-Induced Changes in the Hydrology of the Western United States. *Science* 319, 1080–3.

Barrett, P.M. (2011). The Underground Solution. *Bloomberg Businessweek*, November 7, 66–74.

   (2014). The Phony War on Obama's Plan to Curtail Coal. Bloomberg Business Week (www.businessweek.com/articles/2014-06-03/the-truth-about-obamas-plan-to-curtail-coal-power).

Barringer, F. (2007). A Coalition for Firm Limit on Emissions. *The New York Times*, January 19 (www.nytimes.com/2007/01/19/business/19carbon.html).

Barringer, F. and D. Hakin (2007). New York State Issues Subpoenas to Five Companies Building Coal-Fired Power Plants. *The New York Times*, September 16, 23.

Battisti, D.S. and R.L. Naylor (2009). Historical Warnings of Future Food Insecurity with Unprecedented Seasonal Heat. *Science* 323, 240–4.

Beck, E.-G. (2008). 50 Years of Continuous Measurement of CO_2 on Mauna Loa. *Energy & Environment* 19, 1017–28.

Benson, S.M. (2014). Negative-Emissions Insurance. *Science* 344, 1431.

Benson, S.M. and S.J. Friedman (2014). Carbon Dioxide Capture, Utilization and Storage: An Important Part of a Response to Climate Change. *The Bridge*, National Academy of Engineering, Spring Issue, 42–50.

Bergant, D. (2010). The Bible's Wisdom Tradition and Creation Theology. In *God, Creation and Climate Change: A Catholic Response to the Environmental Crisis*. R.W. Miller, Ed., Orbis Books, New York, 35–48.

Bingaman, J. and L. Murkowski (2011). White Paper on a Clean Energy Standard. Committee on Energy and Natural Resources. United States Senate, March 21.

Blasing, T.J. (2014). Recent Greenhouse Gas Concentrations. Carbon Dioxide Information Analysis Center (http://cdiac.ornl.gov/pns/current_ghg.html).

Bloom, D.E. (2011). 7 Billion and Counting. *Science* 333, 562–9.

Blunden, J. and D.S. Arndt, Eds. (2014). State of the Climate in 2013. *Bull. Amer. Meteor. Soc.* 95, S1–S257. Also, National Climatic Data Center, National Oceanic and Atmospheric Administration (www.ncdc.noaa.gov/bams-state-of-the-climate and www.ncdc.noaa.gov/sotc).

Boening, C., et al. (2012). The 2011 La Niña: So Strong, the Oceans Fell. *Geophys. Res. Lett.* 39(19), 16 (doi:10.1029/2012GL053055).

Bohannon, J. (2008). Weighing the Climate Risks of an Untapped Fossil Fuel. *Science* 319, 1753.

Boswell, R. (2005). Buried Treasure. *Power and Energy*, February, 8–11.

(2009). Is Gas Hydrate Energy within Reach? *Science* 325, 957–958.

Brahic, C. (2006). Price Crash Rattles Europe's CO_2 Reduction Scheme. *Science* 312, 1123.

Brand, S. (2009). Four Sides to Every Story. *The New York Times*, December 15, 2009, A37.

Brandt, A.R., et al. (2014). Methane Leaks from North American Natural Gas Systems. *Science* 343, 733–5.

Breslau, K. (2007). The Insurance Climate Change. *Newsweek*, January 29, 44–6.

Breslau, K. and M. Brant (2006). God's Green Soldiers. *Newsweek*, February 13, 49.

Brock, S. (2012). Faith Organizations and Climate Change. Environmental and Energy Study Institute (www.eesi.org/papers).

Broder, J.M. (2010). Climate Change Doubt is Tea Party Article of Faith. *The New York Times*, October 20 (www.nytimes.com/2010/10/21/us/politics/21climate.html).

(2012). Report Outlines Climate Change Perils: Strains for US Military and Intelligence Agencies Are Predicted. *The New York Times*, November 10, A7.

Broder, J.M. and F. Barringer (2007). E.P.A. Says 17 States Can't Set Emission Rules. *The New York Times*, December 20, A1, A30.

Brook, E.J. (2012). The Ice Age Carbon Puzzle. *Science* 336, 682–3.

(2013). Leads and Lags at the End of the Last Ice Age. *Science* 339, 1042–3.

Brownstein, M. (2013). A New Study Measures Methane Leaks in the Natural Gas Industry. Environmental Defense Fund (http://blogs.edf.org/energyexchange /2013/09/18/a-new-study-measures-methane-leaks-in-the-natural-gas-industry).

Buesseler, K.O., et al. (2008). Ocean Iron Fertilization-Moving Forward in a Sea of Uncertainty. *Science* 319, 162.

Bullis, K. (2014). A Plan B for Climate Agreements. *MIT Technology Review* 117(4), 84–6.

Bunn, M. and O. Heinonen (2011). Preventing the Next Fukushima. *Science* 333, 1580–1.

Burke, M., et al. (2013). Weather and Violence. *The New York Times*, September 1, SR12.

Burkhardt, U. and B. Kärcher (2011). Global Radiative Forcing from Contrail Cirrus. *Nature Climate Change* 1, 54–8 (doi:10.1038/nclimate1068).

C2ES (2012). Climate Change & International Security: The Arctic as a Bellwether. Center for Climate and Energy Solutions (www.c2es.org/docUploads/arctic-security-report.pdf).

(2014). Renewable and Alternative Energy Portfolio Standards (www.c2es.org/node/9340).

CBO (2008). Policy Options for Reducing CO_2 Emissions. United States Congressional Budget Office, Publication 41663 (www.cbo.gov/publication/41663).

(2013). Options for Reducing the Deficit: 2014 to 2023. United States Congressional Budget Office, Publication 44715. (www.cbo.gov/publication/44715).

CCRC (2009). The Copenhagen Diagnosis, 2009: Updating the World on the Latest Climate Science. Climate Change Research Center (www.copenhagendiagnosis.com).

CDIAC (2013). Carbon Dioxide Information Analysis Center. Oak Ridge National Laboratory, US Department of Energy (http://cdiac.ornl.gov/).

Ceres (2014). (www.ceres.org/press/press-releases/world2019s-leading-institutional-investors-managing-24-trillion-call-for-carbon-pricing-ambitious-global-climate-deal).

CIA (2011). The World Fact Book. US Central Intelligence Agency. (www.cia.gov/library/publications/the-world-factbook/geos/xx.html).

(2013). The World Fact Book. US Central Intelligence Agency. (www.cia.gov/library/publications/the-world-factbook).

COEJL (2007). Global Warming: A Jewish Response. Coalition on the Environment and Jewish Life (www.coejl.org/climatechange/gw_jewishresponse.php).

Cornwall Alliance (2011). Evangelical Declaration on Global Warming (www.cornwallalliance.org/articles/read/an-evangelical-declaration-on-global-warming).

CSIRO (2012). State of the Climate 2012. Commonwealth Scientific and Industrial Research Organization (www.csiro.au/Outcomes/Climate/Understanding/State-of-the-Climate-2012.aspx).

(2014). State of the Climate 2014. Commonwealth Scientific and Industrial Research Organization (www.csiro.au/Outcomes/Climate/Understanding?State-of-the-Climate-2014.aspx).

CSSD (2013). Center for Sustainable Shale development (www.sustainableshale.org).

Cala, A. (2012). Growing Doubts in Europe of Future Carbon Storage. *The New York Times*, January 17, B6.

Canadell, J.G., et al. (2007). Contributions to Accelerating Atmospheric CO_2 Growth from Economic Activity, Carbon Intensity and Efficiency of Natural Sinks. *Proc. Natl. Acad. Sci.* 104(47), 18866–70 (www.pnas.org/cgi/doi/10.1073/pnas.0702737104).

Cardwell, D. (2014). NRG Seeks to Cut 90% of Its Carbon Emissions. *The New York Times*, November 21, B3.

Carey, J. (2006a). Senator Inhofe's Climate of Confrontation. *Business Week*, April 3, 43.

(2006b). Business on a Warmer Planet. *Business Week*, July 17, 26–9.

(2009). Carbon Curbs: It's Business vs. Business. *Business Week*, October 5, 66–7.

(2010). Measuring the Gas without the Hot Air. *Business Week*, January 11, 18–19.

(2011a). Storm Warnings: Extreme Weather Is a Product of Climate Change. *Scientific American* (www.scientificamerican.com/article.cfm?id=extreme-weather-caused -by-climate-change).

(2011b). Global Warming and the Science of Extreme Weather. *Scientific American* (www.scientificamerican.com/article.cfm?id=global-warming-and -the-science-of-extreme-weather).

Carson, R. (1962). *Silent Spring*. Houghton Mifflin, Boston, MA.

Carty, T. (2012). Extreme Weather, Extreme Prices. Oxfam International, September 5, (www.oxfam.org/en/grow/policy/extreme-weather-extreme-prices).

Castles, I. and D. Henderson (2003). The IPCC Emissions Scenarios: An Economic-Statistical Critique. *Energy & Environment* 14(2), 159–85.

Chandler, D. (2009). Climate Change Odds Much Worse than Thought. (http:// web.mit.edu/newsoffice/2009/roulette-0519.html).

Chapin III, F.S., et al. (2005). Role of Land-Surface Changes in Arctic Summer Warming. *Science* 310, 657–60.

Chazan, G. (2004). EU Backs Russia's WTO Entry As Moscow Supports Kyoto Pact. *The Wall Street Journal*, May 24, A2, A6.

Chen, J.L., et al. (2006). Satellite Gravity Measurements Confirm Accelerated Melting of Greenland Ice Sheet. *Science* 313, 1958–60.

Chen, X. and K.-K. Tung (2014). Varying Planetary Heat Sink Led to Global-warming Slowdown and Acceleration. *Science* 345, 897–903.

Church, J.A., et al. (2001). Changes in Sea Level. In J.T. Houghton, et al. (eds.), *Climate Change 2001: The Scientific Basis*. Intergovernmental Panel on Climate Change. Cambridge University Press, Cambridge, 639–693.

Clausen, E. (2004). An Effective Approach to Climate Change. *Science* 306, 816.

Clement, A.C., et al. (2009). Observational and Model Evidence for Positive Low-Level Cloud Feedback. *Science* 325, 460–4.

Cohen, J. (2010). Bundle Up, It's Global Warming. *The New York Times*, December 26, SR15.

Collins, J. (2014). Timeline of EPA Actions on Greenhouse Gases. Environmental and Energy Study Institute (www.eesi.org/files/FactSheet_ EPA_timeline_092214.pdf).

Cowtan, K. and R. Way (2014). Coverage Bias in the HadCRUT4 Temperature Series and Its Impact on Recent Temperature Trends. *J. Roy. Met. Soc.* 140(683), 1935–44 (http://dx.doi.org/10.1002/qj.2297).

Creyts, J., et al. (2007). *Reducing US Greenhouse Gas Emissions: How Much at What Cost?* McKinsey & Company.

Crichton, M. (2004). *State of Fear*. Harper Collins, New York.

Cross, J.-M. and R. Pierson (2013). Short-Lived Climate Pollutants: Why They Are Important. Environmental and Energy Study Institute, February 19 (www.eesi.org).

Crutzen, P.J. (2006). Albedo Enhancement by Stratospheric Sulfur Injections: A Contribution to Resolve a Policy Dilemma? *Climatic Change* 77, 211–19.

Cubasch, U., et al. (2001). Projections of Future Climate Change. In J.T. Houghton et al. (eds.), *Climate Change 2001: The Scientific Basis*. Intergovernmental Panel on Climate Change. Cambridge University Press, Cambridge, 525–82.

Cullen, H. (2010). *The Weather of the Future: Heat Waves, Extreme Storms and Other Scenes from a Climate-Changed Planet*. Harper Collins, New York.

Curry, J. (2014). Uncertain Temperature Trend. *Nature Geoscience* 7, 83–4.

DOD (2010). Quadrennial Defense Review Report. U.S. Department of Defense (www.defense.gov/qdr/images/QDR_as_of_12Feb10_1000.pdf).

(2014a). Quadrennial Defense Review 2014. U.S. Department of Defense (www.defense.gov/pubs/2014_Quadrennial_Defense_Review.pdf).

(2014b). 2014 Climate Change: Adaptation Roadmap (www.acq.osd.mil/ie/download/CCARprint.pdf).

DOE (2012). Carbon Utilization and Storage Atlas, 4th Edition. US Department of Energy, National Energy Technology Laboratory (www.netl.doe.gov/technologies/carbon_seq/refshelf/atlasIV).

DOT (2006). *Energy Data Book: Edition 25*. US Department of Transportation.

Dalton, M. (2007). Nymex Plans for Carbon-Emissions Trading. *The Wall Street Journal*, May 10, C2.

Daly, H.E. (1996). *Beyond Growth*. Beacon Press, Boston, MA.

Daly, H.E. and J.B. Cobb, Jr. (1989). *For the Common Good*. Beacon Press, Boston, MA.

Davenport, C. (2013). Large Companies Prepared to Pay Price on Carbon. *The New York Times*, December 5, A1, B4.

(2014a). E.P.A. Staff Struggling to Create Pollution Rule. *The New York Times*, February 5, A12, A15.

(2014b). Threat to Bottom Line Spurs Action on Climate. *The New York Times*, January 24, A1, A21.

(2014c). President's Drive for Carbon Pricing Fails to Win at Home. *The New York Times*, September 28, A8.

(2014d). Optimism Faces Grave Realities at Climate Talks. *The New York Times*, December 1, A1, A12.

Davenport, C. and P. Baker (2014). Taking Page from Health Care, Obama Climate Plan Relies on States. *The New York Times*, June 3, A16.

Davenport, C. and M. Landler (2014). U.S. to Give $3 Billion to Climate Fund to Help Poor Nations and Spur Rich Ones. *The New York Times*, November 15, A5.

Davis, S.C., et al. (2014). Transportation Energy Data Book. Edition 33, Oak Ridge National Laboratory (http://cta.ornl.gov/data).

Davis, S.J., et al. (2010). Future CO_2 Emissions and Climate Change from Existing Infrastructure. *Science*. 329, 1330–3.

(2013). Rethinking Wedges. *Environ. Res. Lett.* 8, 011001 (doi:10.1088/1748–9326/8/1/011001).

Dawidoff, N. (2009). The Civil Heretic. *The New York Times Magazine* (www.nytimes.com/2009/03/29/magazine/29Dyson-t.html)

de Gouw, J.A., et al. (2014). Reduced Emissions of CO_2, NO_x and SO_2 from US Power Plants Due to the Switch from Coal to Natural Gas with Combined Cycle Technology. *Earth's Future* 2(2), 75–82 (doi:10.1002/2014EF000196).

Derksen, C. and R. Brown (2012). Spring Snow Cover Extent Reduction in 2008–2012 Period Exceeding Climate Predictions. *Geophys. Res. Lett.* 39, L19504 (doi:10.1029/2012GL053387).

Dessler, A.E. (2010). A Determination of the Cloud Feedback from Climate Variations over the Past Decade. *Science* 330, 1523–7.

   (2011). Cloud Variations and the Earth's Energy Balance. *Geophys. Res. Lett.* 38, L19701 (doi:10.1029/2011GL049236).

Dessler, A.E. and S.C. Sherwood (2009). A Matter of Humidity. *Science* 323, 1020–1.

Deutsch, C., et al. (2011). Climate-Forced Variability of Ocean Hypoxia. *Science* 333, 336–9.

Dines, W.H. (1917). The Heat Balance of the Atmosphere. *Quart. J. Roy. Meteor. Soc.* 43, 151–8.

Doan, L. (2013). California Carbon Permits Sell above the Clearing Price (www.bloomberg.com/news/print/2013-02-22/california-sells-carbon-allowances-for-13-62-each-in-auction.html).

Dooley, J.J., et al. (2006). Carbon Dioxide Capture and Geologic Storage. Report from the Global Energy Technology Strategy Program, Battelle Memorial Institute.

Doom, J. (2014). RGGI Carbon Prices Jump in First Event with Fewer Permits (www.bloomberg.com/news/print/2014-03-07/rggi-carbon-prices-jump-in-first-event-with-fewer-permits.html).

Dorrepaal, E., et al. (2009). Carbon Respiration from Subsurface Peat Accelerated by Climate Warming in the Subarctic. *Nature* 460, 616–19.

Dowdeswell, J.A. (2006). The Greenland Ice Sheet and Global Sea Level Rise. *Science* 311, 963–4.

Durack, P.J., et al. (2012). Ocean Salinities Reveal Strong Global Water Cycle Intensification during 1950 to 2000. *Science* 336, 455–8.

   (2014). Quantifying Underestimates of Long-Term Upper-Ocean Warming. *Nature Climate Change* 4, 1031–5 (doi:10.1038/nclimate2389).

Dutton, A. and K. Lambeck (2012). Ice Volume and Sea Level During the Last Interglacial. *Science* 337, 216–19.

EC (2008). Climate Action and Renewable Energy Package (http://ec.europa.eu/commission_2010-2014/hedegaard/headlines/topics/package_en.htm).

Economist (2007). Special Report on Business and Climate Change. *The Economist*, June 2.

   (2009a). Last Gasp for the Forest. September 26, 93–5.

   (2009b). Bubbling Under. December 5, 75.

   (2010). Special Report on Forests. September 25, 8–10.

   (2011a). Coming Soon to a Terminal Near You. August 6, pp. 51–3.

   (2011b). A Man-Made World. May 28, pp. 81–3.

   (2011c). A Game of Catch-up. Special Report on the World Economy, September 24, 3–6.

   (2012a). The Melting North. Special Report on the Arctic, June 16.

   (2012b). An Unconventional Bonanza. Special Report on Natural Gas, July 14.

   (2013a). Carbon Trading, ETS, RPI? April 20, 75.

   (2013b). The East is Gray. August 10, 18–21.

   (2013c). Carbon Copy, Companies and Emissions. December 14, 70.

EEnews (2013). E&E Publishing (www.eenews.net/assets/2013/07/15/document_cw_02.pdf).

EIA (2001). Emissions of Greenhouse Gases in the United States: Historical and Projected U.S. Carbon and Total Greenhouse Gas Intensity. U.S. Energy Information Administration, Report DOE/EIA-0573.

(2009). Market and Economic Impact of H.R.2454, The American Clean Energy and Security Act of 2009, SR-OIAF/2009-05.

(2013a). U.S. Energy-Related Carbon Dioxide Emissions, 2012. (www.eia.gov/environment/emissions/carbon).

(2013b). International Energy Outlook 2013. Report DOE/EIA-0484(2013).

(2013c). International Energy Statistics. (www.eia.gov/countries/data.cfm#undefined).

(2014a). Annual Energy Review. (www.eia.gov/totalenergy/data/annual/index.cfm#summary).

(2014b). Annual Energy Outlook 2014. Early Release (www.eia.gov/forecasts/aeo/er).

EPA (2010). EPA to Set Modest Pace for Greenhouse Gas Standards. (http://yosemite.epa.gov/opa/admpress.nsf/).

(2013b). EPA Needs to Improve Air Emissions Data for the Oil and Natural Gas Production Sector. US Environmental Protection Agency, Office of Inspector General. Report 13-P-0161, February 20.

(2014a). Inventory of US Greenhouse Gas Emissions and Sinks: 1990–2012. EPA 430-R-14-003 (http://epa.gov/climatechange/Downloads/ghgemissions/US-GHG-Inventory-2014-Main-Text.pdf). See also (www.epa.gov/climatechange/emissions/usinventoryreport.html).

(2014b). Climate Change Indicators in the United States. Environmental Protection Agency (www.epa.gov/climatechange/science/indicators).

(2014c). Light-Duty Automotive Technology, Carbon Dioxide emissions, and Fuel Economy Trends: 1975–2014 (http://epa.gov/otaq/fetrends.htm).

(2014d). Standards of Performance for Greenhouse Gas Emissions from New Stationary Sources: Electric Utility Generating Units. Proposed Rule, Federal Register Vol. 79, No. 5 (www.gpo.gov/fdsys/pkg/FR-2014-01-08/pdf/2013–28668.pdf).

EPI (2012). Earth Policy Institute. Data Center (www.earth-policy.org/data_center/C25).

Eddy, M. (2014). German Village Resists Plans to Strip It Away for the Coal Underneath. *The New York Times*, February 19, A5.

Eilperin, J. (2012). EPA Imposes First Greenhouse Gas Limits on New Power Plants. *The Washington Post*, March 27 (www.washingtonpost.com/national/health-science/epa-to-impose-first-greenhouse-gas-limits-on-power-plants/2012/03/27/gIQAKdaJeS_story.html).

Evangelium Vitae (1995). Pope John Paul II, §42, March 25 (www.vatican.va/holy_father/john_paul_ii/encyclicals/documents/hf_jp-ii_enc_25031995_evangelium-vitae_en.html).

FEMA (1993). Robert T. Stafford Disaster Relief and Emergency Assistance Act (Amended). Federal Emergency Management Agency (www.fema.gov/robert-t-stafford-disaster-relief-and-emergency-assistance-act-public-law-93-288-amended).

Fairfield, H. (2007). When Carbon Is Currency. *The New York Times*, May 6, A6.

(2014). The Best of Both Worlds in Cutting Emissions. *The New York Times*, June 8, SR5.

Falkowski, P.G. and D. Tchernov (2004). Human Footprints in the Ecological Landscape. *Earth System Analysis for Sustainability*, Dahlen Workshop Reports, H.J. Schellnhuber et al., eds., 211–26.

Fargione, J., et al. (2008). Land Clearing and the Biofuel Debt. *Science* 319, 1235–8.

Farrel, A.E., et al. (2006). Ethanol Can Contribute to Energy and Environmental Goals. *Science* 311, 506–8.

Fernando, H.J.S. and Z.B. Klaić (2011). Addressing Socioeconomic and Political Challenges Posed by Climate Change. *EOS, Transactions of the American Geophysical Union* (doi:10.1029/2011EO350006).

Fialka, J.J. (2004). Russia's Interest in Signing Kyoto Spurs Trading. *The Wall Street Journal*, June 2, C6.

(2006). Big Businesses Have New Take on Warming. *The Wall Street Journal*, March 28, A4.

Fleming, J.R. (2010). *Fixing the Sky: The Checkered History of Weather and Climate Control*. Columbia University Press, New York.

Foley, J.A. (2005). Tipping Points in the Tundra. *Science* 310, 627–8.

Fowlie, M., et al. (2014). An Economic Perspective on the EPA's Clean Power Plan. *Science* 346, 815–16.

Fox, D. (2010). Could East Antarctica Be Headed for Big Melt? *Science* 328, 1630–1.

Freedman, A. (2012). Greenland Ice Sheet Melt Nearing Critical Point. Climate Central, June 29 (www.climatecentral.org/news/greenland-ice-sheet-reflectivity-near-record-low-research-shows).

Friedman, T.L. (2007). The Power of Green. *The New York Times*, April 15 (www.nytimes.com/2007/4/15/magazine/15green.t.html).

Fuller, T. and G. Bowley (2007). Nations Near Agreement on Steps to Revive Climate Treaty. *The New York Times*, December 15, A7.

GAO (2011). Climate Engineering: Technical Status, Future Directions, and Potential Responses. US Government Accountability Office, GAO-11-71, July 28.

GCEC (2014). The New Climate Economy: Better Growth Better Climate. Global Commission on the Economy and Climate (http://newclimateeconomy.report).

GCP (2013). Annual Update of the Global Carbon Budget and Trends. Global Carbon Project, November 19 (www.globalcarbonproject.org/carbonbudget/).

GWEC (2014). Global Wind Statistics 2013. Global Wind Energy Council (www.gwec.net/wp-content/uploads/2014/02/GWEC-PRstats-2013_EN.pdf).

Gallagher, K.S., et al. (2007). Policy Options for Reducing Oil Consumption and Greenhouse-Gas Emissions from the US Transportation Sector. John F. Kennedy School of Government, Harvard University, July 27.

Gardiner, S.M. (2011). *A Perfect Moral Storm – The Ethical Tragedy of Climate Change*, Oxford University Press, New York.

Garvey, J. (2008). *The Ethics of Climate Change*, Continuum International Publishing Group, London.

Gerland, P., et al. (2014). World Population Stabilization Unlikely this Century. *Science* 346, 234–7.

Gertner, J. (2009). Why Isn't the Brain Green? *The New York Times Magazine*, April 19, 36–43.

Giddens, A. (2011). *The Politics of Climate Change*. Polity Press, Cambridge.

Gillis, J. (2010). Reading Earth's Future in Glacial Ice. *The New York Times*, November 14, A1, A14–A15.

(2013). Study of Ice Age Bolsters Carbon and Warming Link. *The New York Times*, March 1, A4, A10.

(2014). Some Scientists Disagree with President's Linking Drought to Warming. *The New York Times*, February 18, A11.

Gillis, J. and F. Barringer (2012). As Coasts Rebuild and US Pays, Again, Critics Stop to Ask Why. *The New York Times*, November 19, A1, A12.

Gleick, P. (2012). Remarkable Editorial Bias on Climate Science at the Wall Street Journal. (www.forbes.com/sites/petergleick/2012/01/27/remarkable-editorial-bias-on-climate-science-at-the-wall-street-journal).

Glenn, D. (2006). A Climate for Not Much Change. *Chronicle of Higher Education*, March 31, A17–A18.

Goodell, J. (2010). *Geoengineering and the Audacious Quest to Fix Earth's Climate*. Houghton Mifflin Harcourt, Boston/New York.

Goodstein, L. (2006). Eighty-Six Evangelical Leaders Join to Fight Global Warming. *The New York Times* (www.nytimes.com/2006/02/08/national/08warm.html).

Greely, B. (2011). The God Clause. *Bloomberg Businessweek*, September 5, 58–67.

Gustafsson, Ö., et al. (2009). Brown Clouds over South Asia: Biomass or Fossil Fuel Combustion, *Science* 323, 495–8.

Hamilton, C. (2013). *Earthmasters: The Dawn of the Age of Climate Engineering*. Yale University Press, New Haven, CT.

Hampton, R. and J.-M. Cross (2014). Climate Adaptation at the Federal Level. Environmental and Energy Study Institute (www.eesi.org/fact-sheet-climate-adaptation-federal-level-16-jan-2014).

Handwerk, B. (2012). Amid Economic Concerns, Carbon Capture Faces a Hazy Future. *National Geographic Daily News*, May 22 (http://news.nationalgeographic.com/news/energy/2012/05/120522-carbon-capture-and-storage-economic-hurdles).

Hansen, J., et al. (2005). Earth's Energy Imbalance: Confirmation and Implications. *Science* 308, 1431–5.

Hansen, J. (2006). Is There Still Time to Avoid Dangerous Anthropogenic Interference with Global Climate? (www.columbia.edu/~jeh1)

(2009). *Storms of My Grandchildren: The Truth about the Coming Climate Catastrophe and Our Last Chance to Save Humanity*. Bloomsbury Press, New York.

Hardin, G. (1968). The Tragedy of the Commons. *Science* 162, 1243–8.

Harris Poll (2011). A Change in the Wind. Americans Increasingly Doubt Global Warming (www.achangeinthewind.com/2011/07/americans-increasingly-doubt-global-warming-harris-poll.html).

Harris, N.L., et al. (2012). Baseline Map of Carbon Emissions from Deforestation in Tropical Regions. *Science* 336, 1573–6.

Hassol, S.J. (2004). *Impacts of a Warming Arctic*. Arctic Climate Impact Assessment (ACIA) Project. Cambridge University Press, Cambridge.

Haszeldine, R.S. (2009). Carbon Capture and Storage: How Green Can Black Be? *Science* 325, 1647–52.

Hawken, P. (2005). *The Ecology of Commerce*. Collins Business, New York.

Hegerl, G.C. and S. Solomon (2009). Risks of Climate Engineering. *Science* 325, 955–6.

Heimann, M. (2010). How Stable Is the Methane Cycle? *Science* 327, 1211–12.

Hoegh-Guldberg, O., et al. (2007). Coral Reefs under Rapid Climate Change and Ocean Acidification. *Science* 318, 1737–42.

Hoegh-Guldberg, O. and J.F. Bruno (2010). The Impact of Climate Change on the World's Marine Ecosystems. *Science* 328, 1523–8.

Hoffman, A.J. (2006). Getting Ahead of the Curve: Corporate Strategies that Address Climate Change. Pew Center on Global Climate Change.

   (2012). Climate Science as Culture War. Stanford Social Innovation Review (www.ssireview.org/articles/entry/climate_science_as_culture_war).

Hönisch, B., et al. (2009). Atmospheric Carbon Dioxide Concentration across the Mid-Pleistocene Transition. *Science* 324, 1551–4.

House, K.Z., et al. (2011). Economic and Energetic Analysis of Capturing CO_2 from Ambient Air. *Proc. Natl. Acad. Sci.* 108(51), 20428–33 (www.pnas.org/cgi/doi/10.1073/pnas.1012253108).

House of Representatives (2007). Political Interference with Climate Change Science under the Bush Administration. (http://oversight-archive.waxman.house.gov/documents/20071210101633.pdf).

Houser, T., et al. (2014). American Climate Prospectus: Economic Risks in the United States. Risky Business Project, Rhodium Group, June 24.

Howarth, R.W., et al. (2011). Methane and the Greenhouse-Gas Footprint of Natural Gas from Shale Formations. *Climatic Change* 106: 679–90 (doi:10.1007/s10584-011-0061-5).

Hoyos, C.D., et al. (2006). Deconvolution of the Factors Contributing to the Increase in Global Hurricane Intensity. *Science* 312, 94–7.

Hsiang, S.M., et al. (2013). Quantifying the Influence of Climate on Human Conflict. *Science* 341(6151) (doi:10.1126/science.1235367).

Hume, M. (2009). *Why We Disagree about Climate Change: Understanding Controversy, Inaction, and Opportunity*. Cambridge University Press, Cambridge.

ICLEI (2011). Local Governments for Sustainability USA (www.icleiusa.org).

IEA (2010). World Energy Outlook 2010. International Energy Agency (www.worldenergyoutlook.org/publications).

   (2011). Technology Roadmap: Electric and Plug-in Hybrid Electric Vehicles (www.iea.org/publications/freepublications/publication/technology-roadmap-electric-and-plug-in-hybrid-electric-vehicles-evphev.html).

   (2013a). Key World Energy Statistics (www.iea.org/publications/freepublications/publication/KeyWorld2013.pdf).

   (2013b). CO_2 Emissions from Fossil Fuel Combustion (www.iea.org/publications/freepublications/publication/co2emissionsfromfuelcombustionhighlights2013.pdf).

   (2014). Snapshot of Global PV: 1992–2013 (www.iea-pvps.org/fileadmin/dam/public/report/statistics/PVPS_report_-_A_Snapshot_of_Global_PV_-_1992-2013_-_final_3.pdf).

IGBP (2013). Ocean Acidification. International Geosphere-Biosphere Program (www.igbp.net/publications/summariesforpolicymakers/summariesforpolicymakers/oceanacidificationsummaryforpolicymakers2013.html).

IMF (2014). Climate, Environment and the IMF (www.imf.org/external/np/exr/facts/pdf/enviro.pdf).

IPCC (1995). *Impacts, Adaptations, and Mitigation of Climate Change: Scientific-Technical Analysis.* Intergovernmental Panel on Climate Change. Contributions of Working Group II to the Second Assessment Report of the IPCC. Cambridge University Press, Cambridge.

(2001a). *Climate Change 2001: Synthesis Report.* Contribution of Working Groups I, II and III to the Third Assessment Report of the IPCC. Cambridge University Press, Cambridge.

(2001b). *Climate Change 2001.* Technical Summary of Working Group I (www.ipcc.ch/ipccreports/tar/vol4/english/pdf/wg1ts.pdf).

(2001c). *Climate Change 2001: The Scientific Basis.* Contribution of Working Group I to the Third Assessment Report of the IPCC. Cambridge University Press, Cambridge.

(2007a). *Climate Change 2007: Synthesis Report.* Working Groups I, II and III Contributions to the Fourth Assessment Report of the IPCC. Chapter 6, Paleoclimate. Cambridge University Press, Cambridge.

(2007b). *Climate Change 2007: The Physical Science Basis.* Contribution of Working Group I to the Fourth Assessment Report of the IPCC. Cambridge University Press, Cambridge.

(2007c). IPCC Assessment of Climate Change and Uncertainties. Contributions of Working Group I to the Fourth Assessment Report. Cambridge University Press, Cambridge.

(2007d). *Climate Change 2007: Impacts, Adaptation, and Vulnerability.* Contribution of Working Group II to the Fourth Assessment Report of the IPCC. Cambridge University Press, Cambridge.

(2014a). *Climate Change 2013: The Physical Science Basis.* Contribution of Working Group I to the Fifth Assessment Report of the IPCC. Cambridge University Press, Cambridge.

(2014b). *Climate Change 2014: Impacts, Adaptation and Vulnerability. Summary for Policymakers.* Contribution of Working Group II to the Fifth Assessment Report of the IPCC. Cambridge University Press, Cambridge.

(2014c). *Climate Change 2014: Synthesis Report.* Fifth Assessment Synthesis Report of the IPCC. Cambridge University Press, Cambridge.

IPSOS (2014). Ipsos Global Trends 2014 (www.ipsosglobaltrends.com/environment.html).

JPL (2013). Study Sheds Light on Arctic Sea Ice Volume Losses. Jet Propulsion Laboratory (www.jpl.nasa.gov/news/news.php?release=2013–057).

Jackson, S.C. (2009). Parallel Pursuit of Near-Term and Long-Term Climate Mitigation. *Science* 326, 526–7.

Jamieson, D. (2014). *Reason in a Dark Time: Why the Struggle against Climate Change Failed – and What It Means for our Future.* Oxford University Press, New York.

Janofsky, M. (2005). When Cleaner Air Is a Biblical Obligation. *The New York Times* (www.nytimes.com/2005/11/07/politics/07air.html?ref=michaeljanofsky).

Johannessen, O.M., et al. (2005). Recent Ice-Sheet Growth in the Interior of Greenland. *Science* 310, 1013–15.

Jones, E.A. (2008). How Texas Struck It Rich Beneath Suburbia. *The Wall Street Journal*, August 2, A9.

Jose, C. (2013). Blessing the Facts: Evangelical Scientists Urge Congress to Reduce Carbon Emissions (www.eenews.net/stories/1059984347/print).

Joughin, I., et al. (2008). Seasonal Speedup along the Western Flank of the Greenland Ice Sheet. *Science* 320, 781–3.

(2014). Marine Ice Sheet Collapse Potentially Underway for the Thwaites Glacier Basin, West Antarctica. *Science* 344, 735–8.

Kahan, D.M., et al. (2012). The Polarizing Impact of Science Literacy and Numeracy on Perceived Climate Change Risks. *Nature Climate Change* 2, 732–5 (doi:10.1038/nclimate1547).

Kahn, S.A., et al. (2014). Sustained Mass Loss of the Northeast Greenland Ice Sheet Triggered by Regional Warming. *Nature Climate Change* 4, 292–9 (doi:10.1038/nclimate2161).

Kanter, J. (2014). E.U. Greenhouse Gas Deal Falls Short of Expectations. *The New York Times*, October 25, A11.

Kanter, J. and J. Mouawad (2008). Pipe Dreams and Politics. *The New York Times*, December 11, B1, B6.

Karion, A., et al. (2013). Methane Emissions Estimate from Airborne Measurements over a Western United States Gas Field. *Geophys. Res. Letts.* 40(16), 4393–7.

Kaufman, L. (2009). Dissenter on Warming Expands His Campaign. *The New York Times*, April 10, A13.

Kaufmann, R.K., et al. (2011). Reconciling Climate Anthropogenic Change with Observed Temperature 1998–2008. *Proc. Natl. Acad. Sci.* 108(29), 11790–3 (www.pnas.org/content/early/2011/06/27/1102467108).

Keith, D.W. (2009). Why Capture CO_2 from the Atmosphere? *Science* 325, 1654–5.

(2013). *A Case for Climate Engineering*. MIT Press, Cambridge, MA.

Kerr, R.A. (2004a). Three Degrees of Consensus. *Science* 305, 932–4.

(2004b). A Bit of Icy Antarctica Is Sliding Toward the Sea. *Science* 305, 1897.

(2005). The Atlantic Conveyor May Have Slowed, But Don't Panic Yet. *Science* 310, 1403–5.

(2006a). Latest Forecast: Stand By for a Warmer, But Not Scorching World. *Science* 312, 351.

(2006b). A Worrying Trend of Less Ice, Higher Seas. *Science* 311, 1698–701.

(2006c). False Alarm, Atlantic Conveyor Belt Hasn't Slowed. *Science* 314, 1064.

(2006d). Pollute the Planet for Climate's Sake? *Science* 314, 401–3.

(2008). Global Warming Throws Some Curves in the Atlantic Ocean. *Science* 322, 515.

(2009). Arctic Summer Ice Could Vanish Soon But Not Suddenly. *Science* 323, 1655.

(2010). Ocean Acidification Unprecedented, Unsettling. *Science* 328, 1500–1.

(2011). Vital Details of Global Warming Are Eluding Forecasters. *Science* 334, 173–4.

(2012). Ice-Free Arctic Sea May Be Years, Not Decades, Away. *Science* 337, 1591.

Kiehl, J.T. and K.E. Trenberth (1997). Earth's Annual Global Mean Energy Budget. *Bull. Amer. Met. Soc.* 78, 197–208.

Kildow, J. and J. Scorse (2012). End Federal Flood Insurance. *The New York Times*, November 29, A23.

Kintisch, E. (2006). Evangelicals, Scientists Reach Common Ground on Climate Change. *Science* 311, 1082–3.

(2007). Tougher Ozone Accord Also Addresses Global Warming. *Science* 317, 1843.

(2008). Roads, Ports, Rails Aren't Ready for Changing Climate, Says Report. *Science* 319, 1744–5.

(2009a). New Push Focuses on Quick Ways to Curb Global Warming. *Science* 324, 323.

(2009b). Projections of Climate Change Go from Bad to Worst. *Science* 323, 1546–7.

(2010). *Hack the Planet: Science's Best Hope – or Worst Nightmare – for Averting Climate Catastrophe*. John Wiley & Sons, Hoboken, NJ.

(2014). Is Atlantic Holding Earth's Missing Heat? *Science* 345, 860–1.

Kintisch, E. and K. Buckheit (2006). Along the Road from Kyoto. *Science* 311, 1705–6.

Klein, N. (2014). *This Changes Everything: Capitalism vs. The Climate*. Simon & Schuster, New York.

Kosaka, Y. and S.-P. Xie (2013). Recent Global-Warming Hiatus Tied to Equatorial Pacific Surface Cooling. *Nature* 501, 403–7 (doi:10.1038/nature12534).

Kramer, A.E. (2005). In Russia, Pollution Is Good for Business. *The New York Times*, December 28, C1, C5.

Krauss, C. (2008a). There's Gas in Those Hills. *The New York Times*, April 8 (www.nytimes.com/2008/04/08/business/08gas.html).

(2008b). Drilling Boom Revives Hopes for Natural Gas. *The New York Times*, August 25, A1, A16.

(2013). U.S. Coal Companies Scale Back Export Goals. *The New York Times*, September 14, B1, B6.

Krauss, C., et al. (2005). As Polar Ice Turns to Water, Dreams of Treasure Abound. *The New York Times*, October 10 (www.nytimes.com/2005/10/10/science/10arctic.html).

Lazard (2013). Lazard's Levelized Cost of Energy Analysis-Version 7.0. August.

LTDL (1993). Bad Science: A Resource Book. Legacy Tobacco Documents Library. University of California, San Francisco (http://legacy.library.ucsf.edu/tid/snc52c00).

Lal, R. (2004). Soil Carbon Sequestration Impacts on Global Climate Change and Food Security. *Science* 304, 1623–6.

Leiserowitz, A. (2010). Project on Climate Change Communication. Yale University. (http://environment.yale.edu/climate/news/global-warmings-six-americas-june-2010).

Levin, K., et al. (2010). Playing It Forward: Path Dependency, Progressive Incrementalism, and the "Super Wicked" Problem of Global Climate Change. (http://environment.research.yale.edu/documents/downloads/0–9/2010_super_wicked_levin_cashore_bernstein_auld.pdf).

(2012). Overcoming the Tragedy of Super Wicked Problems: Constraining our Future Selves to Ameliorate Climate Global Climate Change. *Policy Sciences* 45(2), 125–52.

Lewis, V.B. (2006). The Common Good in Classical Political Philosophy. *Current Issues in Catholic Higher Education* 25(1), 25–46.

Li, J. and R. Stone (2011). China Looks to Balance Its Carbon Books. *Science* 334, 886–7.

Lipton, E., et al. (2012). Flood Insurance, Already Fragile, Faces New Stress. *The New York Times*, November 13, pp. A1, A18.

Liu, Z., et al. (2009). Transient Simulation of Last Deglaciation with a New Mechanism for Bølling-Allerød Warming. *Science* 325, 310–14.

Llovel, W., et al. (2014). Deep-Ocean Contribution to Sea Level and Energy Budget Not Detectable over the Past Decade. *Nature Climate Change* 4, 1031–5 (doi:10.1038/nclimate2387).

Loarie, S.R., et al. (2009). Boosted Carbon Emission from Amazon Deforestation. *Geophys. Res. Lett.* 36, L14810.

Lobell, D.B., et al. (2008). Prioritizing Climate Change Adaptation Needs for Food Security in 2030. *Science* 319, 607–10.

Lomborg, B. (2006). Stern Review. *The Wall Street Journal*, November 2, A12.

Lovins, A.B. (2005). More Profit with Less Carbon. *Scientific American*, September, 74–83.

Lustgarten, A. (2009). The Hidden Danger of Gas Drilling. *Business Week*, November 24, 76–9.

Lutchke, S.B., et al. (2006). Recent Greenland Ice Mass Loss by Drainage System from Satellite Gravity Observations. *Science* 314, 1286–9.

(2012). Coal Power Drops to New Lows. *Mechanical Engineering*, June, 10.

MIT (2006). Report of the Energy Research Council, Massachusetts Institute of Technology (http://web.mit.edu/mitei/about/erc-report-final.pdf).

(2007). *The Future of Coal: Options for a Carbon-Constrained World*. Massachusetts Institute of Technology. (http://web.mit.edu/mitei/research/studies/coal.shtml).

(2011). *The Future of Natural Gas*. Massachusetts Institute of Technology (http://web.mit.edu/mitei/research/studies/naturalgas.html).

(2014). Carbon Capture and Sequestration Technologies (http://sequestration.mit.edu).

Mongabay (2011). REDD (http://rainforests.mongabay.com/redd).

Mahasneh, H.I. (2003). Islamic Faith Statement. Faiths and Ecology. Alliance of Religions and Conservation (www.arcworld.org/faiths.asp?pageID=75).

Malakof, D. (2014). The Gas Surge. *Science* 344, 1464–7.

Mankiw, N.G. (2007). One Answer to Global Warming: A New Tax, *The New York Times*, September 16, BU6.

Mann, M.E., et al. (1998). Global-Scale Temperature Patterns and Climate Forcing over the Past Six Centuries. *Nature* 392, 779–87.

(1999). Northern Hemisphere Temperature during the Past Millennium: Inferences, Uncertainties, and Limitations. *Geophys. Res. Lett.* 26, 759–62.

(2008). Proxy-Based Reconstruction of Hemispheric and Global Surface Temperature Variations over the Past Two Millennia. *Proc. Natl. Acad. Sci.* 105, 13252–7.

Marcott, S.A., et al. (2013). A Reconstruction of Regional and Global Temperature for the Past 11,300 Years. *Science* 339, 1198–201.

Martin, C. (2013). US States Turning Against Renewable Portfolio Standards as Gas Plunges. (www.renewableenergyworld.com/rea/news/article/2013/04/u-s-states -turning-against-renewable-portfolio-standards-as-gas-plunges).

Martin, C., et al. (2013). Why the US Power Grid's Days Are Numbered. *Bloomberg Businessweek* (www.businessweek.com/articles/2013-08-22/homegrown-green -energy-is-making-power-utilities-irrelevant).

Marzeion, B., et al. (2014). Attribution of Global Glacier Mass Loss to Anthropogenic and Natural Causes. *Science* 345, 919–21.

McCright, A.M. and R.E. Dunlap (2011). The Politicization of Climate Change and Polarization in the American Public's View of Global Warming, 2001–2010. *The Sociological Quarterly* 52(2), 155–94.

McGinn, D, (2009). The Greenest Big Companies in America. *Newsweek*, September 28, 34–48.

McIntyre, S. and R. McKitrick (2005). Hockey Sticks, Principal Components, and Spurious Components. *Geophys. Res. Lett.* 32, L03710 (doi:10.1029/2004GL021750).

McKinley, G.A., et al. (2011). Convergence of Atmospheric and North Atlantic Carbon Dioxide Trends on Multidecadal Timescales. *Nature Geoscience* 4, 604–10 (doi:10.1038/ngeo1193).

Meehl, G.A. and C. Tebaldi (2004). More Intense, More Frequent, and Longer Lasting Heat Waves in the 21st Century. *Science* 305, 994–7.

Meehl, G.A., et al. (2011). Model-Based Evidence of Deep-Ocean Heat Uptake during Surface-Temperature Hiatus Periods. *Nature Climate Change* 1, 360–4.
  (2013). Externally Forced and Internally Generated Climate Variability Associated with Interdecadal Pacific Oscillation. *J. Climate* 26, 7298–310.

Miller, S.M., et al. (2013). Anthropogenic Emissions of Methane in the U.S. *Proc. Natl. Acad. Sci.* 110(50), 20018–22. (www.pnas.org/cgi/doi/10.1073/pnas.1314392110).

Mills, E. (2005). Insurance in a Climate of Change. *Science* 309, 1040–4.

Moberg, A., et al. (2005). Highly Variable Northern Hemisphere Temperatures Reconstructed from Low- and High-Resolution Proxy Data. *Nature* 433, 613–17.

Monastersky, R. (2003). Storm Brews over Global Warming. *The Chronicle of Higher Education*, September 5, A16.
  (2005). Demand for their Data on Climate Chills Scientists. *The Chronicle of Higher Education*, July 15, A1, 23, 24.
  (2006). House Panel Heatedly Debates Research behind Global Warming Theory. *The Chronicle of Higher Education* (https://chronicle.com/article/House-Panel-Heatedly-Debates/9801/).

Monnin, E., et al. (2001). Atmospheric CO_2 Concentrations over the Last Glacial Termination. *Science* 291, 112–14.

Moon, T., et al. (2012). 21-st Century Evolution of Greenland Outlet Glacier Velocities. *Science* 336, 576–8.

Morello, L. (2011). NOAA Makes It Official: 2011 Among Most Extreme Weather Years in History (www.scientificamerican.com/article.cfm?id=noaa-makes-2011-most-extreme-weather-year&print=true).

Mouawad, J. (2006). The Greener Guys. *The New York Times*, May 30 (www.nytimes.com/2006/05/30/business/30carbon).

Mouawad, J. and C. Krauss (2009). Dark Side of Natural Gas Boom. *The New York Times*, December 8, B1, B4.

Muller, E. (2011). Cooling the Warming Debate. (http://berkeleyearth.org/Resources/Berkeley_Earth_Summary_20_Oct.pdf). Accessed November 3, 2011.

Muller, R.A. (2012). The Conversion of a Climate-Change Skeptic. *The New York Times*, July 30, A11.

Munich Re (2012). North America Most Affected by Increase in Weather-Related Natural Catastrophes (www.munichre.com/en/media_relations/press_releases/2012/2012_10_17_press_release.aspx).

Myers, S.L. and N. Kulish (2013). Growing Clamor about Inequities of Climate Crisis.*The New York Times*, November 17, A1, A12.

Myrhe, G. (2009). Consistency between Satellite-Derived and Modeled Estimates of the Direct Aerosol Effect. June 18 (www.sciencexpress.org).

NAS (2012). The Effect of Solar Variability on Earth's Climate: A Workshop Report. National Academy of Sciences Prepublication, National Academies Press (www.nap.edu/catalog.php?record_id=13519).

NASA (2006). Global Effects of Mount Pinatubo. National Aeronautics and Space Administration (http://visibleearth.nasa.gov/view_rec.php?id=1803).

(2014). The Sunspot Cycle. National Aeronautics and Space Administration (http://solarscience.msfc.nasa.gov/SunspotCycle.shtml).

ND-GAIN (2014). Notre Dame Global Adaptation Index (www.nd-gain.org).

NOAA (2010). Arctic Report Card: Region Continues to Warm at an Unprecedented Rate. National Oceanic and Atmospheric Administration. (www.noaanews.noaa.gov/stories2010/20101021_arcticreportcard.html).

(2011). Extreme Weather 2011 (http://www.noaa.gov/extreme2011/).

(2013c). Billion-Dollar Weather/Climate Disasters. National Climatic Data Center, National Oceanic and Atmospheric Administration (www.ncdc.noaa.gov/billions).

(2014a). Trends in Atmospheric Carbon Dioxide. Earth System Research Laboratory,(www.esrl.noaa.gov/gmd/ccgg/trends).

(2014b). The NOAA Annual Greenhouse gas Index (AGGI) (www.esrl.noaa.gov/gmd/aggi/aggi.html).

(2014c). El Niño Theme Page. National Oceanic and Atmospheric Administration (www.pmel.noaa.gov/tao/elnino/nino-home.html).

NPCC (2010). Climate Change Adaptation in New York City: Building a Risk Management Response. New York Panel on Climate Change. *Annals of the New York Academy of Sciences*, Volume 1196 (http://bit.ly/9N78gI).

NRC (2002). *Abrupt Climate Change: Inevitable Surprises*. National Research Council. National Academies Press, Washington, DC (www.nap.edu/download.php?record_id=10136).

(2006). Surface Temperature Reconstructions for the Last 2,000 Years (www.nap.edu/catalog/11676.html).

(2008). Potential Impacts of Climate Change on US Transportation. Transportation Research Board Special Report 290 (http://onlinepubs.trb.org/onlinepubs/sr/sr290.pdf).

(2009). America's Energy Future: Technology and Transformation.

(2010). Adapting to the Impacts of Climate Change (www.nap.edu/catalog .php?record_id=12785).

(2011). America's Climate Choices (www.nap.edu/catalog.php?record_id=12781).

(2012a). A National Strategy for Advancing Climate Modeling (www.nap.edu/ catalog.php?record_id=13430).

(2012b). Climate Change: Evidence, Impacts, and Choices. (http://dels .nas.edu/resources/static-assets/materials-based-on-reports/booklets/ Climate-Change-Lines-of-Evidence.pdf).

NSEE (2014). Public Views on a Carbon Tax Depend on Proposed Use of the Revenue. National Surveys on Energy and the Environment (www.closup .umich.edu/files/ieep-nsee-2014-spring-carbon-tax.pdf).

NSF (2011). National Science Foundation, Office of Inspector General, Closeout Memorandum A09120086.

NYT (2004). Subverting Science. *The New York Times*, October 31 (http://query .nytimes.com/gst/fullpage.html?res=940DE1D6103DF932A05753C1A9629 C8B63).

(2006a). Global Warming and the Courts. July 8, A24.

(2006b). Muzzling Those Pesky Scientists (www.nytimes.com/2006/12/11/ opinion11mon3.html).

(2011). Drilling Down (www.nytimes.com/interactive/us/DRILLING_DOWN _SERIES.html).

Nakicenovic, N., et al. (2000). *Special Report on Emission Scenarios (SRES): A Special Report of Working Group III of the Intergovernmental Panel on Climate Change*, Cambridge University Press, Cambridge (www.grida.no/ climate/ipcc/emission/index.htm).

National Academies (2008). Understanding and Responding to Climate Change (www.southernclimate.org/documents/resources/Understanding_and_ Responding_to_Climate_Change_2008.pdf).

(2010). On Thin Ice. *IN FOCUS* (http://infocusmagazine.org/9.2/thin_ice.html).

Nghiem, S.V., et al. (2012). The Extreme Melt Across the Greenland Ice Sheet in 2012. *Geophys. Res. Lett.* 39, L20502 (doi:10.1029GL053611).

Nisbet, E.G., et al. (2014). Methane on the Rise – Again. *Science* 343, 493–5.

Nordhaus, W. (2007). Critical Assumptions in the Stern Review on Climate Change. *Science* 317, 201–2.

(2013). *The Climate Casino: Risk, Uncertainty, and Economics for a Warming World*. Yale University Press, New Haven, CT.

Norris, F. (2013). Population Growth Forecast from the U.N. May Be Too High. *The New York Times*, September 21, B3.

Null, J. (2014). El Niño and La Niña Years and Intensities (www.ggweather.com/ enso/oni.htm).

Octogesima Adveniens (1971). Pope Paul VI, Apostolic Letter §21, May 14 (www .vatican.va/holy_father/paul_vi/apost_letters/documents/hf_p-vi_apl_19710514 _octogesima-adveniens_en.html).

Olivier, J.G.J., et al. (2011). Long-Term Trend in Global CO_2 Emissions. PBL Netherlands Environmental Assessment Agency and Institute for Environment and Sustainability of the European Joint Research Centre (http://edgar.jrc .ec.europa.eu/news_docs/Co2%20Mondiaal_%20webdef_19sept.pdf).

et al. (2012). Trends in Global CO$_2$ Emissions. (www.pbl.nl/en/publications/2012/trends-in-global-co2-emissions-2012-report).

Oppenheimer, M. and R. Alley (2004). The West Antarctic Ice Sheet and Long-Term Climate Policy. *Climatic Change* 64, 1–10.

Oreskes, N. and E. Conway (2010). *Merchants of Doubt: How a Handful of Scientists Obscured the Truth from Tobacco Smoke to Global Warming*. Bloomsbury Press, New York.

Osborn, T.J. and K.R. Briffa (2006). The Spatial Extent of 20th-Century Warmth in the Context of the Past 1200 Years. *Science* 311, 841–4.

Oxford (2014). Oxford Principles. Oxford Geoengineering Programme, University of Oxford (http://www.geoengineering.ox.ac.uk/oxford-principles/principles/).

PCAST (2013). Climate Change Strategy and Policy (www.whitehouse.gov/sites/default/files/microsites/ostp/PCAST/pcast_energy_and_climate_3-22-13_final.pdf).

Pacala, S. and R. Socolow (2004). Stabilization Wedges: Solving the Climate Problem for the Next 50 Years with Current Technologies. *Science* 305, 968–71.

Pan, Y., et al. (2011). A Large and Persistent Carbon Sink in the World's Forests. *Science*, 333, 988–93.

Parkinson, C.L. (2010). *Coming Climate Crisis? Consider the Past, Beware the Big Fix*. Roman & Littlefield Publishers, Lanham, MD.

Parrenin, F., et al. (2013). Synchronous Change of Atmospheric CO$_2$ and Antarctic Temperature During the Last Deglacial Warming. *Science* 339, 1060–3.

Parson, E.A. and D.W. Keith (2013). End the Deadlock on Governance of Geoengineering. *Science* 339, 1278–9.

Patashnik, I.V., et al. (2011). Water Vapor Self-Continuous Absorption in Near-Infrared Windows Derived from Laboratory Measurements. *J. Geophys. Res.* 116, D16305 (doi:10.1029/2011JD015603).

Paulson, H.M. Jr. (2014). The Coming Climate Crash. *The New York Times*, June 22, SR 1,12.

Pedro, J.B., et al. (2012). Tightened Constraints on the Time-Lag between Antarctic Temperature and CO$_2$ during the Last Deglaciation. *Clim. Past* 8, 1213–21.

Peikoff, L. (1993). *Objectivism: The Philosophy of Ayn Rand*. Penguin Books, New York.

Peña, N. and G. Grünbaum (2001). Facts and Figures: Global Climate Data. (www.c2es.org/docUploads/ff_chapter.pdf).

Peterson, P.G. (2004). ~~Running on Empty: How the Democratic and Republican Parties are Bankrupting our Future and What Americans Can Do About It.~~ Farrar, Strauss and Giroux, New York.

Pew (2013a). Who's Winning the Clean Energy Race? The Pew Charitable Trusts. (www.pewenvironment.org/uploadedFiles/PEG/Publications/Report/-clenG2o-Report-2012-Digital.pdf).

(2013b). Deficit Reduction Rises on Public's Agenda for Obama's Second Term. Pew Research Center for the People and the Press (www.people-press.org/2013/01/24/deficit-reduction-rises-on-publics-agenda-for-obamas-second-term).

Pfeffer, W.T., et al. (2008). Kinematic Constraints on Glacier Contributions to 21st Century Sea-Level Rise. *Science* 321, 1340–3.

Pfeil, M.R. (2006). Social Sin: Social Reconciliation? *Reconciliation, Nations and Churches in Latin America*. I. Maclean, Ed., Ashgate, Burlington, VT, 171–89.

Phillips, O.L., et al. (2009). Drought Sensitivity of the Amazon Rain Forest. *Science* 323, 1344–7.

Piedra, A.M. (2006). Economics and the Common Good. *Current Issues in Catholic Higher Education* 25(1), 75–93.

Pielke, R.A. Jr. (2007). *The Honest Broker: Making Sense of Science in Policy and Politics*. Cambridge University Press, Cambridge.

  (2010). *The Climate Fix: What Scientists and Politicians Won't Tell You About Global Warming*. Basic Books, New York.

Pierson, R. (2013). Landfill Methane. Fact Sheet, Environmental and Energy Study Institute (www.eesi.org/fact-sheet-landfill-methane-26-apr-2013).

Pigou, A. C. (2013). *The Economics of Welfare*. Palgrave Macmillan, London.

Pindyck, R.S. (2013). Climate Change Policy: What Do the Models Tell Us? National Bureau of Economic Research, Working Paper 19244.

Plumer, B. (2012). CBO: Carbon Capture Efforts Aren't Going So Well. *The Washington Post* (www.washingtonpost.com/blogs/wonkblog/wp/2012/07/02/cbo-carbon-capture-efforts-arent-going-so-well).

Pooley, E. (2010a). *The Climate War: True Believers, Power Brokers, and the Fight to Save the Earth*. Hyperion, New York.

  (2010b). America Sits Out the Race. *BloombergBusinessweek*, August 2, 6–7.

Porter (2013). Rethinking How to Split the Costs of Carbon. *The New York Times*, December 25, B1, 2.

  (2014). China's Hurdle to Fast Action on Carbon. *The New York Times*, July 2, B1, B9.

Powell, C.A. and B.D. Morreale (2008). Material Challenges in Advanced Coal Conversion Technologies. *MRS Bulletin* 33, 309–15.

Prentice, I.C., et al. (2001). The Carbon Cycle and Atmospheric Carbon Dioxide. In J.T. Houghton et al. (eds), *Climate Change 2001: The Scientific Basis*. Contribution of Working Group I to the Third Assessment Report of the Intergovernmental Panel on Climate Change. Cambridge University Press, Cambridge, pp. 183–237.

Prestowitz, C. (2003). *Rogue Nation: American Unilateralism and the Failure of Good Intentions*. Basic Books, New York.

Putnam, R.D. and D.E Campbell (2010). *American Grace: How Religion Divides and Unites Us*. Simon and Schuster, New York.

REN (2014). Renewables 2014: Global Status Report. Renewable Energy Policy Network (www.ren21.net/portals/0/documents/resources/gsr/2014/gsr2014_full%20report_low%20ores.pdf).

RGGI (2014a). Regional Investment of RGGI CO_2 Allowance Proceeds, 2012. (www.rggi.org/docs/Documents/2012-Investment-Report.pdf).

  (2014b). Auction 24 (www.rggi.org/market/co2_auctions/results/auction-24).

Rabe, B.G. (2006). Race to the Top: The Expanding Role of U.S. State Renewable Portfolio Standards. Pew Center on Global Climate Change.

Rahmstorf, S., et al. (2012). Comparing Climate Projections to Observations up to 2011. *Environ. Res. Lett.* 7(4), 044035 (http://iopscience.iop.org/1748-9326/7/4/044035).

Ramaswamy, V., et al. (2001). Radiative Forcing of Climate Change. In J.T. Houghton et al. (eds), *Climate Change 2001: The Scientific Basis*. Contribution of Working Group I to the Third Assessment Report of the Intergovernmental Panel on Climate Change. Cambridge University Press, Cambridge, pp. 349–416.

Rand, A. (1992). *Atlas Shrugged*, 35th Anniversary Edition, Penguin Books, New York.

Ravishankara, A.R., et al. (2009). Nitrous Oxide (N_2O): The Dominant Ozone-Depleting Substance in the 21st Century. *Science* 326, 123–5.

Redemptor Hominis (1979). Pope John Paul II, Encyclical Letter §15, March 4 (www.vatican.va/holy_father/john_paul_ii/encyclicals/documents/hf_jp-ii_enc_04031979_redemptor-hominis_en.html).

Reed, S. (2013). After Failed Attempt in April, Europe Approves Emissions Trading System. *The New York Times*, July 4, B3.

Regalado, A. (2005). Researchers, Lawmakers Criticize Probe Into Climate Calculations. *The Wall Street Journal*, July 19, A4.

Revkin, A.C. (2003). Into Thin Air: Kyoto Treaty May Not Die. *The New York Times*, December 4 (www.nytimes.com/2003/12/04/international/europe/04CLIM.html).

    (2005). Bush Aide Softened Greenhouse Gas Links to Global Warming (www.nytimes.com/2005/06/08/politics/08climate.html).

    (2014). In Joint Steps on Emissions, China and U.S. Set Aside 'You First' Approach on Global Warming (http://dotearth.blogs.nytimes.com/2014/11/11/in-joint-steps-on-emissions-china-and-u-s-set-aside-you-first-approach-on-global-warming/?emc=edit_ty_20141112&nl=opinion&nlid=16254583&_r=0).

Ricke, K.L. (2010). Regional Climate Response to Solar-Radiation Management. *Nature Geoscience* 3, 537–41.

Rignot, E. and P. Kanagaratnam (2006). Changes in the Velocity Structure of the Greenland Ice Sheet. *Science* 311, 986–90.

Rignot, E., et al. (2013). Ice Shelf Melting around Antarctica. *Science* 341, 266–70.

    (2014). Widespread, Rapid Grounding Line Retreat of Pine Island, Thwaites, Smith and Kohler Glaciers, West Antarctica from 1992 to 2011. *Geophys. Res. Lett.* 41(10), 3502–9 (doi: 10,1002/2014GL060140).

Riihelä, A., Manninen, T. and V. Laine (2013). Observed Changes in the Albedo of the Arctic Sea-Ice Zone for the Period 1982–2009. *Nature Climate Change* 3, 895–8 (doi:10.1038/nclimate1963).

Riley, N.S. (2004). On Thrift. *In Character, A Journal of Everyday Virtues* (http://incharacter.org/archives/thrift/editors-note).

Rittel, H.W.J. and M.W. Webber (1973). Dilemmas in a General Theory of Planning. *Policy Science* 4(2), 155–69.

Roberts, P. (2004). *The End of Oil*, Houghton Mifflin, New York.

Robock, A., et al. (2009). Benefits, Risks, and Costs of Stratospheric Geoengineering. *Geophys. Res. Lett.* 36, L19703 (doi:10.1029/2009GL039292).

Rohter, L. (2004). U.S. Waters Down Global Commitment to Curb Greenhouse Gases. *The New York Times*, December 19 (www.nytimes.com/2004/12/19/science/19climate.html).

Romm, J. (2004). *The Hype about Hydrogen*, Island Press, Washington, DC.

(2010). Climate Progress (http://thinkprogress.org/romm/2010/06/19/206237/the
-oily-operators-behind-the-religious-climate-disinformation-front-group
-cornwall-alliance).

Rose, D. (2011). Scientist Who Said Climate Change Skeptics Had Been Proved
Wrong Accused of Hiding Truth by Colleague (www.dailymail.co.uk/
sciencetech/article-2055191).

Rosenfeld, D. (2006). Aerosols, Clouds, and Climate. *Science* 312, 1323–4.

Rosenthal, E. (2009). Soot from Third World Stoves Is New Target in Climate
Fight. *The New York Times*, April 16, A1, A12.

(2011). Where Did Global Warming Go? *The New York Times*, October 16, SR1, 7.

Rosenthal, E. and A.W. Lehren (2012). Carbon Credits Gone Awry Raise Output of
Harmful Gases. *The New York Times*, August 9, A1, A10.

Rotman, D. (2009). Natural Gas Changes the Map. *Technology Review*, November/
December, 44–51.

Rubin, E.S. (2001). *Introduction to Engineering and the Environment*. McGraw
Hill, New York.

Russell, L.M. (2012). Offsetting Climate Change by Engineering Air Pollution to
Brighten Clouds. National Academy of Engineering, *The Bridge* 42(4), 10–15.

Russell, M., et al. (2010). The Independent Climate Change E-mails Review.
(www.guardian.co.uk/environment/2010/jul/07/findings-muir-russell-review).

SDSN (2014). Pathways to Deep Decarbonization. Sustainable Development
Solutions Network and Institute for Sustainable Development and
International Relations (http://unsdsn.org/wp-content/uploads/2014/07/
DDPP_interim_2014_report.pdf).

Sallenger, A.H., et al. (2012). Hotspot of Accelerated Sea-Level Rise on the Atlantic
Coast of North America. *Nature Climate Change* 2, 884–8 (doi:10.1038/
nclimate1597).

Sanders, B. (2012). Insurers See Growing Risks and Costs from Climate Change. (www
.sanders.senate.gov/newsroom/news/?id=aad5c0b4-76ed-49b2-8fd7-614f2697e31c).

Sanders, E. (2005). Rebuffing Bush, 132 Mayors Embrace Kyoto Rules.
The New York Times (www.nytimes.com/2005/05/14/national/14kyoto
.html?ref=elisanders&_r=0).

Santer, B.D., et al. (2013). Human and Natural Influences on the Changing
Thermal Structure of the Atmosphere. *Proc. Natl. Acad. Sci.* 110(43), 17235–40
(www.pnas.org/cgi/doi/10.1073/pnas.1305332110).

Savonis, M.J., et al. (2008). Impacts of Climate Change and Variability
on Transportation Systems and Infrastructure, Phase I. U.S. Climate
Science Program Synthesis and Assessment Product 4.7 (http://downloads
.globalchange.gov/sap/sap4-7/sap4-7-final-all.pdf).

Sayre, K.M. (2010). *Unearthed: The Economic Roots of Environmental Crisis*.
University of Notre Dame Press, South Bend, IN.

Schaefer, J. (2010). Environmental Degradation, Social Sin and the Common
Good. In *God, Creation and Climate Change: A Catholic Response to the
Environmental Crisis*. R.W. Miller, Ed., Orbis Books, New York, 69–94.

Scheffran, J., et al. (2012). Climate Change and Violent Conflict. *Science* 336,
869–71.

Schlenker, W. and M.J. Roberts (2009). Nonlinear Temperature Effects Indicate Severe Damage to US Crops under Climate Change. *Proc. Natl. Acad. Sci.* 106(37), 15594–8 (doi:10.1073/pnas.0906865106).

Schmitt, J., et al. (2012). Carbon Isotope Constraints on the Deglacial CO_2 Rise from Ice Cores. *Science* 336, 711–14.

Schmitz, R.A. (2002). The Earth's Carbon Cycle. *Chemical Engineering Education*, Fall, pp. 296–309.

Schneider, S.H. (2009). *Science as a Contact Sport*. National Geographic, Washington, DC.

Schurr, E., et al. (2008). Vulnerability of Permafrost Carbon to Climate Change: Implications for the Global Carbon Cycle. *Bioscience* 58(8), 701–14.
  (2009). The Effect of Permafrost Thaw on Old Carbon Release and Net Carbon Exchange from Tundra. *Nature* 459, 556–9.

Schwartz, H.G. Jr. (2010). Adaptation to the Impacts of Climate Change on Transportation. *The Bridge*, National Academy of Engineering, Fall Issue, pp. 5–13.

Science (2011). A Global Perspective on the Anthropocene. *Science* 334, October 7, 34–5.

ScienceDaily (2011). Multiple Ocean Stresses Threaten 'Globally Significant' Marine Extinction, Experts Warn (www.sciencedaily.com/releases/2011/06/110621101453.htm).
  (2012a). Reducing CO_2: Research Shows Chemical and Economic Feasibility for Capturing Carbon Dioxide Directly from Air (www.sciencedaily.com/releases/2012/07/120724104647.htm).
  (2012b). Pulling CO_2 from Air Vital, but Lower-Cost Technology a Stumbling Block So Far, Experts Say (www.sciencedaily.com/releases/2012/07/120724144532.htm).

Screen, J.A. and I. Simmonds (2010). The Central Role of Diminishing Sea Ice in Recent Arctic Temperature Amplification. *Nature* 464, 1334–7.

Searchinger, T., et al. (2008). Use of US Croplands for Biofuels Increases Greenhouse Gases through Emissions from Land-Use Change. *Science* 319, 1238–40.

Seelye, K.Q. (2003). Environmental Groups Gain as Companies Vote on Issues. *The New York Times*, May 29 (www.nytimes.com/2003/05/29/business/29POLL.html).

Service, R.F. (2004). The Carbon Conundrum. *Science* 305, 962–3.
  (2012). Mountains of Data. *Science* 337, 793–5.
  (2014). Cellulosic Ethanol at Last? *Science* 345, 1111.

Shakhova, et al. (2010). Extensive Methane Venting to the Atmosphere from Sediments of the East Siberian Arctic Shelf. *Science* 327, 1246–50.

Shakun, J.D., et al. (2012). Global Warming Preceded by Increasing Carbon Dioxide Concentration During the Last Deglaciation. *Nature* 484, 49–54.

Shepherd, A., et al. (2012). A Reconciled Estimate of Ice-Sheet Mass Balance. *Science* 338, 1183–9.

Shindell, D.T., et al. (2009). Improved Attribution of Climate Forcing to Emissions. *Science* 326, 716–18.

(2012). Simultaneously Mitigating Near-Term Climate Change and Improving Human Health and Food Security. *Science* 335, 183–9.

Shiva, V. (2002). *Water Wars: Privatization, Pollution and Profit.* Chapter 2, South End Press, Boston.

Siegenthaler, U., et al. (2005). Stable Carbon Cycle-Climate Relationship during the Late Pleistocene. *Science* 310, 1313–17.

Sills, J. (2010). Climate Change and the Integrity of Science. *Science* 328, May 7, 689–90.

Silva, J.M., et al. (2014). Produced Water from Hydrofracturing: Challenges and Opportunities for Reuse and Recovery. *The Bridge, U.S. National Academy of Engineering* 44(2), 34–40.

Silver, N. (2012). *The Signal and the Noise: Why So many Predictions Fail – but Some Don't.* Penguin Press, New York.

Singer, F.S. (2004). The Kyoto Protocol: A Post-Mortem. *The New Atlantis* 4, 66–73.

Skinner, L. (2012). A Long View on Climate Sensitivity. *Science* 337, 917–19.

Socolow, R.H. (2005). Can We Bury Global Warming? *Scientific American*, July, 49–55.

(2011). Wedges Reaffirmed. *Bulletin of the Atomic Scientists*, September 27 (www .thebulletin.org/web-edition/features/wedges-reaffirmed).

Soden, B.J., et al. (2005). The Radiative Signature of Upper Tropospheric Moistening. *Science* 310, 841–4.

Sollcitudo Rei Socialis (1987). Apostolic Blessing, Pope John Paul II, December 30 (www.vatican.va/holy_father/john_paul_ii/encyclicals/documents/hf_jp-ii _enc_30121987_sollicitudo-rei-socialis_en.html).

Soon, W. and S. Baliunas (2003). Proxy Climatic and Environmental Changes of the Past 1000 Years. *Climate Research* 23(2), 89–110.

Spahni, R., et al. (2005). Atmospheric Methane and Nitrous Oxide of the Late Pleistocene from Antarctic Ice Cores. *Science* 310, 1317–21.

Spencer, J. and T. Wright (2007). China, U.S. Spar at Climate Talks. *The Wall Street Journal*, December 6, A10.

Spencer, R.W. and W.D. Braswell (2010). On the Diagnosis of Radiative Feedback in the Presence of Unknown Radiative Forcing. *J. Geophys. Res.* 115, D16109 (doi:10.1029/2009JD013371).

(2011). On the Misdiagnosis of Climate Feedbacks from Variations in the Earth's Radiation Balance. *Remote Sens.* 3(8), 1603–13.

Stainforth, D.A., et al. (2005). Uncertainty in Predictions of the Climate Response to Rising Levels of Greenhouse Gases. *Nature* 433, 403–6.

Stavins, R. (2010). Defining Success for Climate Negotiations in Cancún_What Happened and Why. Belfer Center for Science and International Affairs, Harvard University, December 13.

Stern, N. (2006). *Stern Review on the Economics of Climate Change.* Cambridge University Press, Cambridge. See also Executive Summary (www.wwf.se/source.php/1169157/Stern%20Report_Exec%20Summary.pdf).

(2013). The Structure of Economic Modeling of the Potential Impacts of Climate Change: Grafting Gross Underestimation of Risk onto Already Narrow Climate Models. *Journal of Economic Literature* 51(3), 838–59 (doi:10.1257/jel.51.3.838).

Stine, K.P. and W.T. Sturges (2007). CO_2 Is Not the Only Gas. *Science* 315, 1804–5.

Stocker, T.F., et al. (2001). Physical Climate Processes and Feedbacks. In J.T Houghton et al. (eds), *Climate Change 2001: The Scientific Basis.* Contribution of Working Group I to the Third Assessment Report of the Intergovernmental Panel on Climate Change. Cambridge University Press, Cambridge, 417–70.

Stott, P., et al. (2013). The Upper End of Model Temperature Predictions Is Inconsistent with Past Warming. *Environ. Res. Lett.* 8, 014024 (doi:10.1088/1748–9326/8/1/014024).

Sustainable Business (2014). How California Will Spend the $5 Billion a Year from Cap-and-Trade (www.sustainablebusiness.com/index.cfm/go/news.display/id/25771).

Sweet, W. (2008). Jeremy Bentham (1748–1832). Internet Encyclopedia of Philosophy (www.iep.utm.edu/bentham).

TCEP (2011). Texas Clean Energy Project (www.texascleanenergyproject.com).

TR (2008). A Cheap CO_2 Trap. *Technology Review*, May/June, 26.

(2009). The Cost of Cutting Carbon. *Technology Review*, January/February, 15.

Tabuchi, H. and D. Jolly (2013). Japan Backs Off from Emissions Targets, Citing Fukushima Disaster. *The New York Times*, November 16, A4, A9.

Talbot, D. (2007). Planning for a Climate-Changed World. *Technology Review*, May/June, 63–7.

Tester, J.W., et al. (2005). *Sustainable Energy: Choosing among Options.* The MIT Press, Cambridge, MA.

Thompson, W.G. and S.L. Goldstein (2005). Open-System Coral Ages Reveal Persistent Sea-Level Cycles. *Science* 308, 401–4.

Tilmes, S., et al. (2008). The Sensitivity of Polar Ozone Depletion to Proposed Geoengineering Schemes. *Science* 320, 1201–4.

Timmermann, A. and L. Menviel (2009). What Drives Climate Flip-Flops? *Science* 325, 273–4.

Tollefson, J. (2012). Air Sampling Reveals High Methane Emissions from Natural Gas Field. *Scientific American* (www.scientificamerican.com/article.cfm?id=air-sampling-reveals-high-meth).

(2013). Methane Leaks Erode Green Credentials of Natural Gas. *Nature/News* (www.nature.com/news/methane-leaks-erode-green-credentials-of-natural-gas-1.12123#b3).

Trenberth, K.E. (2005). Uncertainty in Hurricanes and Global Warming. *Science* 308, 1753–4.

Trenberth, K.E., et al. (2009). Earth's Global Energy Budget. *Bull. Amer. Meteor. Soc.* 90, 311–23.

(2011). Issues in Establishing Climate Sensitivity in Recent Studies. *Remote Sens.*, 3, 2951–6.

Trenberth, K.E. and J.T. Fasullo (2010). Tracking Earth's Energy. *Science* 328, 316–17.

Trumbore, S.E. and C.I. Czimczik (2008). An Uncertain Future for Soil Carbon. *Science* 321, 1455–6.

Tyndall (2011). How Do CDM Projects Contribute to Sustainable Development? Tyndall Centre for Climate Change Research (www.tyndall.ac.uk/print/book/export/html/1010).

UCSUSA (2012). A Climate of Corporate Control. Union of Concerned Scientists. (www.ucsusa.org/scientific_integrity/abuses_of_science/a-climate-of-corporate-control.html).

UN (2011). World Population Prospects: The 2010 Revision. United Nations, New York (http://esa.un.org/unpd/wpp/index.htm).
  (2013). World Population Prospects: The 2012 Revision (http://esa.un.org/unpd/wpp/index.htm).

UNDP (2014). Human Development Index (HDI) United Nations Development Programme (http://hdr.undp.org/en/content/human-development-index-hdi).

UNEP (2012). Policy Implications of Warming Permafrost. United Nations Environment Programme (www.unep.org/pdf/permafrost.pdf).

USCCB (1991). Renewing the Earth. A Pastoral Statement of the U.S. Conference of Catholic Bishops, November 14 (http://nccbuscc.org/sdwp/ejp/bishopsstatement.shtml).
  (2001). Global Climate Change: A Plea for Dialogue, Prudence and the Common Good. A Pastoral Statement of the US Conference of Catholic Bishops, June 15 (http://old.usccb.org/sdwp/international/globalclimate.shtml).

USGS (2008). News Release. US Geological Survey (www.usgs.gov/newsroom/article.asp?ID=1980).

Unruh, B. (2008). 31,000 Scientists Reject Global Warming Agenda. *WorldNetDaily* (www.worldnetdaily.com/index.php?pageID=64734).

Urbina, I. (2011). Regulation Lax as Gas Wells' Tainted Water Hits Rivers. *The New York Times* (www.nytimes.com/2011/02/27/us/27gas.html).

Van den Broeke et al. (2009). Partitioning Recent Greenland Mass Loss. *Science* 326, 984–6.

Venkataraman, C., et al. (2005). Residual Biofuels in South Asia: Carbonaceous Aerosol Emissions and Climate Impacts. *Science* 307, 1454–6.

Vidic, R.D., et al. (2013). Impact of Shale Gas Development on Regional Water Quality. *Science* 340(6134) (doi:10.1126/science.1235009).

Vonk, J.E., et al (2012). Carbon Release from Collapsing Coastal Permafrost in Arctic Siberia. *Nature* 489, 137–40.

WCED (1987). *Our Common Energy Future.* World Commission on Environment and Development, Oxford University Press, Oxford.

WDPM (1990). World Day of Peace Message. Peace with God the Creator, Peace with All of Creation. Message of His Holiness Pope John Paul II, January 1 (http://conservation.catholic.org/ecologicalcrisis.htm).

WMO (2011). Provisional Statement on the Status of Global Climate_2011. *World Meteorological Organization* (www.wmo.int/pages/mediacentre/press_releases/gcs_2011_en.html).

WRI (2014). China FAQs – Short Take. World Resources Institute. (www.chinafaqs.org/files/chinainfo/ChinaFAQs_short_take.pdf).

WSJ (2004). Kyoto Capitalists. *The Wall Street Journal*, December 13, A16.
  (2006). Hockey Stick Hokum. *The Wall Street Journal*, July 14, A12.
  (2012). No Need to Panic About Global Warming. *The Wall Street Journal*. (http://online.wsj.com/article/SB10001424052970204301404577171531838421366.html)

WWF (2013). The 3% Solution: Driving Profits Through Carbon Reductions. World Wildlife Fund (http://worldwildlife.org/publications/the-3-solution).

Wald, M.L. (2011). Coal Project Hits Snag as a Partner Backs Off. *The New York Times*, November 11, B1, B8.

(2012). Court Backs E.P.A. Over Emission Limits Intended to Reduce Global Warming. *The New York Times*, (www.nytimes.com/2012/06/27/science/earth/epa-emissions-rules-backed-by-court.html).

(2014). The Potential Downside of Natural Gas. *The New York Times*, June 4, B3.

Wald, M.L. and M.D. Shear (2013). Challenges Await Plan to Reduce Emissions. *The New York Times*, September 21, B1, B2.

Warner, K., et al. (2009). In Search of Shelter: Mapping the Effects of Climate Change on Human Migration and Displacement. (www.ciesin.columbia.edu/documents/clim-migr-report-june09_final.pdf)

Washington, H.A. (1861). *The Writings of Thomas Jefferson*. Letter to Thomas Law, June 13, 1814, H. W. Derby, New York (www.yamaguchy.com/library/jefferson/1814b.html).

Watanabe, T., et al. (2011). Permanent El Niño during the Pliocene Not Supported by Coral Evidence. *Nature* 471, 209–11.

Waxman, H. (2009). American Clean Energy and Security Act of 2009, H.R. 2454, 111th Congress (www.govtrack.us/congress/bills/111/hr2454).

Weart, S.R. (2003). *The Discovery of Global Warming*, Harvard University Press, Cambridge, MA. Also available at www.aip.org/history/climate/.

Webster, P.J., et al. (2005). Changes in Tropical Cyclone Number, Duration, and Intensity in a Warming Environment. *Science* 309, 1844–6.

Wells, K. and B. Elgin (2011). What's Killing Carbon Capture? *Bloomberg Businessweek*, July 31, 68–71.

Wessel, D. (2004). How the U.S. Deficit Is a Moral Issue. *The Wall Street Journal*, November 4, A2, A6.

Westerling, A.L., et al. (2006). Warming and Earlier Spring Increases Western US Forest Wildfire Activity. *Science* 313, 940–3.

White House (2002). Global Climate Change Policy Book: A New Approach (http://georgewbush-whitehouse.archives.gov/news/releases/2002/02/climatechange.html).

Wigley, T.M.L. (2006). A Combined Mitigation/Geoengineering Approach to Climate Stabilization. *Science* 314, 452–4.

(2011). Coal to Gas: The Influence of Methane Leakage. *Climatic Change* 118, 601–8.

~~Wild, M., et al. (2013). A New Diagram for the Global Energy Balance.~~ American Institute of Physics, AIP Conference Proceedings (http://dx.doi.org/10.1063/1.4804848).

Willis, J.K. and J.A. Church (2012). Regional Sea Level Projection. *Science* 336, 550–1.

Wilson, E.O. (2002). The Bottleneck. *Scientific American*, February, 83–92.

Wines, M. (2013a). E.P.A. Is Expected to Set Limits on Greenhouse Gas Emissions by New Power Plants. *The New York Times*, September 14, A11.

(2013b). A Push Away from Burning Coal as an Energy Source. *The New York Times* (www.nytimes.com/2013/11/15/us/a-push-away-from-burning-coal-as-an-energy-source.html?adxnnl=1&ref=us&adxnnlx=1384537303-V47XoJd10odOWwWTOVvisw&_r=0).

Winfield, K. (2007). Europe's Carbon Con Job. *The Wall Street Journal*, August 21, A14.

Wong, E. (2014). In Step to Lower Carbon Emissions, China Will Place a Limit on Coal Use in 2020. *The New York Times*, November 21, A4.

WorleyParsons (2009). Strategic Analysis of the Global Status of Carbon Capture and Storage_Report 5: Synthesis Report. Global CCS Institute (www.globalccsinstitute.com).

Yohe, G. (2010). Risk Assessment and Risk Management for Infrastructure Planning and Investment. *The Bridge*, U.S. National Academy of Engineering, Fall Issue, 14–21.

Young, R., et al. (2014). 2014 International Energy Efficiency Scorecard. American Council for an Energy-Efficient Economy (www.aceee.org/research-report/e1402).

Yuan, H. (2014). *Bloomberg News* (www.bloomberg.com/news/2014-09-01/china-seeks-pollution-cut-with-national-carbon-market.html).

Zeller, T. Jr. (2010). Is It Hot in Here? Must Be Global Warming. *The New York Times*, August 1, A4.

  (2011). Studies say Natural Gas Has Its Own Environmental Problems. *The New York Times*, April 12, B1, B6.

Zeng, N., et al. (2008). Climate Change-the Chinese Challenge. *Science* 319, 730–1.

Zhang, X., et al. (2013). Attributing Intensification of Precipitation Extremes to Human Influence. *Geophys. Res. Lett.* 40, 5251–7.

Zimov, S.A., et al. (2006). Permafrost and the Global Carbon Budget. *Science* 312, 1612–13.

Zoback, M.D. and D.J. Arent (2014). Shale Gas Development: Opportunities and Challenges. *The Bridge*, U.S. National Academy of Engineering, Winter Issue, 16–23.

Index